The
Darkening
Land

William Longgood

SIMON AND SCHUSTER
NEW YORK

Published by Simon and Schuster
Rockefeller Center, 630 Fifth Avenue
New York, New York 10020

First printing
SBN 671-21217-6
Library of Congress Catalog Card Number: 72-78544
Designed by Irving Perkins
Manufactured in the United States of America
Printed by The Murray Printing Company, Forge Village, Mass.
Bound by The Book Press, Brattleboro, Vt.

Many hands took part in the preparation of this book. Included were experts in the environmental field, conservationists, and other writers who gathered material that I used. My wife Peggy read copy for me, made helpful suggestions, and offered encouragement during a long and difficult task. Specialists read the manuscript for accuracy. My many friends at Simon and Schuster, among them Peter Schwed, the publisher, and Charlotte Seitlin, who edited the manuscript, and others who worked on it at various stages, performed heroically in helping translate an idea into a book. To single out individuals from the small army of contributors would be an injustice to those left unnamed. To all who helped in any way, directly or indirectly, I extend my deepest thanks.

*In memory of a dear and
gentle friend who loved the
earth and its creatures—*
Y. K. SMITH
1885–1969

Contents

Three federal agencies, consolidating some thirty departments and interdepartmental offices, have primary responsibility for the nation's antipollution effort. Their abbreviations and basic functions are as follows:

CEQ: The President's Council on Environmental Quality. Recommends policy for the environment and carries out provisions of the National Environmental Policy Act of 1970.

EPA: Environmental Protection Agency. Responsible for controlling and preventing pollution of air, water, and land by issuing guidelines and enforcing them.

NOAC: National Oceanic and Atmospheric Administration. Responsible for long-range fundamental research on atmospheric and marine sciences.

Significant antipollution legislation:

National Environment Policy Act of 1970. Creates guidelines for the government to abate and clean up pollution at all federal installations. Requires all government agencies to submit statements on the environmental impact of any proposed action.

Water Quality Improvement Act of 1970 (also known as Water Pollution Control Act). Enhances quality and value of water resources and establishes a national policy for the prevention, control, and abatement of water pollution.

Refuse Act of 1899. Prohibits dumping of harmful material into navigable waters or their tributaries. Major weapon against industrial water polluters. Act revived by 1970 Presidential directive requires all industries to apply for a permit if they discharge into such waters.

Clean Air Amendments of 1970 (known as Clean Air Act). Stresses cooperative activities by the states and local governments for prevention and control of air pollution. Encourages the enactment of improved and uniform state and local laws relating to prevention and control of air pollution.

Amendments to the Solid Waste Disposal Act of 1965, including the Resource Recovery Act of 1970. Purposes of act: (1) to promote the demonstration, construction, and application of solid-waste management and recovery systems which preserve and enhance the quality of air, water, and land resources; (2) to provide technical and financial assistance to state and local governments and interstate agencies

13

in planning and development of resources, recovery, and solid-waste disposal programs.

Government agencies and private organizations mentioned in text and their abbreviations:

ADC:	Aid to Dependent Children
AEC:	Atomic Energy Commission
AMA:	American Medical Association
Amtrak:	National Passenger Railroad Corporation
ARA:	Air Resources Administration (New York City)
ARS:	Agricultural Research Service of USDA
ASR Center:	Atmospheric Sciences Research Center
CCA:	Citizens for Clean Air (New York City)
CRF:	Citizens' Research Foundation (Princeton, N.J.)
EDF:	Environmental Defense Fund
FAA:	Federal Aviation Administration
FAPCA:	Federal Air Pollution Control Administration
FBLM:	Federal Bureau of Land Management
FDA:	Food and Drug Administration
FOA:	Friends of Animals
FWPCA:	Federal Water Pollution Control Administration
FWQA:	Federal Water Quality Administration
HE&W:	Health, Education and Welfare, Department of (also USDHE&W)
IRS:	Internal Revenue Service
NCI:	National Cancer Institute
NTRA:	National Tuberculosis and Respiratory Association
PRD:	Pesticide Registration Division (a unit of USDA, now under jurisdiction of the EPA)
USDA:	United States Department of Agriculture
USDHE&W:	United States Department of Health, Education and Welfare
USDI:	United States Department of the Interior
USPHS:	United States Public Health Service

Introduction

This book attempts to explain the world's intricate interrelationships and how pollutants interfere with the life-support systems. Its scope embraces not only environmental damage and the peril it poses to life but also political, social, and economic influences. It seeks most of all to bring together many isolated fragments into a meaningful whole.

In researching the subject I made many startling discoveries. The most unsettling, perhaps, is how little is actually known about what we are doing to the world and how little real effort and money are being devoted to finding solutions for so many pressing problems. There is more rhetoric than action. More money is being spent on public relations and advertising to deceive and beguile than to cleanse the waters and air and restore the soil.

This is not a definitive work on the threat to the environment. No single book can fill that epic role. The destruction is proceeding too quickly. With so many untested and unknown substances being dumped into the environment, with such massive engineering changes under way, and so much biological dislocation taking place, it is inevitable that more unpleasant surprises await us, like the recent disclosures about mercury, arsenic, and pesticides that tear at the fragile fabric of life. Reassurances by government and industry ring hollow against the reality of what is taking place.

The simple fact is that the environment no longer can absorb all the pollutants being injected into it; the system itself has been damaged by the changes wrought by reckless and shortsighted men. A slow worldwide buildup of these pollutants is taking place. When the accumulation becomes great enough, the life-support systems start to break down. This is happening now in isolated places.

No one knows when breakdowns will start to occur on a national or international scale. The atmospheric envelope that surrounds earth is universally polluted. Even as the average weight of airborne particles over American cities decreases owing to emission controls, the opacity of the atmosphere goes up through the introduction of more and smaller particles, creating a haze that blots out sunlight. Studies show that the opacity of the atmosphere from these minute particles has risen 100 percent during the past year (1971). New York City is often cited as a barometer of the nation. After making heroic efforts to control air pollution, the amount of sulfur oxides was reduced, but in 1972 the air was dirtier than in the previous year and 7 percent dirtier than in 1968. Officials could offer no explanation for this increase. Nor can anyone predict the eventual effect of the buildup of dirt in the air, but weather and climate are changing.

Virtually all the world's waters are now contaminated. The soil is being progressively washed away through erosion, ruined from salting as a result of irrigation, and poisoned by agricultural chemicals. Species of animals are being wiped out. The carpet of breathing green is being smothered by a ruthless creeping blanket of concrete and asphalt. Rampant technology devours the earth's dwindling supply of resources. Men with little understanding or knowledge of the ultimate consequences of their acts are rapidly rearranging the universal ecosystem that has functioned for billions of years in delicate balance. Soaring, uncontrolled populations and their wastes are putting increased stress on an already overburdened earth. Politics and economics, rather than environmental necessity, dictate public policies. One biologist wryly noted that pollution is 90 percent politics and only 10 percent objectionable matter.

It was once suggested that I focus my research on how a private individual can survive in a polluted world. But from the start it was obvious that none of us can go it alone. There is no private air, water, or food supply. Every garden is at the mercy of contaminants in the air and water. Even if we have our own wells, they are subject to pollution from ground waters, sometimes with bizarre effects, as will be shown in these pages.

For most of us it is difficult or impossible to pick up and move to a less polluted environment. If we continue to live as we have, soon there will be no unspoiled areas left to flee to. When many people crowd into a relatively clean area, it is quickly despoiled and the quality of life deteriorates. Our present economy and technology are based on consumption, waste, and environmental degradation. Smokestacks, sewer outfalls, and diminished wildlife are symbols of a bankrupt vision. Unlike the ancients who had little knowledge of nature and much reverence for its mysteries, we have much knowledge and little reverence.

Our salvation depends primarily on a new morality based on respect for our dependence on nature and an understanding of our own limited role in nature. But the whole effort of environmental reform becomes a mockery if people are urged to put candy wrappers in a trash basket, while industry and private cars are permitted to spew millions of tons of deadly matter into the atmosphere; while we continue to degrade the soil that is the base of life, and dump our wastes into the rivers and ocean; while we pass environmental protection laws and do not enforce them; while we appropriate money toward corrective ends and then the funds are not forthcoming; or while comfort and wealth continue to be the sole criteria of progress. It is, as this book demonstrates, a lethal game to continue playing politics with survival, to preserve the economy at the cost of sacrificing the world. The problem is simple. The solution is complex and costly.

This was a fantastically well-endowed nation. Most of our prosperity has been gained at the land's expense. But the days of reckless exploitation, waste, and despoliation must end. The bill for past abuses is due. We must clean up the mess without further delay or perish in our own filth.

Everything we do, everything that must be done, has one common starting point: awareness of what is at stake and where we now stand. We cannot continue to see this planet as so many fragments or isolated parts to be treated separately. We must understand that any insult to any part of the environment is a global offense against the whole complex interacting web of life.

Since pollution is now a recognized fact of life, this book was not conceived as merely another catalogue of outrages against nature. Nor was it intended to blame any segment of society for acts that were, when originally performed, agreeable to all or most people. It was to be about life and death and survival in a polluted environment. Most of all it was to be about people—

the changes they have wrought in the world, by chance and design, and what these changes mean to us and our descendants. To gather such information I had to find experts who were not only technically trained but capable of a broader view of life than a tainted commercial orientation, people who could weld the fragments into a meaningful whole. This was the underlying concept that guided the preparation of the manuscript.

The evidence I collected often made it difficult to be optimistic about the future. Many scientists of stature have already given up on our feeble attempts to reverse the destruction. An editorial in the Sierra Club publication *The Argonaut* (January, 1970), reflects this despair: "Leaders of the scientific community foresee the end of life on earth within fifty to one hundred years. Doomsday predictions are now so common within the academic world that they bring about fatalistic acceptances instead of the action that would turn the tide."

A few people are, however, struggling to turn the tide. Their efforts are sometimes heroic and well publicized. More often they are low key, on a community level, and go unrecognized. These public and private individuals try to do what they can against great odds, sometimes risking their own welfare. But theirs is a lonely and futile struggle without full commitment by government and the society it regulates.

I salute every attempt to alleviate pollution, however modest. But I do not think individual acts will solve so big a problem. Because of the astronomical numbers involved, the problem is basically economic and political.

Government agencies have flunked their obligation to halt corporate abuses. Most pollution-control efforts have been feeble, nonexistent, confusing, or when occasionally bold they have generally been watered down. Government has at times added to the confusion. For years federal officials warned against the hazards of phosphate-bearing detergents and urged the use of substitutes. Suddenly in September, 1971, the same officials did an about-face, recommending the use of phosphate detergents because their replacements might be more harmful. Communities that had banned or were about to ban detergents were confused. Consumers were bewildered. The reversal reflected the lack of strong leadership and a strong environmental philosophy. Are we to believe that a clean white shirt is more important than a clean river?

In the end, only new attitudes rooted in strong laws can save us. A small select group of individuals give society its directions and goals. Democracy never leads; it merely affirms or denies, taking its cues from the strong and resolute. A minority in government now recognize their obligation and have dedicated themselves to change. Some of the proposed changes could be revolutionary to our comfortable way of life. Gradually a few farsighted leaders are helping to impose corrective measures, but reform is resisted by powerful despoilers and their allies in government. That too is part of the story that unfolds here.

During my research I met many individuals who are struggling to reverse pollution by striking at its insidious roots rather than waving high-sounding self-serving press releases about its dramatic effects. They included scientists, professionals, and knowledgeable laymen, men and women in government and private life. Many were extraordinarily generous in giving of their expertise for what they considered the worthy cause of public enlightenment. There are too many to list here or acknowledge my debt to them. Nor can I pay full tribute to all the written sources used in compiling this book. I am especially indebted to *The New York Times* as my primary source for the running story on environmental developments. The *Times* and its correspondents have done valiant public service, both reporting new developments and taking a strong editorial stand for corrective action.

Many individuals, however, are mentioned in these pages. They not only gave me information but also broadened my perspective by interpreting technical data. If any factual errors slipped through, despite the vigilance of the experts who checked the manuscript for accuracy, I take full responsibility for such lapses. My only defense is human error and the magnitude of the task attempted. I tried hard to be accurate and to present an honest picture of where we now stand. The viewpoint expressed is mine. It is based on evidence gathered from thousands of sources.

Deadlines of publishing make it impossible to discuss developments that took place after the manuscript went into final production.* But the reader will have a framework to better evaluate future developments. I considered concepts more important than events. Early in my research I learned that the hour is late. Very late. Practically everyone I interviewed agreed that there is little time to spare if we are to preserve any kind of meaningful life.

---

* Though a few last-minute developments are appended at the back of the book as numbered notes.

The primary condition I imposed on myself in embarking on this task was to learn the truth firsthand in every case possible. My quest opened with the sea—a logical place to start. The sea is believed to be where all life began, and where, according to some, all life will end if we persist in marching to the muffled beat of a funereal drum that we fail to hear over the roar of the sonic boom and the reverberations of the internal combustion engine, or if our blunted senses fail to perceive the menace in the toxic haze that casts its pall over what the Indians, retreating from the white man's greed, prophetically called the "darkening land."

And so I went to the sea, and there I looked upon the face of death. . . .

WILLIAM LONGGOOD
New York, N.Y.

Pollution is a crime compounded of ignorance and avarice.
                                    —Lord Ritchie-Calder

The outlook is dark. Just as this beautiful region . . . once became for its Cherokee inhabitants a "darkening land" when, because of the encroachments of white pioneer settlers, the copper-colored owners of this fertile land were pushed ever farther westward toward the land of the setting sun, until at last there was no place to go, except to death and oblivion, to slumber in the bosom of the hills. The triumphant settlers called it "progress." The decimated Indians called it the "darkening land."

Some shortsighted Americans call our industrial growth and development "progress." Many wiser ones call it, too, the "darkening land."

                          —Representative John D. Dingell,
                          speech before
                          Fontana Conservation Roundup
                          May 16, 1969

The
Darkening
Land

I

The
Earth's
Waters

All roads lead to the ocean. Rivers are prime movers of from there to here. Some water and solids move in from bare cliffs or salt marsh. Except in a few comparatively limited areas which, in this geological era, are in a Dead or Caspian Sea complex every drop of rain and flake of snow, each loose speck of soil is a potential immigrant to the sea, the moisture that goes back up into the sky by evaporation and then transpiration is the good stuff. At least it starts clean. . . .
—*The National Fisherman*, August, 1969

THE SEA

The clam-shaped bucket was dropped over the stern of the *Challenger*, and the winch whined as the steel cable played out into ninety feet of water. Then the cable stopped abruptly as the bottom-grab struck bottom. The winch was reversed, and the clamlike device retrieved and guided onto a wooden stand at the back of the ship. Dr. Jack Pearce forced the steel jaws open and there was the treasure we sought: a vile-smelling mass of greasy, black organic matter, the residue of wastes that had flushed down millions of toilets and sewers, a lifeless muck that formed part of the ocean floor only a few miles off the New Jersey shore where thousands of families swim, sports fishermen catch game fish for their tables, and commercial fishing boats drop their nets to supply world markets.

This expedition to take samples of the ocean bottom was one small part of the vast mosaic that scientists are slowly putting together to find out what is happening to the environment and what it means to man. The task is incredibly difficult and complex. Reliable information is pitifully scarce. Research funds and efforts to learn more are dangerously small. Progress is painfully slow. And yet, a picture is emerging. Biologists warn that in the shadowy outline they see a shape—the face of Death, a progressive and relentless deterioration of water, air, and soil. Many fear that by the time the mosaic is complete it will be too late to do anything about the unfolding disaster.

The sea is part of this disaster. In a way it has played us false. It is not so large as we once thought. Only forty years ago, when large-scale dumping began off the New Jersey coast, scientists believed the ocean was too big ever to be polluted. Now they know better. The assault is so great that the sea has almost lost its ability to purify itself. It is estimated that, if left untouched from today, it would take at least five hundred years to recover.

Into the sea go the wastes of civilization. Anything that cannot be burned, buried, or dragged into some corner of land is hauled off to the deep. There are 210 dumping sites in our coastal waters, 61 on the populous East Coast, 95 on the Great Lakes, the remainder in the Gulf of Mexico and along the Pacific shore. In addition, there is much unrecorded dumping. From United States cities, industrial and other polluters pour some 48 million tons of wastes into the marine environment every year. Predictably, says Senator Gaylord Nelson of Wisconsin, "our mass-consumption, mass-disposal society is responsible for one-third to one-half the world's population input into the sea."

These wastes include every product known to man: garbage, trash, used oil, grease, dredge spoils (the contaminated muck removed from inland waterways and harbors); powerful industrial chemicals and wastes—acids, caustics, cyanides, dyes, alkalies, phenols, ammonia compounds, chlorine, waste liquor; airplane parts, junked automobiles, radioactive discharges; poisons including arsenic, barium, boron, beryllium, selenium, dissolved gases, potassium; obsolete ordnance—nerve, mustard and tear gas, explosives; toxic heavy metals that never break down or disappear—mercury, nickel, lead, chromium, cadmium, copper, silver, etc.; DDT, herbicides, fumicides, defoliants, and all the thousands of other pesticides; detergents, asbestos, cannery wastes, antibiotics and hormones. It's all dumped into the ocean, pouring in from all over the world, every day of every year.

In a recent year New York City alone dumped into its adjacent sea 6.6 million tons of dredge spoils, 4 million tons of sewage sludge, 2.6 million tons of industrial waste acids, 573,000 tons of cellar dirt. This guck includes viruses, germs, and filth from humans and animals. From industry comes a bewildering complexity of chemicals. The Food and Drug Administration estimates that of the half million such substances now in use, most are so recent that the earth's living forms have never before encountered them. Each year hundreds of new chemicals are produced. Many are highly toxic. Many can cause cancer and other disease. The effects of most are unknown. They have never been tested alone or in combination with other chemicals.

There is virtually no supervision over what goes into the ocean. The Army Corps of Engineers is responsible for issuing dumping permits. But Senator Nelson points out that few applications for offshore waste dumping permits are ever denied, "even when environmental agencies strongly oppose the dumping." A recent report, he says, found no instance where the Corps of Engineers ever rescinded a disposal permit, "even when the polluter had clearly violated it."

Jurisdiction is so confused that agencies concerned with ocean dumping rarely have a comprehensive picture of what is going on or in. Some of the dumping areas are not even marked. One mining operator had to turn back his seabed phosphate lease when he found it was in an old Defense Department ordnance dump. Regular monitoring of ocean dumping is almost nonexistent, says the senator, thus leaving the way wide open for abuse of already inadequate permit terms. "Guidelines to determine how dumping will affect fragile ocean ecology and the marine food chain do not exist."

Every day, from New York and New Jersey, some 7 million gallons of raw sewage spill into the harbor and fan out along the coastal area. On one 30-mile stretch of the New Jersey coast there are 15 sewer outfalls discharging directly into the ocean, and more are planned.*

* In early 1972 it was revealed that for more than forty years 15 New Jersey shore communities have been storing thick sewage wastes in holding tanks during the summer and autumn and then dumping them into the ocean through pipelines when the seashore is not used by bathers. Government investigators said that the four million gallons of sludge,

Already there is grotesque and frightening evidence of what this pollution is doing: "grim portents for the future of much of the U.S. marine environment if the practice is allowed to continue," says Senator Nelson. Beaches have been closed to swimming because of contamination and those mysterious blooms of organisms called red tides that some marine scientists believe are caused by pollution. Fish have died and developed strange diseases, their tails and fins rotting away. Massive growths of nuisance organisms, such as seaweed and jellyfish, are now prevalent. The once-abundant oyster beds in New York Harbor are all but eliminated. Clamming has been banned since people eating them developed hepatitis. Lobsters and crabs exposed to the pollutants develop fouled bronchial chambers and gills. Lobsters' gills were found covered with spots of oil and debris; the fine skin covering of the gills was perforated with small openings left by dead tissue.

For several years marine biologists have been alarmed about the deterioration of the sea. The French oceanographer Jacques-Yves Cousteau says that when he explored the Sargasso Sea in 1942 he could see underwater for about 300 feet. Today, the visibility has shrunk to barely 100 feet, he reports. When he began diving in the Mediterranean twenty-five years ago it was filled with life. Today, "you can hardly see a fish three inches long." He estimates that the vitality of the seas, in terms of fish and plant life, has declined some 30 to 50 percent in the past twenty years.

The noted Swiss marine scientist, Dr. Jacques Piccard, an adviser to the United Nations Conference on the Environment held in Stockholm last summer, estimates that at the current rate of pollution there will be no life in the world's oceans in twenty-five years. He said the shallow Baltic Sea, which has no tides, would be the first to die, and that the Adriatic and the Mediterranean, which also have no tides to carry away pollution, would be next.

---

which is pumped out an average of 1000 feet from the shore, contained "countless forms of harmful bacteria and viruses that produced such diseases as paralytic polio, hepatitis, meningitis and salmonella." The dumping practice was ordered discontinued by a federal judge in February, 1972, but the order was appealed by 14 of the communities. Local officials seemed more concerned about the cost of disposing of the sludge and the effect on business than what was happening to the ocean. One was quoted in *The New York Times* as saying that there was so much public protest that the adverse publicity threatened to "ruin our resort business."

Dr. Jack Pearce, an oceanographer connected with the government's Sandy Hook Marine Laboratory, believes that unless we change our ways soon we will so damage the ocean that it will no longer support life. "There is a very real threat that the end of the world, as we have known it, is on the way," he says. "If we continue the way we're going, we will so pollute the ocean that it will preclude life. Man will no longer be able to exist." He believes that such a catastrophe could occur by the end of the century.

Dr. Paul Ehrlich, also a biologist, in a fanciful article in *Ramparts*, "The Death of the Ocean," projected an earlier demise:

> The end of the ocean came late in the summer of 1979, and it came even more rapidly than the biologists had expected. There had been signs for more than a decade, commencing with the discovery in 1968 that DDT slows down photosynthesis in marine plant life. It was announced in a short paper in the technical journal, *Science*, but to ecologists it smacked of Doomsday. . . .

Similar warnings come from other scientists. In 1969 a panel of marine experts, meeting at the Scripps Institution of Oceanography, held an inquiry into "Man's Chemical Invasion of the Sea." They concluded that the effects of making the ocean an endless sink could be as insidious as cancer. Cousteau says there is an urgency about it, because "the sea as a biological concept is in great danger of being damaged or destroyed."

How universal the pollution of the ocean has become is suggested by Thor Heyerdahl, the Norwegian explorer. Of his celebrated voyage on the balsa raft in 1947, he recalls that "we on *Kon Tiki* were thrilled by the beauty and purity of the ocean." Twenty-two years later he and his six-man crew were astonished and depressed by the jetsam bobbing hundreds of miles from land. According to *Time* magazine:

> Almost every day plastic bottles, squeeze tubes and other signs of industrial civilization floated by the expedition's leaky boat. What most appalled Heyerdahl were sheets of "pelagic particles." At first he assumed that his craft was in the wake of an oil tanker that had just cleaned its tanks. But on five occasions he ran into the same substances covering the water so thickly . . . that "it was unpleasant to dip our toothbrushes into the sea. Once the water was too dirty to wash our dishes in."

"Because the oceans of the world have become one gigantic septic tank in which poisonous wastes are accumulating at rates many consider alarming, an international effort is being planned to identify the most dangerous changes before it is too late," *The New York Times* reported. In the spring of 1971, representatives of thirty-five nations met in London to discuss such a program. One stated purpose was to identify pollutants that are not now recognized. Another was "to uncover the more subtle effects of such substances on marine life—those that do not kill, but may, for example, impair reproduction."

Scientists have become increasingly concerned about the indirect effects of pollution, which can be perplexing and bizarre. Pollutants sometimes seem to play a role in producing lethal marine poisons. Some fish, traditionally safe to eat, suddenly have become poisonous. The person who eats them may confuse heat and cold, can ultimately die of respiratory failure.

In the spring of 1971 there was a disastrous failure of shrimp production off Britain's northwest coast. Officials speculated that along with a natural low period for shrimp, pollution was accentuating the effects. One said it "may be that pollution has killed off the weeds in the grass that the shrimp feed on. It may be that pollution has killed off a certain amount of oxygen."

The difficulty in checking on ocean pollution was noted by Dr. Bruce W. Halstead, head of the World Life Research Institute in California and a leading authority on poisons in the marine environment. He estimated that in southern California alone 50,000 industrial plants are contributing to the "gross contamination" of the sea. There is not even "remotely adequate" surveillance of such discharges, he said.

The New York Times report added:

> The discharges come from plating plants, plastic factories, petroleum centers and many other installations, including Norton Air Force Base which Dr. Halstead says yearly dumps 1.8 million gallons of chromium and cyanide compounds, as well as other poisonous wastes.

Specialists warn that the pollution already found in the sea is but a foretaste of what could follow. Pearce sees the dumping ground off Sandy Hook as a microcosm of what the entire ocean might deteriorate into if it continues to be used as a burial place for our wastes.

The dumping area off New Jersey has probably been studied more intensively than any other. Since 1966 oceanographers from the Sandy Hook Marine Laboratory, in a former hospital building at Fort Hancock, overlooking the coast, have been sampling the bottom and its overlying water. They are watching for changes, gradually filling in a profile of the area as part of the environmental mosaic that scientists throughout the world are constructing. The Sandy Hook researchers describe the dumping area as a vast "dead sea."

Their studies began after several important species of game fish began to disappear off the Atlantic Coast. At first it was believed that this was due to "overharvesting" or possibly some cosmic change in the ocean itself. Gradually the culprit was identified as pollution, and attention focused on the wastes that were smothering and poisoning the life of the sea.

The dumping ground is twelve miles off the coast of Sandy Hook, just beyond Ambrose Lighthouse which vigilantly marks the shipping channel into New York Harbor. Dumping takes place over an area of twenty square miles, or so the permit states.

Shortly after dawn I went out on the *Challenger* on a cold, gray day in early December. We set out from the Sandy Hook Coast Guard station, where the ship was moored, and headed for the open sea. Lining the shore were high-rise apartments delineating Coney Island and the Rockaways, speaking silently of still another threat.

During the outbound voyage Dr. Pearce talked about the studies he is making, their implications and background. He spoke of the disappearance of the fish, the mysterious disease that afflicted them, and the startling truth that began to emerge —the extent to which man is altering that part of the ocean adjacent to the shore, the most vulnerable and sensitive part of the sea.

It is commonly believed that the entire sea abounds with fish, but actually most marine life is found in the coastal waters. Fish tend to live along the continental shelf, a gentle slope that extends off the North American coast from approximately 30 to 200 miles. The far reaches of the ocean are almost devoid of life. Out beyond the continental shelf, the ocean floor dips sharply to become an irregular and often violently thrusting series of ranges; frequently their peaks are higher than the tallest moun-

tains, separated by huge gorges and basins. In some places they
are more than seven miles deep, so deep that the sun's rays cannot penetrate the water to produce the oxygen necessary to the
more common forms of marine life we are acquainted with, and
here live grotesque serpentlike monsters out of a child's night-
mares. Scientists are just starting to explore this realm. "From
birth, man carries the weight of gravity on his shoulders," says
Jacques-Yves Cousteau, the French oceanographer. "But man
has only to sink beneath the surface, and he is free. Buoyed by
water, he can fly in any direction—up, down, sideways—by
merely flipping his hand. Underwater, man becomes an arch-
angel." For Cousteau the sea represents beauty, freedom, the essence of life itself.

The dumping ground off the Jersey shore is divided into three
major areas, Pearce explained. One is for sewage wastes, the
solids collected from treatment plants. The second is for dredge
spoils, including those from the Arthur Kill, an "oil highway"
leading into Port Newark. The third, and smallest of the three, is
for acid wastes. More than 12 million tons of the combined pollutants are dumped here every year, transported by scows.

The dumping ground, Pearce says, is divided by the Hudson
Canyon, a huge undersea V-shaped valley that is an important
pathway used by fish to migrate from deep water to their feeding
and breeding areas near the shore. Dredge spoil is dumped on
the west side of the canyon, sewage sludge on the east. Originally it was thought that when the wastes were dumped in the
sea they would pile up, much as they would on land, and remain
more or less stationary. Now it is known that nature is not so
passive or predictable as those scientists believed.

There are many currents and countercurrents in the ocean.
There are surface currents and underwater currents that flow
like rivers. They combine, join, split, and oppose. Surface currents and underwater currents may go in opposite directions.
They agitate the shallows, mixing in the oxygen essential to
marine life; they lift food and debris from the bottom, putting
them in suspension, setting the particles adrift on the ever restless tides and currents. In the abyss the waters are ancient and
upwelling is slow; often they take a half century or more to replace themselves.

In the dumping ground near the shore the wastes have settled
out in a blanket more than three feet deep. They destroy in two
major ways: suffocating bottom life and poisoning all life out-

right. The wastes ferment, interact, and refuse to stay put. Pearce says there are indications that the sewage sludge and dredge spoil are gradually sliding down the steep slopes of the Hudson Canyon. In time they may fill it up, blocking it to the migration of fish or contaminating the water to such an extent that fish can no longer use it. Recent studies indicate that fish respond to smells; abnormal odors tend to drive them away. Many species of fish must return to their ancestral home to reproduce, and they count on their sense of smell to guide them. "If fish in transit come to a polluted area, it seems to affect them like an invisible screen, blocking the migration that is necessary to their survival, almost as effectively as a solid wall," according to Pearce.

Studies of the dumping area have tried to fix the borders of this concentration of pollution. It extends from Asbury Park, about two miles east of Ambrose Lighthouse, as far out as the Atlantic Beach inlet. This is an important area in terms of both fish productivity and human recreation. Due to underwater drift —a phenomenon only recently discovered—the wastes appear to be shifting from south to north, where the most heavily used bathing beaches are.

The day I was on board the *Challenger*, Pearce was trying to determine how much the heavy metals in the sludge are spreading beyond the dumping area. It is important to find this out. Many of these metals, such as arsenic, mercury, nickel, lead, and cadmium, are highly toxic to marine life as well as to man. They are difficult to detect in water because they appear only in minute amounts. Often their presence can be determined only after they build up in living tissues in dangerous quantities.

This process, which came as an unpleasant surprise a few years ago, is now well known: a contaminant appears in the water; microscopic organisms are unable to distinguish it from a nutrient and absorb it in their tissues; the tiny organism is eaten by shellfish or small fish; these, in turn, are eaten by progressively larger fish or birds; each time one creature eats another the amount becomes greater; at the top of the food chain is man, who absorbs the largest concentration of all.

Along with studying the movement of heavy metals, the expedition wanted to see if the same animals present in the area a year ago were still alive. If the population in a sample varies from one time to the next you do not know whether it is because of natural changes in the sea or not, Pearce said. "But if there is no life, you know it's due to pollution, as there is always life when there is no interference by man. By sampling the same places outside the dumping area again and again, we can tell what is going on inside it."

During the day we took samples from six stations, ranging from the north edge of the sewage dumping area, into the dead center, and then to the southern perimeter. Each time the same tests were made. Each time the amount of life recovered was in direct proportion to the amount of pollution present.

Bottom waste was collected by dropping a spring-triggered clam-shaped device with steel jaws into 90 feet of water. The bottom-grab was attached to a steel cable on a winch. When the device struck bottom its steel jaws automatically snapped shut, encapsulating about a cubic foot of sludge. Some of the sludge was put in plastic bags and tubes (providing a profile of how the successive layers were deposited) and immediately frozen for later analysis. The rest of the debris was washed through progressively finer screens to filter out living organisms and any objects present.

On the edge of the dumping area we brought up black oily muck that produced only a few empty shells, a few worms that can tolerate pollution, and some man-made artifacts. The worms were popped into a chemical solution. This relaxed them so, as a girl student biologist on the ship explained, "they can't scrunch up so they can't be identified." Later the worms were preserved in formaldehyde.

We reached the center of the dumping area in midafternoon. It was like cruising around in a turbulent septic tank. The muck from the bottom was thick, black, greasy, and had a sickening smell. There was no life present, not even the pollution-tolerant worms. The muck was hosed off in an oily black stream. Behind were left the usual man-made objects: Band-Aids, plastic and metal-foil wrappers, labels, sanitary pads, medicinal products, cigaret filters.

Pearce valiantly picked up the muck in his bare hands and plopped it onto the screens to hose it down. "You can smell it real nice now," he said, and laughed at my squeamishness.

There was something enormously sad about seeing this stuff being brought up from the ocean, like some terrible betrayal. I thought of a passage from *Moby Dick:* "There is, one knows not what sweet mystery about this sea, whose gentle awful stirrings seem to speak of some hidden soul beneath."

In addition to collecting samples of sludge and its residues, water samples were taken to measure temperature and salinity at various depths. The salt content in the open water usually runs about 35 parts of salt to 1,000 parts of freshwater. If it varies much from this, fish will sicken or die, as they depend on salt content to regulate the water retention in their flesh.

Water samples from the surface and ocean bottom were also tested for the vital oxygen content. Oxygen in seawater usually varies from about 6 to 11 parts per million (1 part per million is often described as a shot of vermouth to a railroad tankcar full of gin). The colder the water, the more oxygen it holds. Most marine animals require from 3 to 5 parts of oxygen just to exist. In heavily polluted areas organisms use up the oxygen needed to break down the wastes. When the oxygen is gone these organisms die. Then the wastes begin to putrify and give off foul-smelling gases.

In the sewage dumping area the oxygen generally runs less than 1 part per million, and life is virtually nonexistent. In a relatively unspoiled area we sampled outside the dumping ground, the ocean bottom was sandy and vibrant with life. Pearce said a cubic foot of unspoiled ocean bottom would produce around 200 species of life. But where the wastes were concentrated in a three-foot blanket, marine life was disrupted, suffocated, or poisoned outright. Only a few scavenger worms were found. It is a frightening possibility that these worms fore-shadow some wretched form of evolution that could replace the life that now exists on earth.

In addition to the direct effects of pollution, there are many indirect effects and subtle warnings of possible disasters in store. In 1968 thousands of fish mysteriously died on the Jersey shore. The water was tested and found to contain less than 1 part per million of oxygen. For reasons unknown the dead water over the dumping ground apparently moved in toward the shore to suffocate the fish.

The threat of lifeless water is now local there. It could be universal. As water is corrupted by pollution it becomes more opaque. This cuts down the amount of sunlight that can penetrate the water, thus interfering with the ability of plants to use sunlight to manufacture their food and, in turn, to give off oxygen. Without oxygen the sea must die.

Pearce says the pollution of the ocean is already so advanced that algae no longer will grow in water 60 feet deep where it was once found. In water only 30 feet deep, where luxuriant growths formerly bloomed, there is now nothing.

Recently some of the water from the acid dumping ground off the Jersey coast was studied at the Woods Hole Oceanographic Institution. The acids were found to inhibit the growth of diatoms, the basis of the ocean's food chain. These microscopic one-celled plants are called the grasses of the sea; they float in the water pastures and help feed many grazing animals, among them fishes and whales. If they are not present, there is no basis for marine life. Without plant life there can be no animal life.

Dr. Roger L. Payne, a New York zoologist who is an expert on the sea, believes that the damage is so great that there may be no reversing it:

> We're putting the kind of stuff in the water now that is destructive to the small organisms that create oxygen for our increasingly polluted atmosphere. We are attacking our biosphere from two levels simultaneously.
>
> This is the way to destroy life on earth. Probably hydrogen bombs wouldn't do it. If you want to look for the thing to do the job, destroy the algae life in the ocean. When that happens it will be the end of life on earth.
>
> It doesn't matter where you run. You can't hide. The sad part is that most people don't realize to what a phenomenal extent it has gone already. My feeling is that you could make a pretty tight argument, based on some very good data, that it's too late already. Maybe we've passed beyond the point of no return.

Like Ehrlich and many other scientists, he is concerned about the DDT in the ocean: "People think that when they spray DDT, for example, it's a local application. You don't spray your neighbor's yard, you spray your own. But in fact you spray the whole world. When the rain comes along, the stuff eventually washes down and is carried to the ocean, the final cesspool."

For many years it was believed that the microscopic diatoms in the sea were unaffected by DDT. Then in 1968 came another of those surprises that nature keeps handing us. There was unsettling evidence that DDT does indeed slow down photosynthesis in the ocean. No one knows what the ultimate effect will be, but it has catastrophic implications.

The engines slowed. We were back at the dock in Sandy Hook. The samples collected during the day were carried ashore. Later the samples would be taken to the laboratory for analysis. The disaster we had witnessed would be given scientific names and precise dimensions, and it would be fitted into the emerging environmental mosaic that ultimately will spell out the fate of man in his imperiled world.

THE COASTAL ZONE

It is near the shore, where the assault is the greatest, that the sea is the most vulnerable. This transitional area where land and sea meet is neither land nor sea but part of each. It is, according to the Department of the Interior, "part of a vital zone that includes not only the mouths of rivers but vast reaches of the continental shelf and, by congressional definition, the Great Lakes." An estuary can take many forms, as salt marsh, tidal flat, lagoon, bay, shallow sound, slough, and tidal river, like the lower reaches of the mighty Hudson that spills into New York Harbor.

These soggy borders of land that lead into water are collectively known as wetlands. The estuary is a coastal wetland. It is part river, part ocean, part freshwater, part salt, a marvelously intricate web of life, a constant interaction of water, land, wind, and sun, pulsating with life, marine animals, waterfowl, songbirds, scavengers, wild creatures, and the ultimate microorganisms that make possible that unending chain of beginning and end, death and renewal, on both sea and adjoining land.

We know little about these irregular stretches of coastland that fringe the earth and almost nothing about the invisible submicroscopic relationships and interactions that occur within estuary and marsh. There is, however, much evidence that they go far beyond the normal values we give them, even as nurseries and granaries for the sea. They are the essential first step in the chain of existence that looks to the ocean as the mother of all life.

About 70 percent of the world's population lives within 50 miles of the ocean. In the United States approximately one third of the people live and work in this region, and 40 percent of our industry is concentrated there.

Estuaries are frequently spoken of by scientists as "ecosystems." Approximately 850 of them bracket the 53,000-mile coastline of the continental United States. Between 80 and 90 percent of the Atlantic Ocean and Gulf of Mexico coasts and 10 to 20 percent of the Pacific coast are composed of estuaries.

These natural treasures are distributed among 26 coastal states and territories of the nation.

Unfortunately, we have failed to appreciate our good fortune. We have ruthlessly destroyed and altered many of these resources. What is gone is gone forever. What remains is often damaged. Already we have squandered half of our estuarine heritage, and that which still remains is threatened. The disaster is greater because we so often fail to recognize it until too late.

Some of the finest salt marshes in the world were in New England. The long spartina grass, green in the summer, brown in the winter, provided a cover for small animals and birds, a perpetual pantry for the sea. Many of these marshes are now long gone.

The early settlers lived near the marshes, which sustained them just as they did plant and animal life. The farmers took from the marsh small game for their table. From the adjoining sea they took a variety of fish. They took only what they needed and no more, living harmoniously with nature.

For generations their families grew up and continued to live on the old homestead; they worked the soil and replenished its fertility with manure from domestic animals. Then, gradually, change came. The children became restless and went to live in the city. Hunters visited the marsh. Often they came at night to catch birds, stalking their sleeping quarry with lights, twisting their necks and popping them in bags by the hundreds. City milliners bought the feathers; the carcasses were sold to butchers. Finally the choice birds were wiped out.

Then the plunder turned against the small animals. They were killed for their skins, for meat, for sport. They were slaughtered by the thousands: raccoon, opossum, skunk, fox, marmot, rabbit, marten, squirrel, and others. Animals that had depended on them for food or as predators to keep the delicate balance in the marsh also disappeared, and the spartina grass no longer marked their trails.

The old house was empty. The present owners sold the land to speculators. Soon the bulldozers came and filled in the marsh. A housing development sprang up. New families moved in. No one even knew about the salt marsh that had once graced the land where they now lived.

Then, as now, to the speculator and the developer, the marsh or estuary is an unsightly and unusable property begging for "land reclamation." This means filling it in for commercial use.

The complex role of an estuary is neither understood nor appreciated. If it bears no dollar sign, no concrete cover or buildings, it is considered worthless.

The coastal wetland is, in fact, one of nature's masterpieces, subtle, intricate, marvelously complex, and essential in maintaining equilibrium between sea and land. It is never static but ever changing, always in motion, tempered by the delicate blending of salt and freshwater, temperature change, the flow of nutrients and silt, the multiple and profound influences of nature and men. Dr. John Clark, a marine biologist who was formerly assistant director of the Sandy Hook Marine Laboratory at Sandy Hook, New Jersey, explains the "biological magic" of an estuary:

> . . . The coastal wetlands receive phosphates, nitrates and other minerals washed down from the land. Through their warm saline waters sunlight penetrates to the bottom, providing lush plant growth for feeding, for attachment of small invertebrate organisms, and for fertilizing the estuary. Plant life occurs in the wetlands not only as marsh grasses but also as algae to the bottom and much smaller algae which grow on the mud or floating algae in the water as tiny cells or phytoplankton. The larger plants die, are detached, then decomposed and broken down by bacteria. The particles, called organic detritus, together with phytoplankton make of the water a rich nourishing broth.
>
> . . . the gentle action of the tide keeps the rich mixture of nutrients stirred and moving through the shallow water. Nothing is wasted, with bacteria and crabs helping to recycle the nutrients through the system. Bait fishes, worms, shellfish, shrimp, and plankton all thrive in this fertile environment. They in turn provide a ready supply of nourishment for the growth and health of salt water fishes along the whole coast.

The estuary becomes a trap for the nutrients. As the river slows upon entering the estuary, the suspended material settles out, forming tidal mud or sand flats along the fronts and margins of the delta. The new land, formed of mud and sand, rises; it is colonized by algae and flowering plants that can tolerate salt and brackish water. The plants catch the sediments, building up higher ground on which new plants take root and grow. Eventually the whole estuary becomes soft springy earth; it is channeled with meandering creeks that soak up water, releasing it slowly into the earth to replenish underground water tables. The estuary absorbs the fury of storms and hurricanes, protecting the land better than any man-made levee or dyke; it is nature's own first line of defense against the unpredictable sea.

Tides continue to wash the land, decomposing the thick salt grasses, ever adding to the wealth of the estuary, making it a rich fare for the sea, a "cold vegetable soup." The marsh is the richest land in the world, a half-dozen times more productive than the most fertile wheatland. By day and night, in rhythm set to light and darkness, the rising and falling salt tides, the weather and seasons, the estuary throbs with life and its imperative of death.

There are urgent mating calls, shrill cries of alarm, the undulation and movement of the brackish water as small creatures swim, and others crawl over moist land or burrow in the mud, stalking and hiding. This stage is brilliant with colorful berries and flowers, marine life and land life. They all live off one another. At high tide mussels eat plankton and crabs scavenge detritus. A few hours later, at low tide, raccoons eat the mussels, herons feast on crabs, and killifish gobble up mosquito larvae. Baby fish eat plankton. Larger fish eat the smaller fish. Each organism, each plant, plays its own role in the economy of the marsh.

In the summer, life is active and intense. In the winter it is deceptively still.

Nothing is lost or wasted in the marsh; all material, plant and animal alike, is retained and recycled within its closed circuit. The tides continue to mix and redistribute the nutrients in their various forms throughout all parts of the system, taking nourishment from the sea and returning a multiplied bounty. Most of the life cycles of the marsh and estuary—plankton, those wandering rootless microscopic plants of the sea, microorganism and small animal—are short and end violently; turnover is rapid; nutrients rarely remain locked in unusable form for long. The system is magnificently efficient, ever self-sustaining.

Among the important visitors and inhabitants of the estuary are the fish. Two out of every three of the most valuable commercial and game fish on the Atlantic coast depend on estuaries in some way. Fish, like people, have evolved their own individual life-styles and strategies of survival. Different species take advantage of the estuary in the way they have evolved. The spotted weakfish, for example, cleverly uses the grass beds for protection; its mottled pattern of coloring enables it to be perfectly camouflaged to inhabit the bottom vegetation so abundant in estuaries. Without the grass beds, the camouflage pattern would be useless as protection from predators, and without the essential grass the small fish and shrimp on which the weakfish feasts would be gone. Dr. Clark explains:

> Although many dangers threaten all fully grown fish, biologists find that the most critical time in the lives of fish is the period of early development—when they are an inch or less long. It is while they are so tiny that fish are most susceptible to starvation and to attack by multitudes of predatory creatures. In the first few months of their lives, fish have little defense and they must escape from all dangers or perish.
>
> For a majority of coastal fishes the young are best fitted for survival in tide waters, not in the open ocean. Once the floating eggs and hatchlings of offshore specimens have survived the jellyfishes and small predatory fishes, the baby fish in the sea are now vulnerable to multitudes of predatory fishes. But if they drift and swim back to the coast, they can find refuge. In the shallow, brackish waters, the young fish are safe because the large predators are rarely found there. . . . Here in the shallows, the young can escape from any attack that might occur by dashing for a shallow beach or by hiding in a weed bed. There are millions of acres of coastal marsh and shallow waters along our coasts where juveniles abound but where the large fish are rarely found. These are our precious and irreplaceable baby fish sanctuaries. They support the whole cycle of life of the estuarine-dependent coastal fishes.

Some fish live most of their lives within the protected waters of the estuary. Others, primarily oceanic, are critically dependent on the shallow waters for their nursery areas. Still others spawn in the estuaries, and then deposit their young, remaining behind to find the protection and abundant food they need at this critical stage of their lives. Each species becomes a specialist in feeding, spawning, rearing its young, and self-protection. Such specialization ties it ineluctably to its environment; if this habitat is damaged or altered, the species must perish because it has become too specialized and dependent to change to a new environment. In this way animals are akin to man; he is similarly locked into his environment. We alter it at our peril.

The strategies of survival worked out by various species of fish are awesomely ingenious. The story of the salmon struggling back to its ancestral home to spawn and die has often been told. Less well known are the habits of some ocean fish. Many of these have come to light only in recent years. Some fish spawn a thousand miles from where their young will live. The fluke (or summer flounder) reproduces at sea. Its eggs drift south on certain tides that carry them into shore inlets; when the baby fluke reach one centimeter they turn on their right side to lie flat, and the eyes migrate to the top, the skin pigment changing to offer camouflage on top and bottom. One race of bluefish spawns far off New Jersey, setting its larvae adrift on southerly currents; the eggs hatch and, at a certain time, about 800 miles from the starting point, the young strike out for the estuaries of South Carolina. At the same time a race of bluefish from those waters are sending their larvae adrift on currents that will deposit them in the estuaries of New Jersey. Thus nature guards against local disasters that could wipe out a species.

But nature, apparently, did not count on the greatest scourge of all—man, who in his ignorance and arrogance has imperiled the life of the sea. Half of the estuaries we were blessed with are gone forever. Those that remain are endangered by either outright destruction or pollution. The National Estuarine Study, completed in the spring of 1970, is a melancholy three-volume recital of this mass destruction and degradation. Page 497 of the report states:

> It seems clear that most, if not all major estuarine areas in the continental United States are now or soon will be affected by disturbances of more than one identifiable type. These systems are characterized by heterogeneous patches of chemicals, fertilized waters, waters low in available oxygen, turbidities, acids and other conditions alien to normal life of estuarine ecosystems.

The coastal wetlands have been filled in to make housing developments and industrial complexes, parking lots, airports, shopping centers, refineries, roads, garbage dumps, landfills, and recreation areas. They have been dredged and gouged out for marinas and harbors, crisscrossed by bridges, jetties, piers, and causeways. They have been mined for valuable gravel, sand, and shellfish deposits, minerals and oil. The inflow of freshwater has been curtailed or eliminated entirely by dams, reservoirs, and hydroelectric plants. Bold engineering projects turned out to be trial-and-error experiments ending in ecological disaster. The tampering, usually done without thought or understanding of ecological consequences, upset natural balances, ruining the harmony that is the heart of the system.

Of some six million acres of estuarine habitat, more than a half million acres have been destroyed by dredging and filling. Dredging tears up the bottom, plants are ripped out, and the remaining sediment shifts about with the current. Silt settles in a thick mass on the bottom, suffocating animals and plants. Photosynthesis cannot take place in the turbid water. The adjoining marshland or tidal flat is covered with the removed spoil to raise its level and create dry land for commercial purpose. This wipes out all aquatic life. Bulkheads are installed to hold the fill in place for seaside housing; this seals the fate of the marsh by preventing normal tidal flushing action. In this way, Dr. Clark explains, both the marshes and adjoining bays are destroyed as aquatic life zones. Once destroyed, an estuary is gone forever. The effects are not just local; they may echo for thousands of miles, possibly throughout the world.

Stupidity is often magnified by greed and corruption. A popular and profitable scheme has been to dredge huge quantities of gravel and sand from estuaries under the pretext of improving navigation. A New York firm sought a dredging permit in Connecticut to dig a channel to bring small boats to a marina. The trench was to be 300 feet wide, 30 feet deep, and a mile long— large enough for an aircraft carrier. The scheme failed. But many schemes, no less outrageous, succeed. Municipalities often sacrifice or virtually give away estuaries. Developers have been remarkably successful in persuading municipal officials to let them fill in estuaries and marshes for commercial purposes or housing developments; the justification is almost invariably the need for new taxes. In the case of housing this is most usually an illusion at best; essential services generally cost two to three times the increased revenue. A New York community found that it saved money by buying and keeping open land as a park rather than having it developed for housing.

For industry there are many advantages in locating on coastal land. There is water for cooling, to discharge pollutants in, and for cheap transportation. As waterfront property has become progressively scarce it has soared in value. In South Carolina, for example, it has shot up within a few years from $75 to $1,600 a foot. Of the 53,000 miles of coastline in the nation, only about half are considered suitable for recreation or public habitation. Of the remaining 26,000 miles, 1,200 are publicly owned. The rest is open to exploitation or destruction unless protected—and time is fast running out.

A few states have tried various stopgap measures to preserve their remaining coastal lands. Some have attempted, with the help of federal funds, to buy estuaries and marshes. But such funds are in short supply. There is also frequent and strong local resistance to "wasting money on useless land." Others resist any interference with the private development of land.

Newspapers abound with these tragic dramas. One was recently played out on Bald Head Island in North Carolina, one of the few undeveloped sea islands remaining on the Atlantic coast, lying at the mouth of the Cape Fear River. It consists of 9,000 acres of salt marsh and 3,000 acres of high ground, a habitat for fish and birds. The owner, Frank O. Sherrill, sold the land for $5.5 million to real estate developers who intend to build a convention center, golf course, yacht club, motels, and house lots. The state of North Carolina was interested in trying to preserve the area and was ready to meet the owner's price with funds advanced by the Nature Conservancy. But Sherrill rejected the state's offer, in the words of a *New York Times* editorial, "and in effect extended a loan to the developers by granting them a mortgage on the property:

> There is no defense for this unconscionable deal . . . a major salt marsh can no longer be regarded as one man's asset to be bartered for private profit. It is part of every man's natural heritage to be held in trust for the public good.

But such ethical considerations mean nothing unless underwritten by law. While talk goes on, irreplaceable coastal wetlands are being gobbled up at a rate of about 1 percent a year. Along the North Atlantic coast, from Maine to Delaware, 45,000 acres of tidal wetlands were destroyed between 1955 and 1961. North Carolina is losing its wetlands at the rate of one acre a day. Maryland's loss is more rapid: 1,000 acres a year. Connecticut has lost half of its tidal marshes. Almost a third of Long Island's are gone, and there is no assurance that the remainder can be saved. New Jersey has sacrificed some of the finest salt marshes in the world.

Several states have passed laws to try to save what remains of their coastal lands. These legislative acts vary in effectiveness. Massachusetts and Rhode Island were among the pioneers, empowering the state to zone privately owned lands for ecological reasons. New Jersey, after losing 36,000 acres of estuarine land since 1953, passed the nation's strongest law late in 1970. It gives the Commissioner of Environmental Protection the power to regulate all wetlands (those covered by one foot of water at mean high tide), and if necessary to forbid any activities that would alter their character. Delaware, under the leadership of Governor Russell W. Peterson, in June, 1971, passed landmark legislation barring heavy industry from Delaware's coast. The measure preserves the state's Delaware Bay and Atlantic Ocean coastlines for recreation and tourism. It was strongly opposed by the Delaware Chamber of Commerce, the oil and natural gas industries, international cargo shippers, and the United States Commerce and Treasury Departments.

To date most state statutes have not proved adequate. Often they were fragmentary, cumbersome, vague, or failed to provide for enforcement or funds to buy imperiled properties. The courts have also undermined some of the laws by favoring developers. Even New Jersey's new strong law has not proved adequate. Developers are trying to devour valuable estuaries before they can be charted and protected by the state.

The need is obvious: to put the coastal wetlands under federal protection with a strong, uncompromising law. At this time there is no central power to act when the states or local communities fail or refuse to meet their responsibilities. Confusion and bureaucratic red tape immobilize the government; no fewer than seventeen different agencies and bureaus have a finger in the estuaries. One agency may be trying to save a wetland while another is destroying it.

Several bills are pending in Congress, but all are cast in the same weak mold. They offer grants to the states as incentives to develop or implement management plans for coastal areas, but do not make any state action mandatory. The government's role would be primarily that of passive referee. This is not only meaningless, but virtually an invitation to continue the exploitation and destruction of the vital wetlands.

One serious effort was made at the federal level to save the wetlands. Representative John D. Dingell of Michigan tried to get through a measure that would have enabled the Secretary of the Interior to veto any project that was damaging to the natural

value of all wetlands. But it suffered the unhappy fate of practically every piece of environmental legislation: it was emascu-
lated by commercial interests and those who believed that the
Almighty created nature for the convenience and profits of
private enterprise. As it emerged, the bill provided simply for a
national inventory of estuaries. This follows a familiar scenario:
when a private interest is threatened, order another study. Delay
follows delay. Finally, the studies become little more than post-
mortems.

The best of laws have not touched the second great threat
against estuaries: pollution. Most of the chemicals and sewage
wastes pouring into the ocean are dumped by rivers emptying
into coastal waters. The full impact is concentrated in these
shallows where most of the life of the sea exists. It takes only a
relatively small amount of insecticides, petrochemical wastes, or
oil to knock out or disrupt an entire ecosystem. Usually the deadliest substances are invisible and unsuspected until the damage
is done. We hear a great deal about the spectacular accident—a
big oil spill that fouls a beach, or a massive fish kill—but it is
not these publicized local disasters that are significant, says Dr.
Peter Korringa, director of the Netherlands Institute for Fisheries Investigation. Rather, it is "the general trends, the stealthy
deterioration of environmental conditions in sections of the sea
of vital importance for its living resources."

This slow degradation continues, usually invisible, day by
day. Often it can be noted only by indirect evidence—sick and
dying fish, foul odors, putrid miasma rising from the waters. Of
nearly 23 billion gallons of industrial wastes dumped annually,
only 29 percent receive any treatment. Sewage sludge, as noted
before, contains significant amounts of pesticides and poisonous
heavy metals. There are also radioactive wastes. Dr. Edward D.
Goldberg, of the Scripps Institution of Oceanography, is quoted
in the *National Fisherman* as saying that "chemists can isolate
man-induced radioactivity in any 50-gallon sample taken anywhere in the sea." No one, he says, has the vaguest concept of
what this man-made chemical experiment will produce. Another
investigator was unable to find an uncontaminated sample of
surface seawater within 300 miles of the California coast that
did not contain added lead.

Into the coastal waters pour millions of pounds of phosphorus
and nitrogen, from detergents and fertilizer runoff and other
products and chemical uses. These compounds encourage algal

growth. They use up oxygen in the water and destroy marine life, bringing death to the water and gradually filling it with accumulating solids. From outfalls pour sludge and silt; this vomited suspended matter adds to the problem, reducing sunlight penetrating the water, inhibiting and killing microscopic plants that are the base of the food chain for the entire sea.

The disaster is greatest in industrial areas, where the tragedy of one is the story of all. There is a difference in degree only in the volume and composition of effluents pouring into Boston Harbor, the Houston Ship Channel, the radioactive wastes Great Britain dumps into the Irish Sea, the billion gallons of unprocessed sewage and chemicals that Russia annually pours into the Ob River.

All this degradation of water, land, and air has come with devastating swiftness. Only forty years ago Raritan Bay, off New Jersey, was used for bathing, surf fishing, commercial clamming, and aesthetic pleasure for people. Today the bay is fouled and ugly—an industrial horror—ringed with refineries, factories, and smokestacks; it is beset by noxious odors and the sky is blotted out by a suffocating blanket of fumes. The water has an oily scum and a deadly gray pallor. A 1951 analysis of Raritan Bay revealed high concentrations of phenols, formaldehyde, copper, and arsenic. Flora and fauna were wiped out by sewage wastes and oil refinery effluents. The bottom spoil that would hinder shipping was hauled off and dumped into the ocean beyond Sandy Hook. Since then the problem has worsened with intensified populations and multiplying chemicals.

Elsewhere it is much the same. Industrial pollution has now reached staggering proportions, as *Newsweek* reports:

> The Reserve Mining Company, near Silver Bay on Lake Superior's ruggedly beautiful north shore, each day discharges an average of 67,000 tons of waste iron-ore grindings, slowly but surely building up ever-thicker layers of sludge.* Nationally Federal Government experts estimate

* The Environmental Protection Agency asked the Justice Department, on January 20, 1972, to file suit against the Reserve Mining Company to force it to stop polluting Lake Superior. EPA stated that for fifteen years the company has been dumping 67,000 tons of taconite tailings a day into Lake Superior at Silver Bay, Minn. Reserve is owned on a 50-50 basis by the Republic Steel Corporation and the Armco Steel Corporation. For two and a half years the federal government "tried unsuccessfully to persuade the company to dispose of the fine tailings on land," reported *The New York Times*. EPA officials said the decision to bring suit against Reserve was difficult because of political complications. "Willis Boyer,

the food, textile, paper and chemical, coal, oil, rubber, metals, machinery, and transportation industries spill a staggering total of 25 trillion gallons of waste water annually. Just four steel mills in the area south of Chicago pollute Lake Michigan each day with one billion gallons of waste water containing 429,000 pounds of suspended solids, 650,000 pounds of oil, and 66,000 pounds of suspended iron filings.

What this can mean to a body of water is suggested by the fate of Lake Champlain. Since the late nineteenth century a paper mill located on the New York side has been dumping its wastes into the water. The wastes now form a 300-acre mass of sludge that oozes across the bottom from the mouth of Ticonderoga Creek; in some places the sludge bed is 20 feet thick. Vermont is starting a legal action to force New York, the mill, the government, or somebody to clean up the lake that it shares with New York. It charges that the mass not only robs the water of oxygen needed to support plant and animal life; in hot weather, huge chunks break away from the bottom and are washed to Vermont beaches. A new paper plant is replacing the old one that did the damage, but the mess remains; the cost of the cleanup is put at over $2 million. A company spokesman said, "There are other towns and companies discharging into that area and we do not believe it is our doing alone." The suit will set a precedent of one state suing another to clean up an environmental mess. At this time there is no law requiring a polluter to clean up existing wastes. Up to now efforts were aimed mainly at stopping current pollution; little attempt was made to assess the responsibility for past actions.

The greatest assault of waterborne pollutants is against the ocean. "All pollution, including that of motor vehicles, ends up

---

chairman of Republic, is vice chairman of the Ohio Republican Finance Committee, and C. William Verity, head of Armco, is a Republican fundraiser." Armco is said to have contributed $12,000 to the 1968 Republican campaign. Last September, the *Times* stated, "a Federal District Court in Texas ordered Armco to stop dumping into the Houston Ship Channel 975 pounds of cyanide a day." Verity, according to the account, gave a speech soon afterward relating how, after the decision, he had written to the President and had made contact with Secretary of the Treasury John B. Connally. Subsequently the company, on November 4, 1971, was given until the following July to cease the discharges. "Mr. Verity was quoted by *The Houston Business Journal* as telling the Houston Rotary Club, 'I pray every night for John Connally.'"

in the sea," says Captain Jacques-Yves Cousteau. One estimate holds that mankind has added a half-million different substances to the ocean, many of them harmful to ocean life.

The impact is largely concentrated near the shore in about 0.1 percent of the ocean's total area. From this small portion comes about half the world's catch of fish. Many of these areas are now closed because of contamination. More than a tenth of the 10.7 million square miles of shellfish-producing waters bordering the United States have been declared unusable. Shellfish are excellent barometers of what is in the water. An oyster pumps large amounts of water through its system in a day; as it takes in oxygen, nutrients, and pollutants, it concentrates various poisons in its flesh until it becomes a biological time bomb (see "Common Environmental Poisons," Appendix). Like clams, oysters often thrive on tainted fare, but they can kill and maim birds and fish that are foolish enough to eat them. Oysters sometimes have an oily taste; their flesh may be greenish rather than white from copper wastes that marinate them in the bay. Clams, concentrating organisms, have caused hepatitis (a liver infection) in humans. Shrimp are highly vulnerable to DDT and certain other pesticides; often they are first to disappear from polluted waters.

In 1908 the production of shellfish in 19 coastal states was two and a half times what it is now. Chesapeake Bay, which formerly produced up to 12 million bushels of oysters per year, is now down to 1 million and the amount is getting smaller each year. Production of shellfish in Connecticut dwindled from $40 million in 1950 to $3 million in 1966. Forty years ago Jamaica Bay in New York had a major oyster, clam, and commercial fishing industry which today is almost nonexistent. San Francisco Bay, which once proudly boasted of the Pacific Coast's biggest fishing fleet, producing millions of pounds of shellfish, is now 90 percent closed to their harvest; its once iridescent waters have been debased into a vast cesspool, a receiving tank for garbage, wastes, and landfill—one third of it filled in, the rest made barren by thousands of tons of greases, oils, acids, and sewage that pour into it daily.

The fortunes of fish have also gone downhill. This is particularly true of the more valuable game fish. On the Atlantic Coast a dozen of the most prized varieties have almost disappeared. In the past seven years alone, commercial fishing off the New York–New Jersey coast (where 88 percent of the country's ocean dumping occurs) has dropped from 673 million pounds to 133 million, an 80 percent decline. In the same period oyster production has fallen off 43 percent.

Some fish cannot tolerate pollution. Others that must go into freshwater to spawn, such as the Atlantic salmon, often are unable to continue instinctive breeding patterns because they no longer can sniff out their home waters owing to changes brought on by pollution. The disappearing species include the spot, sheepshead, and Spanish mackerel. One biologist asks, "Are saltwater fish in the United States on the way to being exterminated?" Seventeen species of fish caught off the Atlantic Coast have been found to be suffering with the bacterial disease "fin and tail rot." Up to 80 percent of the bluefish and 50 percent of the fluke suffer from it.

The financial loss to commercial fishermen and sports fishing interests has been severe. Between 1955 and 1965 the catch of estuarine-dependent fish off the Atlantic and Gulf states dropped from 393 million pounds to 291 million. In the last half of that decade the catches of 18 Atlantic coastal species fell by as much as 50 percent. The loss from reduced sports fishing is in the billions. Florida, which has suffered heavily, says fishing alone brings the state more than $1 billion annually. Several states have tried to place a dollar value on their coastal lands for purposes of compensation due to loss, but it is an arbitrary figure. How can a price be placed on the priceless?

The loss of marine life may be only an omen of what is in store as the chemical assault continues against the oceans. A three-year study of Long Island Sound disclosed a rising incidence of birth defects in a colony of terns, believed by researchers to be caused by pollutants in the Sound's waters. The abnormalities were first noticed in 1969 among young roseate and common terns on Great Gull Island at the eastern end of the Sound. There were chicks born without down feathers or that later failed to develop adult feathers, a four-legged chick, birds that lost their primary and secondary feathers (which are needed for flight), one born with stumps for legs, and others with crossed mandibles and too-small eyes. The defects were likened to the birth deformities caused by the sedative thalidomide, used by pregnant women in the late 1950s.

The researchers, Dr. Robert Risebrough, an ecologist at the University of California at Berkeley, and Helen Hays, an ornithologist at the American Museum of Natural History, called Long Island Sound one of the more polluted areas of the world's oceans, "a distinction it shares with the Baltic, the Mediterranean, and the Southern California coastal waters." They urged that a study be made to determine what pollutants are in the Sound.

In researching this book I visited many estuaries and marshes. I saw them in many seasons and in many moods. I recall a marsh on Long Island. My host was Tony Taormina, regional supervisor for the New York State Conservation Department. Many conservationists are dedicated. Taormina is impassioned. He showed me the good and the bad. The good was Mt. Sinai Harbor, about halfway out on Long Island, off the north shore, near Setauket. The bad was Port Jefferson, a few miles south.

Port Jefferson Harbor is one of Long Island's busiest ports. It is a microcosm of the problems of modern industrialized society. It is full of boats, expensive yachts, and more modest pleasure craft. The water is skimmed with oil and adrift with litter. There are barges and dredges. On the shore are restaurants, shops, huge oil storage drums, enormous mounds of traprock and sand to feed the area's unending construction. On one slope is the Long Island Lighting Company plant, an ancient red brick building, its smokestack spewing dark smoke; from the plant run cables supported on huge steel towers that slash across the land.

The plant discharges its effluent and heated water into the harbor, raising the temperature of the receiving waters, according to Dr. Charles Schnee, a retired physician who lives in Setauket, now a public-spirited campaigner for a better environment. Dr. Schnee is also a critic of an adjacent sewage plant; it was built in 1939 and is long antiquated and overburdened, providing only primary treatment. He says an underwater pipe from the plant is broken and the effluent is discharged only 50 feet from the shore, adding to the progressive deterioration of Long Island Sound. Port Jefferson is blasted by whining roar of jets going to and from Kennedy Airport. "Here you see the unnatural world," said Taormina. "Crowded, smoking, joined together, noisy."

When we got out of the car he had locked the door. "It's all part of the same thing," he said wistfully. He stared moodily at the harbor. "There's a correlation between human density and degradation. You can't preserve your values when you stand jowl to jowl. . . ."

We drove the short distance to Mt. Sinai. The difference was startling. Here was water that sparkled brilliantly in the sun.

Mt. Sinai begins with a culvert that spills freshwater from the land into the salt estuary formed by the sound. We looked out over the water that sparkled brilliantly in the sunlight. Far from shore, on a tiny patch of green covered by the long marsh grass of late spring, was a solitary white spot. "Swan," said Taormina. "She's been there all spring, nesting. But she's wasting her time. She didn't get back far enough, the water rose and chilled her egg and killed it. Now she just keeps sitting there, week after week, incubating away and waiting for the egg to hatch." It seemed incredible that we were only about twenty miles from Manhattan, with its asphalt covering, its canopy of dirty air and moat of contaminated water. Here there was only the blue water ruffled by a gentle breeze, the azure, unbroken sky, the swan hovering over her dead egg. Nature too makes its mistakes. Taormina interrupted my reflections. He spoke of compromise. "Here you see the great compromise," he mused.

Compromise? "Yes," he said. A company had mined gravel from the estuary for fifteen cents a square yard. Everyone had a vested interest. The company wanted to make money. The local officials wanted to be reelected. The boat owners wanted the channel dredged so they could get in and out at any time without waiting for the tides. The fishermen did not want silt from the gravel mining to kill the fish. The property owners along the shore put in fill to extend their land; each one justified what he did by saying, "What difference does my little bit make?"

"And this is my vested interest," he said, taking in the estuary with a sweeping gesture. "I want to keep it like it is."

"One reason I came here was to enjoy Mount Sinai. I would come here with my two young sons, and in twenty minutes we would get enough shellfish for two meals—hard clams, soft clams, blue mussels, razor clams, periwinkle, even oysters. But all we saw, day by day, was the dredges gobbling up sand and gravel and destroying the basic values that were prevalent here —the shellfish and wildlife—and no end in sight."

He said there were a series of confrontations. In the end there was "a great cry from the people." The company was forced to leave. "Now what happens to a place like this? The old guard says, 'Let it stay as it is.' The big property owners want to make money out of it; they say, 'How can you convert it to dollars?' So we tried to make a semblance of organization. We asked ourselves: 'What are the amenities of the harbor? How do we distribute them—give something to everybody?' More compromise . . ."

Compromise ultimately produced a plan—an attempt to find a point of equilibrium—that would permit people to exist harmoniously within a "natural" environment. The plan was designed to "protect the environment from the ever-encroaching development and commercialization of the island. As the miles of blacktop and concrete stretch over the horizon, will there be corresponding areas of quiet relief to calm the senses and refresh our minds? Or will urbanism completely dominate the environment and consume man's essential natural heritage?"

You have to define your values, says Taormina. The plan provides for trying to keep the estuary as it is, one of the finest in the world, unpurchasable and irreplaceable, under public ownership. It would preserve the salt meadow where the swan nested. The relatively pure water and tidal flats would continue to produce edible shellfish and fish, waterfowl and other wildlife. It provides for a limited number of boats ("Eight hundred, not thousands, and those on a first-come basis"). Fishing is permitted. There is an area for bathing. More important, the plan leaves untamed and unmanaged a large portion of the estuary that is bordered by the undulating sand dunes and irregular beach area; there will be wild patches covered by red cedar, bayberry, beach plum, black cherry, and even poison ivy. In areas that have already been filled in, healing plants are beginning to take over: saltbush, saltwort, miscellaneous weeds and grasses, and patches of salt meadow grass. Together it provides a sweet diversity that will attract and protect wildlife and delight nature lovers and children in their explorations. There will be the open horizons that are necessary for the eyes and the soul.

The plan, signed by Taormina, his boss, William G. Bentley, Regional Supervisor of Fish and Game for New York State, and members of the committee that helped formulate it, concludes with a statement of philosophy and optimism rare in these troubled and often despairing times:

> It should not be necessary for the present generation to feel that it must plan for the ultimate development and use of the area. Let us plan only that there will be hundreds of

generations ahead of us, and that to each succeeding generation, uncluttered and undeveloped landscape will become increasingly valuable and vital. It is in the hands of this generation to see that such land will be left for them to be used according to their wisdom.

RIVERS, STREAMS, AND LAKES

The Passaic River in northern New Jersey is not known for its length, and has never been celebrated in song or fable. Other streams have been immortalized for their breadth, beauty, majesty or because they stirred some deep spiritual sense. The Passaic's only claim to fame is that it has been called the dirtiest stream in the United States.

Old-timers can recall the Passaic as the scene of family outings, a place for swimming, gay boating regattas. They lament its loss of innocence, which also may be theirs. Its most eloquent biographer, perhaps, is Sid Moody, an Associated Press correspondent. He traced the Passaic's corruption from its gentle upland birth, through a series of disasters and attempted rejuvenations, to the ultimate catastrophe when it dumps into the cesspool known as Newark Bay—a body of water so hopelessly polluted that a wag said of it, "You don't sink in Newark Bay; you walk on it."

"Upland ridges. Wood. Red barns. Forget-me-nots. Cool freshets. Ponds. Ducks."

At the end: "Lime flotsam. Garbage. Oil scum. A flowing desert, a Styx. It takes but 90 miles to be born and die. It dies from pollution. And it dies from ignorance, from indifference, from illegality, from narrowness of mind, from parsimony, from the many who do nothing and despite the many who do much."

Men have always had a strong emotional attachment to water. Possibly this is because we are vitally dependent on it. Perhaps we retain some mystic attachment to water because it is our common ancestral home; the oceans that are believed to have given us birth and from which we took the juices for our bodies are the same oceans we now go to for physical recreation and spiritual restoration. Is it significant that the oceans cover some two thirds of the earth, and a man's body is made up in almost the same proportion of solids to liquids, as if reflecting a cosmic balance?

It is fashionable nowadays to talk about the endless riches of the sea, says Cousteau. "The ocean is regarded as a sort of bargain basement, but I don't agree with the estimate," he told *The New York Times:* "People don't realize that water in the liquid state is very rare in the universe. Away from earth it is usually a gas. This moisture is a blessed treasure and it is our basic duty, if we don't want to commit suicide, to preserve it."

Our allotment is 317 million cubic miles of saltwater, enough to fill a giant cube with each dimension slightly less than 700 miles, just enough to cover Utah, Colorado, Arizona, and New Mexico, according to the *National Fisherman*. It is marvelous stuff, this water, odorless and colorless, two parts hydrogen and one of oxygen bonded together, the molecule so infinitely small that it takes a trillion trillion to fill a one-ounce glass.

Water is necessary to all life. Its distribution determines civilizations and life patterns. To maintain this life the water must be delivered in an even, steady flow. Throughout history centers of commerce have tended to settle on major waterways. Most rose and fell with the fortunes of the adjacent water. Often this reflected the way they treated it. In the United States, as in most countries of the world today, every major city is on a major waterway.

We know much about water, its composition and effects, but behind our superficial knowledge is profound mystery. What formed water? What brought about the bonding of two molecules of hydrogen and one of oxygen? There are many theories. The story, as we infer it, is a dramatic one. As the molten rock hurled off by the sun began to cool and solidify, thrusting and heaving, forming the great depressions and peaks that were to become valleys and mountains, gases bubbled out of the flaming interior of the globe to surround it. These gases formed a veil of air, a vapor that cooled and formed as clouds. The clouds chilled, water condensed, the first rains fell. The falling water struck the seething rock surface, rose as great clouds of steam and fell again. For millions of years it rained steadily, or so it is believed, and this downpouring of water leveled mountains and ranges, chiseling and gouging out the deep irregular basins that were ultimately to become our oceans.

Water obeys laws that it alone recognizes. It answers to its parent, the sun. The sun beats on the earth. Water evaporates from the great ocean reservoir into the air. While the amount of water is finite, it is ever in motion, circulating about the globe in an endless cycle between earth, sky, and ocean; the same mole-

cule is on a perpetual journey that takes it through soil, rock, ice,
rivers, vegetation, animals, people, and the sea.

Once airborne, water tends to form droplets around dust from
spent stars, volcanoes, and other matter in transit. The liquid
droplets condense to form a protective shield against the searing
rays of the sun. As the filtered sun strikes the earth, the reflected
heat warms the adjoining mass of air. The vapor thickens as
clouds form. The clouds are driven by the winds, churned by the
spinning of the earth and gravitational forces. A mountain ob-
trudes. The moving cloud strikes it and is deflected upward, or it
otherwise collides with a mass of cooler air. Now incapable of
holding its excess moisture, the chilled cloud spends itself as
rain, snow, sleet, or some other form of precipitation. The droplets tumble through the atmosphere, picking up nutrients. A recent theory holds that microscopic life from the atmosphere
gathers to live in clouds, which become floating ecosystems, and
there they manufacture vitamins and other nutrients.

The rain strikes the earth, enriching it. Some immediately
evaporates back into the atmosphere. The rest runs off or is absorbed by organic matter on top of the soil, which acts as a huge
sponge. Some of the water is returned to the atmosphere locally
through the complicated process of transpiration; it is drawn up
by plants and trees, and the green vegetation serves as so many
fountains that "spray" water into the air, cooling, refreshing,
affecting local climate. The rest of the water slowly percolates
through the purifying filter of soil and rock, replenishing ground
waters, seeping into buried lakes, streams, springs or freshets,
eventually joining rivers that empty into bays or estuaries. Ultimately all water returns to the open sea, where once more the
endless cycle begins, the water continually used and reused, intermittently purified by the process of evaporation powered by
the parent sun.

Water has the extraordinary ability to dissolve a greater
variety of substances than any other liquid. Falling through the
air it collects atmospheric gases, salts, nitrogen, oxygen and
other compounds, nutrients and pollutants alike. The carbon dioxide it gathers reacts with the water to form carbonic acid.
This, in turn, gives it greater power to break down rocks and soil
particles that are subsequently put into solution as nutrients and
utilized by growing plants and trees. Without this dissolving
ability, our lakes and streams would be biological deserts, for
pure water cannot sustain aquatic life.

Water dissolves, cleanses, serves plants and animals as a carrier of food and minerals; it is the only substance that occurs in all three states—solid, liquid and gas—and yet it always retains its own identity and emerges again as water. This is a key to its indestructibility.

It cleans itself. Some pollutants are bleached out by sunlight. Others sink to the bottom to settle there; they may be sealed off by sediment or scale. The rest are consumed by beneficial bacteria. These organisms, using oxygen in the water, convert organic matter (carbon, nitrogen, and phosphorus) to inorganic forms (carbon dioxide, nitrate, and phosphate). This inorganic material is absorbed by plants. The plants, using sunlight and carbon dioxide in the water, produce new oxygen for the bacteria, fish, and other marine life. At the same time they purify the water. The plants are eaten by the animals. The animals produce carbon dioxide and wastes that are attacked by the bacteria of decay. These wastes support new growths of algae, thus completing the cycle. The cycle is perpetual, an interwoven chain that revolves around the four basic elements of life: carbon, oxygen, hydrogen, and nitrogen. Each organism is dependent on the rest. The system cannot be broken down or any part isolated. If one part breaks down the entire system collapses. The point is illustrated by John Stewart Collis in *The Triumph of the Tree* with a simple experiment. He tells us to place a bowl of water so that the sun can strike it, and put in a snail, a fish, and a waterplant; the population of three in this cosmos will thrive for months by virtue of mutual exchange:

> The fish lives on the plant. The waste of the fish is prepared by the snail so that it can be manufactured by the plant which uses the waste of both fish and snail for its own purposes, and in so doing releases oxygen that purifies the water and guards the animals from suffocating. When the physical and chemical equilibrium is thus maintained by the action of different creatures, we get what ecologists call a balanced environment. Remove the snail, and the plant will drop and the fish will fail. Remove the fish, and there can be no further exchange between the plant and the snail. Remove the plant, and neither fish nor snail can pasture. It looks like three in one and one in three. But of course it is more than that. We must not forget the water. We must not forget the sun. Take their action away from the fish, the plant, and the snail, and again this cosmos will totter to its foundations.

The secret is balance. Each element in the right proportion; the system must not be overstressed by too much or too little of each. Large quantities of wastes with their overabundance of nutrients can cause an explosion of plant growth. This causes the greedy bacteria to overmultiply. Plants proliferate even more wildly. Now the sunlight cannot penetrate the thick plant mass in the water. Bacteria use up the available oxygen faster than it can be replenished. Crisis deepens into disaster: the fish and other animals die, plants die, and the bacteria suffocate. The dead matter decomposes and sinks to the bottom. Putrefaction and death ensue. A miasma of gases and foul odors rises.

In the same way poisons injected into a stream can kill the microorganisms, plant or animal life, disrupting or killing the system outright. A severely damaged stream or body of water can take a long time to recover. Biologists are not sure if, after the most severe assaults, the original condition can ever be restored. Numbers may be as great as ever, but the variety is reduced, the richness is gone. In badly polluted water diversity is replaced by a few rough species that can live with little or no oxygen. It is the difference between an estuary and a cesspool. Those who have never known the richness and diversity of life accept this poverty of variety as normal; for them the loss may be greater as they have never known better.

There are said to be 12,000 different toxic chemical compounds in industrial use today, and more than 500 new chemical compounds are developed each year. More and more of these are finding their way into the nation's water supply. *The New York Times* reports:

> Thus water pollution is getting more complex each day and includes a growing roster of weedkillers, pesticides, fungicides, phosphates from detergents and fertilizers, trace metals such as mercury, lead and cadmium, acid from mine drainage, antibiotics, hormones and perhaps carcinogens— cancer-producing agents. Water treatment facilities simply cannot cope with this increasing chemical crush.

Pollution is now so extensive that it affects virtually every stream in America, according to the government. The Mississippi River, which Mark Twain wrote of with lyrical beauty, is now known as "the colon of mid-America." Water samples taken from the river near St. Louis were so toxic that when diluted with 10 parts of clear water, fish placed in the mixture died in less than a minute. When samples were diluted 100 times, the test fish still perished within 24 hours. This led to the caustic comment: "Fortunately the people living south of Saint Louis have more rugged constitutions than do fish."

A government study released in late 1971 reports that waste dumped into the Mississippi River is not only destroying America's greatest river but also threatening aquatic life in the Gulf of Mexico and human health in southern Louisiana.

Surprisingly little is known about the specific dangers to man and other forms of life from the hundreds of man-made substances dumped into the Mississippi, notes *The New York Times*. "Day by day and year by year, no government agency at any level tries to keep track of the wastes dumped into the Mississippi system and their impact on people, wildlife and plants. Only minimal and erratic efforts are made to measure the amounts of oil, heavy metals, organic chemicals and other toxic substances that may be in the seafood eaten by thousands of people in this area each day."

The Environmental Protection Agency found 46 organic chemicals in the drinking water of New Orleans and Carville, a small community up the river, after the water had been treated and presumably purified, according to the article. "Two of those chemicals are thought to cause cancer. Four others have caused changes in the tissues of experimental animals. What the 40 other chemicals might do is largely unknown." The *Times* quotes the government study as stating:

" 'The health and well-being of 1.5 million people who drink water from water plants using the Mississippi River as the source of raw water may be endangered by the discharge of industrial wastes containing materials known to have toxic, carcinogenic, teratogenic or mutagenic properties.' "

A previous study, in 1959, had found that residents of New Orleans had three times as great an incidence of bladder cancer as those of Atlanta or Birmingham. The New Orleans drinking water, which comes from the Mississippi, was suspected.

New industrial sites are constantly being bought along both banks of the river, carved from land long used for sugar cane plantations. The EPA found that industrial plants already established are contaminating the river with cyanides, phenols, arsenic, lead, cadmium, copper, chromium, mercury and zinc (see

Appendix). The Mississippi not only furnishes nutrients for the Gulf; it has also become its greatest source of man-made poisons.

"The agency has identified 89 organic chemicals in either the industrial waste dumped from the industrial plants or in the drinking water treated by 40 water treatment plants along this part of the Mississippi.

"The study says that 37 manufacturing plants dump at least five pounds a day of at least one heavy metal such as lead or mercury. These concentrations 'may endanger human and the life of the aquatic biota,' the report says. Lead, which is highly toxic in large amounts, causes the greatest concern. The agency has found that 27 industrial plants dump from 5 to 3,700 pounds of lead a day into this section of the river."

Mercury dumping reportedly has been reduced but some still continues. There are also raw sewage and inadequately treated sewage wastes, pesticide and fertilizer runoff.

The Missouri, second only to the Mississippi in length and glory, is written off as an open sewer. The Cuyahoga in Cleveland, an arm of the dying Lake Erie, is so polluted with oil that it frequently catches fire and requires firebreaks and a fire patrol. The Justice Department has filed a civil suit against the Jones & Laughlin Steel Corporation for allegedly discharging into the Cuyahoga more than 900 pounds of cyanides daily. Investigations are reported under way into similar complaints against United States Steel and Republic Steel. The once "Beautiful Ohio" contains hundreds and perhaps thousands of dangerous pollutants. The Potomac River, which threads through the nation's capital, is described in a government report as "a severe threat to the health of anyone coming into contact with it."

The zenith of pollution may be the 50-mile Houston Ship Channel, which runs from downtown Houston to the Gulf of Mexico. In a three-month period in 1971 some 200 workmen were felled by yellowish, noxious fumes that hang over the channel. A Presidential board called the channel "an amazing demonstration of a city polluting itself into extinction." The Texas Water Quality Board has granted 277 permits to dump effluents into the channel, which results in some 315,000 tons of suspended solids being dumped into the waterway each day, according to the Environmental Protection Agency. About half of this finds its way into Galveston Bay, the largest estuary in Texas;

the other half settles to the bottom. Much of the metal and raw sewage is found in shellfish, tests disclosed.

Wastes dumped daily into the waterway are estimated by EPA to total 1,600 pounds of lead, 5,000 pounds of cadmium, 7,900 pounds of zinc, 300 pounds of chromium, 400 pounds of phenols, 100 pounds of cyanide, and 55,000 pounds of oil and grease. In addition, an estimated total of 215 million gallons a day of domestic waste, including 110 million gallons that are inadequately treated or not treated at all, are dumped into the 44-foot-deep, 300-foot-wide ship channel each day.

The Texas board requires that all waters be "substantially free" of oil. Nevertheless it has granted 81 permits to dump the 55,000 pounds of oil and grease that flow daily into the channel. Oil and hydrocarbon residues in oysters range from 23 to 26 parts per million (ppm) in the approved area—where they may be harvested—to 237 ppm in a prohibited zone.

At a hearing on the Bay, called by the EPA, Congressman Robert Eckhardt of Texas said the state had spent $1,401,000 and the federal government $565,000 studying the ship channel and Galveston Bay pollution, "but not one cent for enforcement of pollution standards." He added, "In Texas, studies have been used for one major purpose, to delay enforcement." As soon as one state study was completed, he said, another was started to make sure that the preceding study was correct. The Texas water board was deeply resentful because the government had interfered in what it considered a local problem.

Texas' pollution is not altogether unique. The Merrimack River in New England is so contaminated that drops of spray thrown up by motorboats are considered dangerous, according to Leonard Wolf in *Environment*. The suspended solids settle on the bottom of the Merrimack and decompose, sending gases bubbling up to the top as if the river were cooking:

> These gases often contain hydrogen sulfide, which smells like rotten eggs and can cause expensive damage by discoloring the paint on houses and boats. The river is completely opaque and usually black in color, though it is occasionally gray, milky white or colored, depending on the particular variety of industrial wastes that are currently predominant. Oil slicks, grease balls, black oozing sludge, slimes, fecal debris, and condoms may float by on top of the stream.

How did our waters get to such a point? Why do we permit such conditions to persist? The answer is complex, but primarily it revolves around economics, as most pollution problems do. To clean up the mess would cost billions every year. To spend such huge sums would require a complete change of present attitudes and values. We would have to realize that flushing a toilet is not just a sanitation measure. It becomes a moral issue. It is not the problem of the person flushing the toilet but of the one who lives downstream and drinks the water. The only ones willing to spend money to clean up the rivers to date are those who live farthest downstream. But suddenly a horrible realization is taking root: we have created a world in which we all live downstream.

The problem itself is simple enough: we are putting more filth into the water than we are taking out. We are building treatment plants that are inadequate and outdated before they are built. We have set minimum standards of cleanliness and are failing to meet even them. We have long forfeited on the time schedules. Many states and communities have refused to spend the necessary monies. There is more arguing than action. The federal government has failed to keep its commitments; it is spending pennies where it should be spending millions. Much of the entire program has broken down in politics, squabbling and corruption.

In many ways New York is typical of the nation. The state approved a $1 billion Pure Waters bond issue in 1965 and the people were told that New York's waters would be clean enough to swim in by 1970. In 1967 Governor Nelson Rockefeller asked the federal government for a delay until 1972. According to former Representative Richard L. Ottinger, in a 1969 congressional address, "there appears to be no date. Instead we have glowing reports of the success of the program."

Ottinger, in a speech before the Sierra Club on February 2, 1970, recalled a Rockefeller speech only a few days earlier in which the governor noted that the Hudson River, "this majestic waterway," has been slowly suffocating from 757 million gallons of raw or poorly treated sewage poured into it daily from 108 polluters along its shores. He noted that Rockefeller declared that every major polluter is now under an abatement order and following a timetable for completing sewage treatment facilities. Does that include municipalities? asks Ottinger. Does it include New York City, which alone pours 365 million gallons of untreated sewage into the river every day. Does it include the hundreds of towns and villages along the Hudson that use the "majestic river" as an open sewer. What *is* the timetable? Will it mean a clean river by 1972, the date promised by the governor three years ago (1967).

Then this thrust by Ottinger:

> If everything is going so well, why is our water so dirty? As a lifelong resident of the Hudson Valley, I invite the Governor to join me in what used to be a great institution—a friendly swim in the river. He can choose his date—1970, as he first promised; 1972, the last date he's mentioned, or any future date. The only condition is that he get in first, and I mean *in;* it's no fair just walking on the water.

The tragedy of the situation we face today is that "pollution suddenly makes good politics," says Ottinger, "but politics, so far, only makes for more pollution. And the danger is that the rhetoric and promises will assuage the demand until it is too late."

To put the problem in context, a little technical background information is necessary. We start with sewers. Septic tanks and cesspools cannot work in densely populated communities; in many rural areas the soil is now so saturated from these facilities that groundwaters have become contaminated, creating serious sanitation and health problems. Sewers are expensive to install. Only about 140 million Americans are served by any kind of sewers. Often these are outmoded, nonfunctioning, or faulty. Some 2,139 of the nation's 11,420 communities have sewer systems that flush raw sewage directly from the toilet into the nearest stream or lake.

There are two kinds of sewer systems: combined and separate. Older communities, such as New York City, almost invariably have combined storm and sanitary sewers. This consists of a single pipe network that carries both sewage and street runoff. Separate systems have double pipes, one for sewage and a second for street and storm runoff. A combined system is generally adequate in dry weather, but in storms the runoff is greater than most treatment plants can handle. The excess then bypasses the plant, pouring directly into the receiving waters, carrying filth flushed from streets as well as an estimated 65 billion gallons of raw sewage into lakes and rivers every year. All new sewer systems must now be separated, but to convert the old combined systems would mean tearing up every street in a city. The American Public Works Association estimates the cost of replacing all of the nation's combined systems at $48 billion. To get around this obstacle many plans have been suggested. One is to have holding or settling tanks to retain excess water until it can be processed. A more ingenious proposal, for New York, was to lay enormous plastic bags along the bottom of the Hudson to hold the excess runoff until it could be treated.

Once sewers are installed, another problem emerges: what is at the end of the sewer lines? Is there a treatment plant, or does it merely dump into a stream or lake? With the water in a single stream now being reused eight and ten times, treatment becomes increasingly important. Half of all the nation's wastes goes into streams either completely untreated or inadequately treated, and the other half is questionable. *The New York Times* states:

> Every second, day and night, about 2 million gallons of sewage and other fluid waste pour into the nation's waterways. No way has been devised of measuring the resultant totality of pollution, but nobody has yet ventured any contradiction of President Johnson's 1967 statement that "every major river system in the country is polluted."

There are three main sources of pollution: 'community sewage, industrial wastes, and agricultural runoff. Little has been done about agricultural pollution because it is so diffuse, the *Times* points out. "The abatement of both municipal and industrial sewage has been palpably insufficient."

An estimated 7 percent of the municipal sewage dumped into our waterways is completely untreated, and wastes of only about one third of the population get secondary treatment. Of 600 industrial plants sampled, only 25 percent provided adequate treatment of their own facilities. Of the 280,000 manufacturing businesses in the United States, all but 25,000 discharge into municipal sewers. These wastes may or may not be treated, or they may be inadequately treated. Industry pays absurdly small fees to use municipal sewers, as it pays small fines for polluting air and water. This amounts to a huge public subsidy to industry. Even worse are the firms that discharge wastes with no treatment whatsoever directly into streams. In 1969 a single steel company discharged into Lake Michigan every day, via the Calumet River, approximately 13,750 pounds of ammonium nitrate, 1,500 pounds of phenol, 1,700 pounds of cyanide, and 54,000 pounds of oil. No one knows the amount or composition of most industrial wastes. Efforts to get such information have been successfully blocked by industrial pressures. The only way an enforcement official can find out what is being dumped is by paddling in a boat to an outfall pipe and taking a sample for analysis. Even then he cannot be sure of success. Many of the new compounds being discharged into public waters have not been identified. "The poor water plant operator really doesn't know what's in the stream—what he is treating," says James H. McDermott, director of the Bureau of Water Hygiene in the Public Health Service.

Conservationists say the threat becomes progressively greater because of a loophole in the law. To protect waterways the federal government depends primarily on the Refuse Act of 1899 that prohibits the discharge of injurious substances into waterways. But the act specifically exempts liquid discharges from municipal sewers. At a 1971 enforcement conference conservationists expressed concern that industries could avoid prosecution because more and more of them were connecting their waste pipes to municipal systems.

Many municipalities encourage industries to tie in to their treatment facilities to help pay the huge costs. Toxic wastes not only could overwhelm the municipal systems, say the conservationists, but also make it difficult to trace and prosecute specific industrial polluters. The effect on the waterway, of many toxic substances, would be little different whether the wastes went through a treatment plant or were dumped directly into the waterway.

There are two basic methods of treating wastes: primary and secondary treatment. Half the population in the United States is served by primary and secondary sewage treatment; some 45 percent get only primary treatment; 5 percent receive no treatment. Primary is only screening and settling out the larger solids and in some cases chlorinating the effluent to kill harmful bacteria. Secondary treatment goes a step farther: after the solids are removed, the organic matter is filtered through a bed of rocks or charcoal, aerated and digested by bacteria, much in the way a stream works naturally in breaking down wastes. The effluent may be chlorinated, especially during the bathing season. Good secondary treatment reduces organic matter by 90 percent. The remaining sludge is usually buried or barged out to sea.

The biggest shortcoming of even the best secondary treatment is that it barely touches the enormous problem of toxic chemicals in water. Practically all of the deadly substances that go into a treatment plant will emerge in the same amount. Into drinking water go the pesticides, heavy metals like mercury, lead and zinc, and other powerful toxins. These can be removed only with the most advanced technology—the so-called tertiary treatment, and even it is not wholly effective. Plants of this type cost at least twice as much as a modern secondary treatment plant. Many experts say the present technology is too limited to make the outlay worthwhile.

There is another shortcoming of secondary treatment. Basically it is designed to protect the aquatic life in the receiving waters by reducing the organic wastes to inorganic form. But Dr. Barry Commoner, one of the country's leading environmental scientists, points out that this is a deception we play on ourselves. The organic matter that is transformed to inorganic form is returned to the water, and in this state it is more available to algae and other growth, which quickly reconvert it to new organic matter. This can lead to even more problems than it sets out to correct, according to Dr. Louis T. Kardos, in *Environment*. He explains that all organic waste that eventually returns to the air, the land, or water is broken down by microorganisms and nourishes new growth. Although man cannot escape this cycle, he can violently unbalance it. "Instead of the slow, regular, daily and seasonal cycle, he can separate the cycle's various segments widely in space and time":

> In a sewage-treatment plant, the end products of the cycle, the inorganic minerals, are produced by a highly controlled microbial system and then discarded into the nearest body of water. This stimulates a sudden, more profuse growth of green plants (algae) and the entire subsequent cycle of feeding, waste, death, and regrowth fluctuates wildly. The result may be the crowding out and the eventual disappearance of entire species that once were part of the natural system.

The more secondary treatment plants that go up, the greater the mountain of sludge that must be disposed of. Present methods do not correct the problem as much as they transfer the site of pollution from immediate receiving streams to the universal ocean.

No one really knows what to do about the problem of sludge. New Jersey passed a stringent law in 1971 requiring most of the state's sewage sludge and industrial wastes to be dumped into the Atlantic Ocean 100 miles offshore rather than the present 12 miles. But New Jersey has no legal control over these deeper waters. The Army Corps of Engineers, in charge of ocean dumping, the previous year rejected a New Jersey request to increase the dumping grounds to a 100-mile limit. The Corps of Engineers said it would be too expensive to take the sludge 100 miles to sea and that it would cost at least $50 million more. The problem is further complicated because Philadelphia and Delaware also dump their wastes off the Jersey shore. A Philadelphia water official said his city turned to ocean dumping when no more land area could be found for the sludge. He said that nutrients in the sludge "should be helpful to the fish and other aquatic life."

A federal study suggested that the sewage sludge might be used to restore land ravaged by such practices as strip mining. But Richard J. Sullivan, commissioner of the New Jersey Department of Environmental Protection, said it could not be used for this purpose. "They [the wastes] contain such a quantity and variety of industrial chemicals that they would probably kill everything in sight."

The nation's waters testify to the ineffectiveness of present programs of water control. A 1969 study by the General Accounting Office found that despite an expenditure of more than $5 billion for new sewage treatment facilities there has been no improvement in the quality of the nation's waterways. The study covered eight rivers throughout the country. In every case it was found that the efforts of government were being overwhelmed by continued outpourings of industrial waste.

On a stretch of the Willamette River in Oregon, $2.1 million was spent on waste treatment facilities. The amount of municipally produced pollution was reduced by 20,000 units on a test scale, but during the same period two paper mills were dumping into the same area of the river between 500,000 and 2 million units. In another waterway $7.7 million was spent to cut municipal pollution by 3 percent; at the same time industrial pollution increased 350 percent.

During the past decade tremendous sums have been spent to clean up Long Island Sound. In that period pollution has doubled, according to Representative Ogden R. Reid. The sound is in danger of becoming another Lake Erie, he warns. "The water is twice as dirty as it was ten years ago and the pollution is spreading out from the shoreline." This is blamed on a booming population, the increasing flotilla of pleasure boats that dump wastes directly into the waters, and new industries. Some 179 municipal sewage facilities now pump wastes into the sound; less than half provide secondary treatment. The influx of new people mock a target date to have all sewage receiving secondary treatment by 1972. Each increase of 10,000 in population adds about one million gallons of sewage effluent a day.

During my early sewage explorations, I was perplexed by the inadequacy of these plants. Several times I heard that there had been no improvement in sewage treatment for forty years. I asked one of the country's leading engineers about this. He confirmed the fact and agreed that it was remarkable. "In what other field do you find no advance in the technology in forty years?" he said. And why was this? He snorted. "Because sewage is a total political venture. That inhibits innovation. To get a contract for a new plant you must grease the palm of the right official. Sewage belongs to the communities. It's a filthy, dirty business, worse than the stuff handled. Politicians have not wanted innovations. There's too much money in the status quo. They're glad to leave it the way it is." He smiled and shook his head. "It should be taken out of the hands of local officials. My old professor said we drink chlorinated pisswater. He was right. It's not even good dilution. We deceive ourselves. In Cleveland the major intake for drinking water is surrounded by two sewage outfalls a mile or so away. It's ridiculous. . . ."

The quality of water deteriorates with use. We now use about 350 billion gallons of water a day. By the end of this century usage will soar to a trillion gallons a day. But there are not a trillion gallons of freshwater to be had each day. This means that the same water must run through more digestive tracts. It must be recirculated through more industrial processes. It will percolate through more soil, wash off of more streets, go through more sewers, carry more filth and disease germs.

The problem of sewage disposal is rooted in outmoded and primitive ideas and techniques. A toilet uses about 6 gallons of water each time it is flushed. This is wasteful and breaks the natural recycling system of nature, while polluting the water. "It is insane to be taking all of this organic matter off the land and dumping it into the water," Dr. Commoner says. And Dr. Donaldson Koons, chairman of Maine's Environmental Improvement Commission, lists the flush toilet along with the automobile as "crimes against humanity." Dr. Koons says, "We have lagged dreadfully behind in devising systems for separating contaminated water from relatively clean water."

Because of the flush toilet, he explains, we have to design water systems to provide 100 gallons of water per person per day. If the flush toilet were not in general use, one person would, on the average, use only about 10 gallons of water a day. A federal official noted that the flush toilet system is inefficient because "it uses a whole lot of water to carry away a little bit of waste." He explained that sewage was actually about 99.9 percent pure water and that the billions being invested in sewage plants were being spent just to remove the small fraction that makes it polluted.

In many cases we already hoist a glass of liquid that only recently went down somebody's toilet. The time between toilet and tap will get shorter as water usage increases. In some parts of the world, the future is already at hand. Windhoek, the capital of South-West Africa, recently had the distinction of becoming the world's first city to recycle its wastes directly into drinking water. This is an engineering euphemism for drinking "chlorinated pisswater." In the United States we are more squeamish than the people of Windhoek. We still prefer to drink other people's wastes. The astronauts could have "recycled their own wastes," but they took along the means of synthesizing their own drinking water to avoid this additional psychological stress.

Fish are often used as a barometer of water quality. In 1968 alone an estimated 15,236,000 fish were killed in 42 states by identifiable pollution sources, according to the Federal Water Pollution Control Administration (FWPCA). This includes only reported kills. Usually the cause of death is some violent poison dumped into the water, perhaps accidentally; the effects make bizarre reading. One government report refers to some 5,000 rainbow and brook trout that "slowly died a few days after the application of carbine herbicide to an adjoining field. The eyes of the fish were white prior to death. The fish gasped and struggled at the surface before succumbing."

Industrial accidents are also becoming an increasing threat not only to fish but to drinking water supplies. In the summer of 1971 the GAF Corporation in Rensselaer, New York, reported that it accidentally dumped more than 7,000 gallons of a poisonous chemical into the Hudson River. The company said the accident occurred when a valve in a line leading from a tank storing aniline was left open by mistake. A state official said the level of aniline in the Hudson as a result of the accident posed "absolutely no danger" to the drinking water of Poughkeepsie, 65 miles downstream, the closest community drawing water from the river, the *Times* reported:

> Figures provided by the Army Corps of Engineers show that 5 parts of aniline for each million gallons of water would be the danger level for drinking water. According to Mr. Stratton [Charles H. Stratton, manager of the plant] the accident resulted in an estimated 1.6 parts for each million.

As water gets progressively worse, health risks increase proportionately. A few years ago the only waterborne viral diseases were hepatitis and polio, according to Harry P. Kramer, director of the Taft Sanitary Engineering Center in Cincinnati. "Today there are over one hundred."

A cupful of water taken at random from the Connecticut River near Hartford was found to contain 26 different infectious bacteria, including typhoid, paratyphoid, cholera, salmonella, tuberculosis, anthrax, tetanus, and *all* the known viruses, including polio and hepatitis, as well as parasitic life forms such as tapeworm, roundworm, hookworm, pinworm, and blood flukes.

The usual test for water safety is the amount of organic matter present—which completely ignores chemical content. This is based on the number of coliform bacteria in a given amount of water. Coliforms live in fecal matter in human intestines, along with any disease germs a person may carry. If coliforms are found in water, this is considered evidence that the water is contaminated with fecal bacteria; therefore, it is likely that disease microorganisms are present. The United States Public Health Service's standard for drinking water is less than 2.2 organisms (coliform) per 100 milliliters of water. This would mean fewer than 414 in an eight-ounce glass of water. For water where shellfish grow, 70 coliforms per 100 milliliters is acceptable. The maximum for bathing is 2,400 per 100 milliliters of water.

Often these coliforms appear in fantastically high amounts. The Potomac, which daily receives wastes equivalent to about 80 million gallons of raw sewage, has counts up to 4,000 times the safe level. The Maumee River in Ohio and Indiana runs as high as 24,000 times allowable maximums. In New York Harbor, bacteria count has risen 10 to 20 times during the past decade, according to Dr. Barry Commoner. In the past most intestinal bacteria died when they left their warm and happy abode in the human gut, but evidence suggests that this is now changing. "The possibility exists that the bacteria, entering the water from sewage or the soil, are now able to grow in the enriched waters of the bay," according to Dr. Commoner.

Water is considered the most effective carrier of disease. It is blamed for recent outbreaks of cholera throughout the world. Worse may be in store. Many scientists warn that if bacteria should develop an immunity to present treatment methods, primarily chlorination, or if treatment systems should break down owing to chemicals that kill beneficial bacteria, there could be a recurrence of the typhoid epidemics of the last century or, worse, the disastrous plagues of the Middle Ages. As long as we continue to dump raw and poorly treated sewage into waters we drink and bathe in, we have a slim margin of safety.

Even the best treatment methods now used are not wholly effective in removing disease organisms. Secondary treatment, with chlorination, can remove up to 99 percent of the bacteria. But viruses pose an even greater threat, according to a report by the American Chemical Society. Indications are that the secondary chlorination, as it is now practiced, does not produce a virus-free effluent. Carbon filtering does remove certain viruses, the report states, but the process "is reversible and the infectious properties of the desorbed viruses are unimpaired." In other words many viruses, even when removed from water, remain infectious: "Viruses could be a serious problem of water reuse, and it appears that improved means of removing or inactivating them will be required to achieve virus-free effluents from advanced waste water treatment." [1]

The threat of viruses becomes more serious because of what is called a "dual infection" mechanism. Dr. F. Kingsley Sanders, a virologist at Sloan-Kettering Institute for Cancer Research in New York, explains that in a dual infection the virus might be present in the body for many years without any symptoms of the disease. For the disease to break out, if it ever did, would require the introduction of another agent that acts as the trigger. That agent, he said, could be another virus or the result of such environmental conditions as chemicals, diet, stress, cigaret smoking or pollution.

This leads to the biggest question of all: how safe is this water to drink? Much of it is not safe at all, according to recent USPHS studies. Millions of Americans are drinking water that is potentially dangerous because of bacterial and chemical contamination. In one study 30 percent of almost 4,000 samples were found to exceed USPHS limits on standards for drinking water. Pesticides appeared in nearly all samples from surface waters. Nine percent of the samples from the tap showed biological contamination. About 0.4 percent contained arsenic above USPHS standards.

Larger cities tend to provide water of "average quality," the report stated, while smaller systems tend to deliver waters of "inferior quality." *The New York Times* stated that at least 340,000 people living in small cities (those with populations of less than 100,000) were being served "waters of a potentially dangerous quality—that is, waters capable of causing disease. Traces of fecal bacteria, lead, copper, iron, manganese, nitrate, arsenic, chromium, and selenium were detected in many such samples."

"We are calling it a crisis," said a government spokesman, "because we know we must act now to improve the situation before there is a major disaster." [2]

By law the USPHS has authority to adopt standards for the control of communicable diseases. It has no legal authority to deal with the kind of noncommunicable chronic disease that may be induced by some chemicals. Under the present law, water might be pronounced safe for drinking because it has no more than permissible levels of bacteria, and at the same time it could be laced with arsenic, nickel, mercury, DDT, and other lethal compounds. The *Times* quotes a "high official" of the Department of Health, Education and Welfare as stating: "Our [drinking water] treatment plants weren't designed for and are not prepared to take the chemical onslaught now coming from our streams."

Permissible limits for 51 chemical components of water are set by the USPHS. Of the 51 components, 39 cannot be removed from water by existing technology. Included are some of the most dangerous pesticides known, including DDT, chlordane, dieldrin, aldrin, and other chlorinated hydrocarbons. Present treatment methods also cannot remove the poisonous metals, such as mercury, cadmium, and radioactive radium-226 and strontium-90. For some of the pesticides there are not even any detection tests. "We don't even know whether they're in the water," said a New York water official.

Throughout the world, water is becoming more dangerous as pollution increases. There is the immediate threat of infectious disease and the long-term hazard from ingesting chemicals. Water is gradually losing its self-cleansing ability as the onslaught increases. In many places public beaches have been closed. In many others they should be but are not. Curiously, closing a bathing beach is often dictated not by sanitation needs but political expediency.

One of the great truths of pollution is that, for most people, pollution does not exist as long as it is not brought to their attention. They protect themselves with ignorance and zealously guard that ignorance. They resent being told that something is unsafe. They resist being upset or having their comfort disturbed. They will swim in the worst filth as long as they are not told it is filth. Close the beach and they are furious—not at the polluter but at the officials who closed the beach. Politicians, wise in the perversions of people, have developed strategies to avoid this dilemma. Many simply refuse to have the water tested. Others order health officials to ignore adverse findings. Many public health officials understand this little game and go along with it; either they do not test the water, or they keep the results to themselves, or insist that there is no hazard, and hope for the best.* Their unspoken motto is: If you can't make it go away, pretend it isn't there.

The same process of denial of risk has permitted other hazards to flourish. Among them are the onslaught of phosphates and nitrates that overenrich waters and cause algae explosions

---

* In Connecticut, according to a Yale professor, beaches were kept open, despite pollution of three quarters of them, to avert slum violence. In New Jersey, where pollution is widespread from nearby ocean dumping and untreated sewage from New York Harbor and local communities without sewage facilities, the state takes no samples to determine whether beaches are unsafe for swimming. "It is mainly a local matter," said an official of the State Department of Environmental Protection. He told the *Times* that he could recall no time in the ten years he had been in the department that a beach had been closed because of sewage pollution. One of the most dramatic battles was carried out in the Rochester area where a group of University of Rochester scientists fought spiritedly to force local health officials to recognize water pollution and enforce state health laws. The story is told in "Water Pollution," a Scientists Institute for Public Information Workbook compiled by Dr. George G. Berg, and a report that appeared in *Scientist-Citizen* (now *Environment*) under the title "Water Crisis—The Rochester Area," by Dr. Berg and Cynthia Brodine.

with the attendant loss of oxygen and eventual eutrophication. Phosphorus is in short supply in nature. The enormous amounts of man-made phosphorus compounds now being dumped into water from detergents, along with nitrogen from fertilizer, add to the burden of sewage. Each pound of phosphorus can propagate up to 700 pounds of algae. The United States alone produces about a billion and a half pounds of phosphorus-bearing detergents, and the phosphorus content has gone up with the introduction of the enzyme presoaks. We have all heard about the woes this enrichment has brought to Lake Erie. It is now virtually dead, and Lake Michigan is dying.

Lake Erie is subject to almost every known insult water can suffer. In the past 50 years it is said to have aged the equivalent of 15,000 years. Some 800 square miles on the western edge is covered with algae slime two feet thick. About 10 million people dump their wastes into the lake or its tributaries. Municipal wastes flowing into the suffering lake are equal to raw sewage from 4.7 million people. In addition, Lake Erie received industrial discharges from 360 sources; more than half are classified as giving inadequate treatment. From surrounding farmland comes a torrent of nitrogen fertilizer runoff; this is rated as the equivalent of sewage nitrogen produced by about 22.5 million people—more than double the actual number in the surrounding communities. The phosphate industry argues that the culprit is the nitrogen from fertilizer. A Canadian study suggests that it is neither phosphate nor nitrogen but the carbon dioxide given off by bacteria while breaking down the algae, but this fails to account for what makes the algae bloom in the first place. Only one absolute truth emerges: we are bringing about profound changes in the environment without knowing what we are doing or even how we are doing it.

The goal has been to clean up the nation's waters during this decade. The target becomes increasingly elusive. The timetable for cleaning up Lake Erie was 1972, but in late 1970 the Federal Water Quality Administration (FWQA) said that 78 of 110 cities on the lake were not meeting their schedule; 49 were more than one year in arrears. Of 130 industrial polluters, 44 were lagging, with 39 more than a year behind schedule.

By 1971 pollution in the lower Great Lakes and the St. Lawrence River was so bad that a joint commission from Canada and the United States said it presented a danger to health and property. In June, 1971, the two countries signed what was described as "a historic agreement" to eliminate water pollution in the Great Lakes by 1975. Gladwin Hill, a *New York Times* environmental writer, noted that the agreement was historic in another sense: "It was, roughly, the 62nd anniversary of the first meeting between the two countries where the same goal was agreed on."

In 1909 a similar commission was set up to prevent pollution in the Great Lakes. The commission has been wholly ineffective, Hill points out. It has done "a lot of work, held many meetings, and produced a five-foot shelf of reports—while the Great Lakes have become more and more polluted." In 1951 the commission reported lamentable pollution in the lakes. Not until 1956 did the United States, "the source of about 90 percent of the pollution then and now," produce a mechanism for coping with the problem: the Water Pollution Control Act of 1956. Six more years passed before federal water pollution officials initiated the first abatement action relating to the Great Lakes—to clean up the Detroit River, a major tributary of Lake Erie. Since then four more abatement actions have been begun.

> In none of these actions has actual abatement been achieved, despite a schedule of pollution-control installations supposed to be completed by 1972. The Detroit River —at last report the repository of 540 million gallons of municipal wastes and one billion gallons of industrial waste daily—is still something no thinking man dips his hand into.

Lake Erie's problems are universal. In many areas of the United States plans have not even been drawn up for water purification programs. There are lack of money, lack of organization, lack of commitment; there are bureaucratic red tape and political maneuvering. The biggest obstacle is cost. Some put the figure at about $26 billion—what the government is now spending in Vietnam in a year. A more realistic figure is $100 billion, says Senator Gaylord Nelson. To reclaim the Great Lakes would cost an estimated $20 billion, up to $10 billion to clean the Hudson, another billion for New York Harbor.

What is actually being spent is a token sum. The Nixon Administration boasts that it is putting out $10 billion to help municipalities build sewage treatment plants over the next five to eight years. But even that inadequate amount is inflated. Actually the federal appropriation for community assistance for the fiscal year 1970–71 was $1 billion, and only about $500 million was forthcoming. Communities and states were supposed to make up the rest. Under the law the federal government is obligated to contribute up to 55 percent of the local cost for treatment plants. In practice the government gives considerably less, and the communities often have a hard time collecting that. Mayor Roman S. Gribbs of Detroit testified before a Senate subcommittee that his city had been served an abatement order by the federal government but that since 1966 the government had made good on only 20 percent of its pledge of $10 million in waste treatment grants. Other mayors complained of the same hypocrisy. The New York State comptroller's office said it had received only 7 percent in actual cash of the 55 percent provided by federal legislation.

For federal aid for construction of sewage plants the government has been less than munificent. In 1970 President Nixon asked Congress for only $4 billion to be spent over four years. The next year he asked for $2 billion to be spent in each of the next three years. This was more realistic, but still far less than the $35 billion the National League of Cities said is necessary for municipal water treatment facilities.*

The entire funding for sewage treatment presents some novel aspects. Treatment plants must be served by sewers; they are useless without them. But the President's Council on Environmental Quality reported in 1970 that less than a third of the country's population was served by both sewer systems and adequate treatment plants. About a third had no sewers at all. The rest had sewers but inadequate plants or none at all. In 1970 Congress authorized a modest $1 billion to build water and sewer lines. But the President apparently considered any amount excessive. He allowed the authorization to become law without his signature "because he recognized that a veto would be overridden," *The New York Times* stated. "But his proposed budget contains no request for any money to carry out this authorization."

These oddities affect only municipal sewage. "The staggering problem of industrial pollution is virtually untouched by our federal antipollution programs," observed Senator Nelson, "even though industry contributes twice as much pollution to our

---

* In a 1971 speech, Representative Paul N. McCloskey Jr. said, "If we can justify increasing the national debt by $11 billion this year for the war in Vietnam, we can justify increasing it by $30 billion over 10 years for clean waterways."

waters as do municipalities." A Ralph Nader task force found that industrial pollution is responsible for four to five times as much pollution as domestic sewage causes.

Legally we take a morbid and corrupt view of our water. The goal is not to make it clean as much as to avoid immediate disaster while exploiting it to the utmost. Under the Water Quality Act of 1965, freshwater (FW) has four classes: FW1, waters in their natural pristine state (these are virtually nonexistent); FW2, drinking water quality; FW3, recreation and fishing; FW4, primarily for industrial use. The country is divided into 22 river basins, and the states within each basin, or region, set standards that must be approved by the federal government. Often the same stream will have several different classifications for different portions.

The higher the rating, the more it costs to clean up the water. Consequently the states tend to seek the lowest standards possible. Water need be no cleaner than necessary for the purpose intended. It can then be degraded to that level but not below it. This does not impose a mandate to clean up the water; rather it is a license to foul it to the designated level of use, despite official denials. Engineers do not call this pollution. In their vocabulary it becomes "assimilative capacity." It means, in effect, that a polluter is free to dump in all of the garbage or filth that the stream can take. Most streams are pushed beyond their "assimilative capacity." They become open sewers.

Many public officials and polluting industries reject effluent standards on principle. They insist that it is "nobody's business how dirty the stuff we discharge is as long as it doesn't demonstrably lower the quality of the waterways you're trying to clean up." In Missouri, pollution experts held that the Missouri River no longer could be expected to perform the conflicting role of removing sewage and providing a source of clean water at the same time. Since a choice had to be made, the experts said, the use of the Missouri for waste removal "has *economic* value far greater than use of the river as a source of municipal and industrial water supply." In other words, abandon it to the use as a sewer. The board did not express fear that the river might cause disease and debase life but that failure to allow the Missouri and other streams in the state to assimilate wastes to utmost capacity would "lower standards of living . . . and decrease employment."

At best, the laws are weak, and such as they are they have not been enforced. Representative John A. Blatnik, chairman of the influential Public Works Committee, and himself instrumental in getting the 1956 and 1961 acts passed, described federal efforts at water pollution control as "absolutely miserable." A Nader task force concluded after 21 months of study that the program has been a "miserable failure." Industry pressure was pinpointed by the Nader panel as the primary reason for the failure, on both the state and federal level.

Federal laws have left to the states primary responsibility for enforcement of standards, and state governments have yielded to industry pressure to procrastinate in setting standards and to set them as low as possible. Four years after the June, 1967, deadline for final submission of standards, 22 states still did not have them finally approved.

On the federal level, industrial pressure weakened enforcement powers. The several water pollution control acts gave the FWQA (now the Water Quality Office under the Environmental Protection Agency) "discretion" to enforce or not. They also provided for tortuous administrative proceedings before a recalcitrant polluter can be taken to court and directed the court to consider whether an abatement order is technically or economically "feasible."

These charges were spelled out by U.S. Comptroller General Elmer B. Staats before the House Public Works Committee. He said the government is virtually powerless to act on water pollution within individual states and cannot act in interstate cases until pollution has already become a problem. Under the 1965 act, he said, "a minimum of 58 weeks" is required between the time EPA calls a conference to identify polluters and decide on corrective measures and the date the agency can refer the case to the Attorney General for court action. The act also provides for a 180-day notice to polluters to take, or agree to take, voluntary action to abate pollution before court action—in effect a license to continue polluting for another six months after already long delays.

Only after the long waiting period can court action be taken. Even then the procedure has various limitations and has been used only about a dozen times. According to the Conservation Foundation:

> For instance, in the case of pollution which is endangering the health or welfare of persons only in the state where the discharge takes place, suit cannot be brought except

with the written consent of the governor. And even if a case gets to court, the court may go beyond the usual issues of law and fact and give "due consideration to the practicability and to the physical and economic feasibility of complying" with established standards.

This provision has provided an effective loophole for many polluters. They often get around the law by going to court and claiming that established standards are not practical, or that they are not physically and economically feasible. The courts are remarkably sympathetic to this appeal.

Abatement procedures have generally proved relatively ineffective; they involve lengthy "enforcement conferences," reports, public hearings, and the like. The government, and most states, have been reluctant to prosecute. They depend primarily on voluntary compliance and it has not worked. The most effective enforcement tool is an 1899 statute that forbids the discharge of polluting materials into navigable waters. This is usually invoked only in emergencies. If strictly enforced it would virtually shut down American industry and most of our municipal communities.

Cleanup is further hampered by bureaucratic confusion. Government agencies often work against one another through ignorance or conflicting interests. A recent enforcement conference held by the Environmental Protection Agency revealed that while EPA was trying to stop the pollution of Long Island Sound, the Army Corps of Engineers was regularly issuing permits to a chemical company to barge its waste three miles into the sound and dump it. The practice had been going on once or twice a week for fifteen years. The waste was described by a spokesman for Pfizer, Inc., of Groton, Connecticut, as a cakelike brown substance left over from the production of antibiotics. He insisted it was not toxic. But the revelation prompted Murray Stein, Assistant Commissioner for Enforcement in the EPA Water Quality Office, to ask, "Are we really asking municipalities to spend millions of dollars to clean up their sewage so we can make Long Island Sound suitable for an industrial dump?" Subsequently the Pfizer Company agreed to stop dumping its wastes into the sound—within a year.

The solution to the water problem is not mysterious. It requires more money, stronger laws, greater cooperation among government agencies, and more vigorous enforcement. Most of all it requires a will to clean up the water and to stop temporizing with pollution. There is a tremendous need for research to find better ways to purify water. Meanwhile, every polluter—industrial, municipal, state, and federal—should be forced to install the most effective devices now available without exception and without delay.

Frequently the suggestion is made to tax industry's effluent. This, in effect, is a license to pollute. It does not solve the problem but gives it economic and legal sanction. This is not a way to conduct an enforcement program, says Richard J. Sullivan, New Jersey's Commissioner of Environmental Protection: "In our judgment, if the technology is available it should be applied and we shouldn't have people taking the position that riparian rights include the right to pollute other people's streams."

Sullivan is responsible for the Passaic River, which a government official bluntly called "a disgrace to the United States." Sullivan does not quarrel with the designation. He is quick to point out the river's shortcomings. Recently he asked the federal government to adopt the Passaic as a model demonstration project with the nation's best minds concentrating on finding more effective ways to control pollution. Conventional methods have failed to solve the Passaic's horrendous problems, he says. "I submit, that nowhere could you find a river with more problems than the Passaic. If we could solve the problems here, we could solve them anywhere."

The Passaic's problems are monumental. In degree, they are universal. Between the upland meadow where it is born to its melancholy terminus 80 miles distant in New York Harbor it suffers progressive indignities. Twenty-nine municipalities and 700 industries dump their wastes into it. The stream is laced with sewage, chemicals, fertilizer runoff, pesticides, and solid waste. Dams impound its waters and interfere with a steady flow, bringing a buildup of pollution and eutrophication. The banks are lined with junkyards, used cars, old refrigerators, and other abandoned appliances. At low flow, in some places, the river is almost half sewage. Coliform counts have soared to astronomical figures. This same stream provides drinking water for more than 700,000 people. The state has tried to clean up the river but has been hampered by the magnitude of the problem and lack of funds.

If the Passaic's problems cannot be solved, there is little hope for us, the people whose lives are so inextricably tied to the waters of the earth.

2

The
Ambient
Air

Very plainly, we are left with but a single choice. Either we stop poisoning our air—or we become a nation in gas masks, groping our way through dying cities—and a wilderness of ghost towns.

—President Lyndon Johnson

IN THE BEGINNING

Scientists believe the earth is about 4, possibly 5, billion years old. In the beginning, as the molten lava cast off by the mother sun, or some other heavenly body, cooled, the rocks gave off gases that provided a hostile atmosphere. This original envelope was made up of toxic fumes, perhaps ammonia, hydrogen, carbon dioxide, and sulfuric dioxide. There was, it is believed, no free oxygen.

Then, some 3 billion years ago, so the theory goes, the simple inorganic compounds in the atmosphere were bombarded by ultraviolet rays from space and warmed by the heat of the sun. These compounds combined to form larger molecules. Under the ultraviolet charge complex forms containing carbon and nitrogen developed. This, many scientists believe, was the beginning of life.

The first life form was simple, we are told. Gradually it developed, utilizing elements formed under the bombardment of cosmic rays. From the toxic atmosphere it took the elements it needed, such as iron and sulfur, and utilized them with the energy from the sun. In this hostile environment, a remarkable event took place; this, according to the theory, was one of the most significant advances in the emergence of life: The primitive organism learned to produce its own food.

Sunlight beat down on its pigment. From the air it took carbon dioxide and merged it with water to form sugars, with which it fed itself. Now life was self-sustaining. In the process of feeding itself it released a by-product, oxygen; the process came to be known as photosynthesis (putting together with sunlight).

The miracle of photosynthesis made possible the cleansing of the air and the Carboniferous Age, a time when the carbon dioxide was taken from the air and locked into the earth in a form that was to become coal. It is difficult to visualize this alien world. It must have been covered by dense vegetation, possibly tree-ferns. These ancient plants probably absorbed carbon from the air, stored it in their tissues, fell into decay and were followed by others that rose and fell—a terrifying world of unspeakable sadness and silence, of growing vegetation, decay, of unceasing rain and an accumulating ooze and slime of rotting matter. One forest grew out of the remains of another. Each, in turn, sank. There were floods. Land rose and fell. The sun beat down. New life forms emerged. They died and others took their place. Many were pressed into the beds of ooze that squashed them into pulp and hardened into the stored black sunshine we call coal. The preserved life forms are our fossils. They write the fable of time.

Photosynthesis provided the oxygen that sustains life. Great beds of coal, oil, and gas have been formed in the earth from living matter. Nature is a conscientious bookkeeper. Everything is in equilibrium. Each growing thing produces just enough oxygen for its own consumption by fire, bacteria, or other natural forces that could recycle the elements into their original inorganic form. The oxygen we breathe is a form of life capital. When we burn the fossil fuels, we are consuming oxygen and putting back into the air carbon dioxide that plants removed from it more than 200 million years ago.

This has three primary effects: First, it uses up the reserves of fossil fuels, putting plant remains back into circulation as airborne particles of matter, ash and sulfur dioxide gases; all fossil fuels contain some minerals and sulfur, and these contaminate the environment when burned. The particles are breathed into the lungs and the gases absorbed by the blood and carried to the cells of the body. Second, increased fossil fuel combustion leads to an increase in carbon dioxide concentration in the atmosphere. Third, the burning of fossil fuels changes the oxygen–carbon dioxide balance in the air.

No one knows the ultimate effect of these changes. Over millions of years, since emerging creatures developed a lung, life has depended on an atmosphere with about 21 percent oxygen, 78 percent nitrogen, a small amount of carbon dioxide (0.03 percent), gases in trace amounts, including the inert argon which does not react with other gases, and varying amounts of water vapor.

This complex mixture of air is vital to our existence, an extension of our physical being. A person is composed of about 60 percent oxygen. We can go seven to ten days without water, a month or two without food, but we cannot be deprived of oxygen for more than six minutes without suffering fatal brain damage.

No one knows how critical any decrease in oxygen production may be, says ecologist Dr. LaMont Cole of Cornell University:

> The carbon-oxygen relationship is essential to photosynthesis and thus to the maintenance of all life. But should this relationship be altered, should the balance between the two be upset, life as we know it would be impossible. Man's actions today are bringing this imbalance upon us. . . . The carbon-oxygen balance is tipping. When, and if, we reach the point at which the rate of combustion exceeds the rate of photosynthesis, we shall start running out of oxygen.

Should this happen gradually, he says, the effect would be "approximately the same as moving everyone to a mountaintop. . . . However, the late Lloyd Berkner, director of the Graduate Research Center of the Southwest, thought that atmospheric depletion might occur suddenly."

A few scientists warn that we are in danger of running out of oxygen. Most deny it; they insist that supplies are almost unlimited. A 1970 study by government agencies maintained that oxygen in the earth's atmosphere has remained essentially constant over the past 60 years. The greater threat is that the existing air will be so corrupted that it will become unbreathable. It makes little difference whether we are strangled to death or poisoned; that is for pathologists and record keepers to determine. The important point now is the contradictions among trained scientists. Again and again we discover that life hangs by the narrow thread of what they think is true—and often they turn out to be wrong. By the time the mistake is discovered they are generally gone from the scene and we are left to live with the unfortunate result of their miscalculation.

Many shadows hang over us. One of the more ominous is not that we will run out of this or that commodity, but that we fail to respect the equilibriums of nature. All life rests on a fine point of balance. We disturb this balance at our peril, says Dr. Cole. It has existed and sustained life for millions of years. Man himself is a relative newcomer to the earth, a presumptuous upstart in his daring experiments. Biologists sometimes dramatize man's recent origin by quoting a fable that compresses the development of the earth into a single year. Dr. J. H. Rediske, a forestry scientist with the Weyerhauser Company, cited it during a speech in Tacoma, Washington:

> By this analogy, the months of January, February and March reveal a desolate, dreary picture of a lifeless planet subject to much geological erosion and change. It is April before single-celled living organisms in the warm sheltered coastal waters appear. In the middle of July come the first land plants; and in September, insects and the first dinosaurs are seen. The Grand Canyon is dug on December 25; and about noon on December 31—a few short hours before the end of the year—man appears on the scene. About five or six minutes before twelve o'clock New Year's Eve, comes the dawn of western civilization. Compressed within these last few minutes are the golden age of Greece and the birth of European civilization, and Columbus discovers America twenty seconds before the New Year.

All of this adds up to a tremendous impertinence on the part of mankind, a virtual newcomer to the world. Rediske noted that now, for the first time in the history of life, the oxygen process has been reversed. We are returning the poisons to the atmosphere. "If we continue at the rate we are now going, with populations expanding, the result is too drastic to contemplate. Unless we do something about it, the conclusion is very clear," said Rediske.

Behind him was a huge picture window that overlooked Puget Sound. The water was skimmed with oil wastes and debris; it had a gray cast, even in the brilliant sunlight. Ringing it was a vast complex: huge oil storage tanks, construction materials, a paper mill, chemical companies, the usual phalanx of smokestacks emitting long trails of black smoke that drifted into a guileless blue sky. It all seemed slightly absurd. Here we were, an oddly assorted group of men, well dressed and well fed, sitting behind a picture window observing our own self-destruction.

RETURNING THE POISONS

It is no longer realistic to talk about pure air. Most experts do not even talk about clean air. Now they limit themselves primarily to trying to keep the atmosphere from getting worse than it already is. Under present conditions that goal seems unlikely. Gradually we are compromising ourselves into extinction. Our official position seems to be that a lungful of air that does not carry pollutants which can cause cancer and other disease is a luxury we can no longer afford. That is to say that life itself has become a luxury we can no longer afford.

New York is typical of many of the nation's major cities. In a recent interim report to Mayor John V. Lindsay, a special Task Force on Air Pollution stated that "the conclusion is inescapable that most of the large cities in the United States would slide beyond the danger line in air pollution within 7 to 10 years. Putting it in its bluntest form, New York City could be considered uninhabitable within a decade."

There is a basic difference between air pollution and water pollution. Water can be largely cleansed. An individual who wants to avoid contaminants can buy spring or distilled water, as millions of Americans are now doing, or he can dig a deep well that may be relatively free of contaminants. But there is no escaping the air. Each of us must breathe whatever is at hand. We are all at the mercy of the polluters—with whatever relief industry and government provide us.

Our primary defense against air pollution, to date, has not come from industry or government, but rather from the human lung. Against so formidable an enemy as the 200 million tons of pollutants now being hurled into the air every year, it is a fragile protection indeed.

THE CLEVER LUNG

The human lung is an ingenious organ. It enables us to take into our bodies the free oxygen necessary for metabolism—the combustion and utilization of the food-fuel we consume—and to discharge the resulting carbon dioxide. The mechanically minded like to compare a person to a tiny factory. He uses oxygen much as a furnace does, producing heat energy, as well as wastes that must be disposed of. A breathing animal functions exactly the opposite of a green plant in its process of photosynthesis: a plant takes in carbon dioxide from the atmosphere and breathes out oxygen. A human takes in oxygen and breathes out carbon dioxide. Thus man and his environment are linked by a symbiotic weld, each utilizing the other's by-products for mutual survival, each dependent on the other.

The lung has two primary lines of defense. First is a filtering process in the nose and pharynx. Tiny particles of dust and matter that escape entrapment in the nose and pharynx are met by small hairlike projections, called cilia, that grow on the surface of the windpipe (trachea) and bronchial tubes. These cilia are like tiny brooms that sweep the particles, borne in a continuous stream of mucus, back toward the throat; there they are swallowed into the digestive tract or, if large in amount, coughed up.

In the lungs the vital exchange between oxygen and waste gases takes place. Oxygen goes into the bloodstream and hence to the brain, heart, and every minute cell of the body, enabling the specialized cells to utilize the nutrients also brought to them by the blood. On its return to the lungs, the blood carries the waste that is expelled as carbon dioxide.

The average person breathes about 16 times each minute. In a single day he filters through his lungs some 16,000 quarts of air. An adult's daily intake of liquids and food is less than 10 pounds, but the air he breathes may weigh more than 35 pounds. In addition to the atmosphere's natural components, the air is likely to include a mixture of smoke, dust, fumes, mists, radioactive wastes, odors, gases, tiny pieces of carbon, ash, oil, grease, asbestos and other fibers, as well as microscopic fragments of metals and metal oxides. These substances may be sucked in alone or in combination. Along with the known substances are many still unidentified. The effects of only a few have been studied. The majority are still undetermined. Many are unstable chemically and unpredictable biologically. Many are highly poisonous. Others can cause various respiratory difficulties and diseases. These pollutants, individually and collectively, may work directly or with treacherous indirection. United States Surgeon General William H. Steward reported in 1967 that there is "compelling evidence" that air pollution is killing and disabling Americans in every area of the nation.

In addition to health effects, pollutants kill trees and damage crops and other vegetation. They corrode steel, chip and burn paint, eat away stone. A classic horror story often cited is that several inches of granite have disappeared from Cleopatra's Needle, the stately obelisk brought to Central Park from Egypt in 1883. It suffered greater damage in New York City during the past 90 years than in 3,500 preceding years of exposure to the grinding pressures of sand, wind, and sun in Egypt's desert.

DIRTY AIR, DIRTY CITIES

The overall loss from dirty air cannot be computed. No figure can be placed on sickness, suffering, and death. Dollar losses are, at best, arbitrary. Some economists put the figure at $18 billion annually. Others put it much higher. The blight is no longer confined to big cities and industrial centers. Now, according to the Public Health Service, more than 43 million Americans living in 300 cities suffer from "major" air pollution, and every community of more than 50,000 has an air-pollution problem to some degree. The problem is nationwide.

Even this does not tell the whole story. Astronauts returning to earth saw the globe in its finite misery, a black ball hurtling through limitless space, poisonous mists smeared across the entire Eastern Seaboard, a long blackish-brown smudge that almost blotted out land and sea. Foul air has billowed out of the cities, finding those who fled the taint of urban centers, refugees who tried to find relief in the pure air of deserts and mountains. It has invaded the onetime respiratory sanctuaries of the stricken. Phoenix recently was blotted out for eleven days as a thick brown pall filled canyon, gulch and gully, obscuring the desert skyline, grounding planes, causing the afflicted to cough, wheeze and gasp for breath. Airplane pilots have trailed great clouds of poisoned air from eight copper smelters drifting across the sky into Phoenix and Tucson; eight smelters in the state pour nearly 10.5 million pounds of sulfur dioxide into the Arizona air every day. In Montana, Dr. C. P. Brooke, a civic leader in Missoula, said: "We have always believed that we live in the Big Sky country, the Big Blue Sky country, the country with clear, rushing streams, that we had more pure air and more water than we could ever use. Now we know better."

Four Corners, where Utah, Colorado, Arizona and New Mexico meet, is the site of a giant utility known as the Four Corner Plant, "possibly the worst single source of air pollution in the world," according to Virginia Brodine, writing in *Environment*, January-February, 1972. The plant's five operating units emit roughly 300 tons of sulfur oxide and 400 tons of particulates per day; "these emissions can be compared with the daily average emitted in the entire New York metropolitan region, estimated at 633 tons per day in 1966." Four Corners Plant, augmented by five more huge utility plants being built in the Southwest, will increase coal consumption from the present 25,000 tons per day to 130,000 "and unless the present reliance on stack height is given up for genuine control, sulfur, particulates, and nitrogen oxide emissions will rise more or less proportionately."

Smog has turned that Big Blue Sky into a Big Gray Sky, says John T. Middleton, Commissioner of the National Air Pollution Control Administration (NAPCA): In parts of Montana the death rate from bronchitis "has tripled in a decade and lung cancer has more than doubled."

Denver, once celebrated for its sparkling air and scenic beauty, is so afflicted now that the nearby Rocky Mountains are often lost in a brown haze, and wary residents wear breathing filters on the worst days. Los Angeles is the place where smog got its name; when pollution levels soar, children are forbidden to exercise so they won't breathe deeply, and doctors annually advise some 10,000 patients to leave the city to escape the air, but admit they don't know where to send them. New York, a leader in so many things, is said to have the dirtiest air in the nation, second only to London in the world.* People who fled that tainted city to live in the green suburbs complain that the foul air of the metropolis has pursued them there. Even Suffolk and Nassau counties on Long Island, once swept by fresh sea breezes, have begun to wilt under the assault of fumes and aerial garbage. New Jersey does not have a county in the state that is not suffering crop damage due to smog, and some produce no longer can be grown. California estimates its crop damage at $100 million a year. Smog drifts from Los Angeles to kill 1,000 acres of towering ponderosa pines in the San Bernardino Mountains 80 miles west of the city at an altitude of 5,000 feet. It

* Some patriots claim that the distinction for the dirtiest air belongs to Mexico City, Munich, or Tokyo.

threatens vegetation and timber in the great national parks at Yosemite and Sequoia. Trees are dying in California at such a rate from auto exhausts that in some areas the state is now replacing them with plastic trees.

Smog strikes at timber in the isolated Appalachian Mountains. Some experts fear that in a few years pollutants in the air will determine which crops can be grown. One warns that agriculture in certain parts of America will soon cease to exist.

The spread of pollution not only has affected the environment but has also brought a dramatic increase in respiratory ailments, primarily emphysema and chronic bronchitis. Emphysema's mortality rate has shot up 500 percent in the past 10 years and now affects women in about the same proportion as men, according to the New York Tuberculosis and Health Association. An average of more than 1,000 workers a month are said to have been forced to retire prematurely in recent years because of the disease. "Over six percent of those who receive monthly Social Security disability payments suffer from it—a percentage exceeded only by those with arteriosclerotic heart disease. Bronchitis has had a 200 percent increase in the past decade," the association reports. Both emphysema and bronchitis can disable and kill.

Emphysema is the progressive breakdown of the alveoli, tiny balloonlike air sacs in the lung wall. The sufferer is left with· what one researcher calls "a flabby, inadequate bellows with which to breathe." Each breath can become a tortured gasping for air. Bronchitis, an irritation of the bronchial tube leading to the alveoli, causes chronic coughing and is accompanied by frequent acute bronchial infections. Both result from changes in the lung's elasticity, basically a disease of old age; often aging turns out to be due to external factors such as air pollution. Also linked to dirty air are a variety of other respiratory infections, ranging from the mildly annoying to the fatal: asthma, a painful "kind of airway resistance"; colds; chronic cough; pneumonia; and lung cancer, which has almost reached epidemic proportions, especially among smokers in urban areas; nearly 60,000 cases now develop each year. An estimated 15 percent of the population suffer some form of respiratory disease.

All chronic respiratory disease interferes with the body's ability to supply its oxygen needs. This, in turn, forces the heart to work harder to make up the deficiency. The extra burden can lead to heart disease or kill those already suffering from it. Pollutants can also work in a variety of other ways, directly or with undetectable stealth. Airborne germs can directly cause infections and disease. Or they, along with poisons and other irritants, can weaken and destroy the body's normal defenses, enabling viruses and bacteria already present to cause active sickness. Some studies suggest that pollutants commonly found in the air actually alter the structure and function of the respiratory tract.

In various ways pollutants can hamper the lung's self-cleansing powers. They can slow down or even stop the sweeping action of the cilia lining the airways so they can no longer eject mucus and foreign matter, thus leaving the sensitive underlying cells without protection. Irritants can cause the production of increased or thickened mucus. They can bring about the constriction of airways and induce swelling or excessive growth of cells that form the lining of the airways. They can cause a loss of the essential cilia or even of several layers of cells. One or more of these reactions can make breathing more difficult, the tuberculosis association explains, "and foreign matter, including bacteria and other microorganisms, may not be effectively removed so that respiratory infection can more easily result."

THE STEALTHY KILLERS

But the preceding, generally, are long-term effects. Under certain conditions dirty air can suddenly turn into a swift and savage killer. By now the big killer episodes are well known. First was the Meuse River Valley in Belgium, which took 60 lives in December, 1930. Donora, Pennsylvania, followed in October, 1948, when 6,000 of the mill town's 14,000 population sickened and 20 died. In December, 1952, London had an epic five-day attack that left 4,000 dead and double the usual number of patients applying for emergency bed service. New York City has had four major episodes: 1952, 1962, 1963, and 1966 claiming at least 600 lives.

These episodes struck down not only the old, the weak, and the ill, says Dr. Rogers S. Mitchell, a specialist in pulmonary diseases at the University of Colorado: "Air pollutants can kill already sick people, and they can kill healthy people when they're bad enough. . . . Such disasters are going to occur again and again as time goes on, and possibly more frequently than before."

Many investigators believe that people never completely recover from these massive assaults. Evidence suggests that once people undergo a prolonged heavy exposure they become vulnerable to pollution in lesser amounts and other forms of bodily stress, "almost as if they had developed certain allergies." A study made ten years after the Donora episode revealed that people who had been acutely ill were dying at an earlier age than other townspeople who had not been so severely afflicted.

But the greatest danger of pollution is not the dramatic event that kills outright. It is the slow, subtle effects of breathing in these deadly compounds in lesser amounts day by day. Most of the time dirty air is a sly assailant. Rarely does it leave its fingerprints on the corpse. It wears many disguises. We are more likely to think of it as a nuisance than as a threat. It is ugly. It smells bad. It offends our aesthetic senses. We can watch this gray, silent toxic ghost at work without recognizing it. It can be so elusive, in fact, that a killer episode can take place without our being aware that it happened. Ten years passed after the first major pollution episode in New York before statistical studies revealed that excessive deaths had taken place at that time. "The methods we have for detecting excessive deaths are so crude that there has to be a pretty big excess for us to realize that it's there at all," said one investigator. "What we do know is that people get killed by air pollution." We have not even begun to investigate the subtle effects of air pollution. It is suspected that it may increase neuroses and anxieties and influence behavior, along with other effects not generally associated with it.

The biggest problem is the nature of the pollutants themselves. There is overwhelming statistical evidence that air pollution sickens and kills. But there is little solid medical proof that specific pollutants normally found in urban air cause specific diseases. For example, the Public Health Service has found that deaths from lung cancer occur in large metropolitan areas at twice the rural rate, even making allowance for different smoking habits. But no one has specifically identified what makes the difference. It took some twenty-five years "just to prove conclusively that cigarets can cause lung cancer and to identify the tars as the major culprits."

The nature of disease has changed drastically in recent years. In the past we were cut down primarily by infectious diseases—pneumonia, typhoid fever, etc. Today we are assailed by ailments that previously were considered the expressions of old age

—cardiovascular diseases, stroke, cancer, chronic pulmonary diseases including bronchitis and emphysema. These strike not only more often but earlier. Most of us sicken long before we die. Even children are hit by ailments that once were confined to the old. Cancer, formerly a medical rarity in the young, is now second only to accidents as a cause of death in children.

Medical researchers hold that today almost all sickness and disease are caused by environmental factors, possibly attacking some internal weakness. Children, unfortunately, are more vulnerable to most environmental insults than adults. Some aspects of growth and development may be affected by certain air pollutants, warns a committee of the American Academy of Pediatrics. Most studies on air pollution as it affects children were done by foreign scientists, *The New York Times* reports. "The Academy said this fact 'suggests a relative lack of serious consideration of the problem in this country.'"

OUR TANGLED SKEIN OF AIR

The problem is to pinpoint and remove the environmental culprit. In some cases this can be done easily enough. With air pollution it is virtually impossible. Who can sort out and trace the individual threads in so wildly tangled a skein? There are too many variables. Different pollutants work on different people in different ways at different times, says Dr. Stephen H. Ayers, director of the Cardio-Pulmonary Laboratory at St. Vincent's Hospital in New York City. Dr. Ayers, an expert on air pollution, explains that all of us have built-in weaknesses:

> In the daily bombardment of insults we receive an overall stress. This, in turn, imposes an overpowering burden on the body that it is unable to contend with. But different bodies and different organs in different bodies break down at different points. Which substances singly or in combination strike at what weakness? How can you say this did in Joe? And that Smith? In an autopsy you can't tell what caused death; just the result.

A postmortem merely tells where we arrived, not how we got there. The problem is compounded by the complexity of the pollutants themselves, the way they act and interact. This means the researcher is faced with the gargantuan task of trying to separate the elements in the fermenting cauldron of air to determine whether they are harmful, how, and in what quantities and under which conditions. If identified and tested in the laboratory, who can say how they will behave in free atmosphere in combination with thousands of other substances? They may change form. Or they may not change form themselves but cause other compounds to change or speed up normal changes. They may combine to form new and more complex compounds. They may have a greater strength together than the sum of their separate strengths. They might assume a different character altogether when exposed to sunlight, water, or other atmospheric conditions. Other variables can also mask the effect on health: an individual's age, physical condition, daily habits, where he lives, genetic inheritance, the weather.

Instead of recognizing our vulnerability and lack of knowledge as a warning to be wary of what we put into the air and water (and ultimately into our bodies), we have taken ignorance as a license to recklessly and wantonly dump unknown substances into the environment. We do not require industry to prove that these pollutants are safe before they can be used. Rather, once they are in the air or water, the public must prove they are harmful. Even when proved harmful they are not generally removed. At best the dose may be reduced. Usually even that primary precaution is not taken. Many substances that now appear commonly in the environment are known to be harmful. Frequently they appear in amounts known to exceed harmful limits. Most of the air pollutants have never been thoroughly tested. Less than 40 percent of the garbage being pumped into the atmosphere has been identified; a mere 25 percent, according to another estimate. We are still reacting to yesterday's problems, says a federal air official. "We haven't even caught up to today's problems, let alone finding the answers." Each year we fall farther behind as thousands of new products are being developed, releasing into the already overburdened air more new, unknown, and unidentified substances.

The basic problem is much the same as that of water pollution. We are putting more garbage into the air than we are taking out, and control methods are not keeping pace with the ability to pollute. Every day, according to a government estimate, we hurl into the atmosphere more than 140 million tons of contaminants. Only two years ago it was 130 million tons. Next year the figure will be substantially higher than it is today. Each year after that it will continue to rise. Each estimate of the total amount varies, but all agree that the problem becomes progressively more severe and the hazard greater.

The contaminants come from stationary sources and motor vehicles. From industry alone we annually get 2 million tons of carbon monoxide, 9 million tons of sulfur oxides, 2 million tons of nitrogen oxides, 4 million tons of hydrocarbons, 6 million tons of particulate matter. From automobiles come 66 million tons of carbon monoxide, 1 million tons of sulfur oxides, 6 million tons of nitrogen oxides, 12 million tons of hydrocarbons, 21 million tons of particulate matter. From space heating devices come another 8 million tons of assorted poisons. Power plants account for 20 million tons more. Refuse disposal adds another 5 million tons. There is much more. Name it and you will find it in the air. Or, more likely, it does not yet have a name. No one even knows it's there.

What happens to these substances? No one knows that either. Little is known about the flow, dispersion, and degradation of air pollutants, nor the lifetime of the individual contaminants. According to the American Chemical Society, the air is a huge chemical vat, like the water below, where unpredictable and unknown chemical reactions take place. The experts are often baffled by these events. Only recently New York started having Los Angeles type smog. Why? No one knows.

THE UNSOLVED PROBLEM

The essential problem of all pollution is linked to a basic law of physics. Matter is never destroyed. It merely goes elsewhere, often in another form. Frequently we think we are correcting a problem, only to find that we have created another. Often the new is more dangerous than the old because the visible has become the invisible, the annoying has turned sinister. Sometimes, it turns out, not only is the solution wrong, but researchers were working on the wrong problem.

Los Angeles once had the distinction of being the nation's most heavily polluted city.[3] The difficulty was blamed on smokestack emissions and open burning. The city waged an admirable and effective campaign to tame these offenders. Then, to the surprise of all, it turned out that the real source of trouble was the automobile. Pittsburgh had a similar problem. It was virtually buried under a suffocating black cloud that sunlight could not penetrate. The city also waged an effective campaign to get rid of the pall. But behind that pall was a more deadly enemy. "The Golden Triangle is tarnished again by an air pollution problem that is both less obvious and more complex than the old smoke problem," said John W. Gardner, former Secretary of Health, Education and Welfare, referring to automobile emissions.

Until recently air pollution was seen largely as a problem of black smoke pouring out of a chimney, more nuisance and economic liability than threat to health. But this was easily enough controlled; it took only a will and expenditure of funds necessary for existing emissions-control equipment. The particles could be trapped or burned at higher temperatures, or both, so the emission became gray or even invisible. In our innocence we believed that the larger the particulates, the blacker the smoke, the more sinister the threat.

Attitude and the solution alike turned out to be largely an illusion. Experts now realize that the smaller the particulate, the greater the hazard it poses. Many "clean air" devices on smoke-stacks and exhaust pipes are really a subterfuge," says Vincent J. Schaefer, director of the Atmospheric Sciences Research Center, State University of New York, Albany. They only convert smoke into less visible but more harmful gases and particles, he says. "We ought to outlaw chimneys."

Air pollution comes in three forms: solid matter, liquid droplets, and gas. Both solid and liquid matter are considered particulates. They are usually measured in microns. A micron is equal to about 1/25,000 of an inch, according to the National Tuberculosis and Respiratory Disease Association. Particles over 10 microns in diameter can be seen with the naked eye. Viruses are said to range from 0.01 to 0.1 microns in size. Bacteria are between 1 and 25 microns in size; fog droplets 5 to 60 microns; raindrops 400 to 5,000 microns.

Particles settle out of the air at a rate depending on their size and weight. Dust and dirt larger than 10 microns fall quickly and usually near their source. The same is true of fly ash (the impurity remaining after coal is burned). Other particles, smaller and lighter, liquids, solids and gases, take off on the winds. They include aerosols, smoke, fumes, dust, and mist. These fragments come from steel mills, electric power generation stations, iron and steel foundries, cement plants, aluminum ore reduction factories, phosphate fertilizer plants, refineries, burning, evaporating, condensing, impregnating, and complex industrial processes. They come from smoldering dumps, fuel combustion, municipal incinerators, from cars and other vehicles and devices, from debris blown off streets and other paved surfaces, construction and other sources of dirt, dust and minute contaminants. Each ton of coal burned gives off 200 pounds of solids. A ton of refuse burned by ordinary incineration (around 1,600 degrees) releases 25 pounds of solids.

A primary measure for air quality is the number of particulates in a given amount of air. Clean air is classified by the Atmospheric Sciences Research Center as having fewer than 1,000 particulates per cubic centimeter. Country air has 1,000 to 5,000. Suburban air ranges from 5,000 to 50,000. Industrial city air has more than 50,000. Today these airborne particles are traveling farther than ever before. They are penetrating previously untouched areas, often in startling concentrations. Distant rural areas are starting to have the pollution counts of suburban areas. The ASR Center, located atop the Whiteface Mountain, 120 miles from Albany and 269 miles from New York City, has air as pure as any in the country, yet the particulate count there is reaching that of the suburbs, and the picturesque valley below is often obscured by a blue haze from automobiles and industry.

This spread of pollution has ominous health implications. In every breath we take there are more contaminants. The human lung, for all its cleverness of design and tolerance to insult, is ill equipped to deal with the more subtle dangers of air pollution. The larger particles are quite easily trapped and ejected by the vigilant cilia. But the smaller the particle, the more chance it has to slip past these defenders. Highly toxic or even cancer-causing particles can invade deep into the lung and remain in direct contact with sensitive tissue for long periods; gases or other chemicals may cling to these particles. There is virtually no defense for the tiniest particles and gases that invade the lung. If the ejecting mechanism is knocked out of action by other chemicals, there is little or no protection against the larger particles.

How many of these particulates we inhale and how they affect us depends largely on the individual circumstances of our lives: where we live, our biological makeup, state of health, local conditions, nature of the pollutant, concentration of the dose. The most heavily polluted parts of heavily polluted cities are bombarded with an estimated 50 to 100 tons of particulates per month per square mile. Kansas City on an average winter day gets 66 tons per square mile per month. New York City is blasted with 60 tons of soot and debris per square mile every month.

A New Yorker takes in, roughly, about 70 particulates with each breath, according to Simon Conrad, a meteorologist. This works out to something over a million particulates breathed in by each person per day. On the worst days, Conrad said, the number of particulates could be six or seven times that amount. A New York City medical examiner, quoted in the *Times*, states that on the autopsy table it is unmistakable: "The person who has spent his life in the Adirondacks has nice pink lungs. The city dweller's are black as coal." Once I saw one of these urban anatomical exhibits in a jar. It was astonishing not only how dark the lung was, but how large some of the individual particles were.

We concern ourselves with particulates primarily because they can be more easily measured and seen. They can be coped with. We tend to pick the problems that can be handled and ignore the others. But particulates are becoming the lesser problem as technology becomes more sophisticated and control devices reduce the amount of escaping solid matter. Particulates now account for only about 10 percent by weight of the pollution in the air over the United States, according to Commissioner Middleton. "A full 90 percent of air pollution consists of largely invisible but potentially dangerous gases."

THE SULFUR MENACE

Sulfur is found in nature and is almost invariably present as an impurity in coal and fuel oil. For many years it was believed that sulfur in the air was beneficial to health. As late as 1940 physicians in Pittsburgh were still extolling this ancient belief. Now sulfur is recognized as a harmful contaminant. When fossil fuels are burned, the sulfur joins with oxygen in the air (a process known as oxidation) to form sulfur dioxide. Under certain atmospheric conditions it can form sulfurous acid, sulfur trioxide, and the devouring sulfuric acid. Sulfur leads to sulfuric oxides, the villain that disintegrates nylons, dissolves marble, eats away steel and iron, and damages vegetation. Washed into water by rainfall, its salts can increase acidity and destroy aquatic life. It can cause vile odors and affect health. It is believed to be the great killer in the notorious pollution episodes, such as that at Donora. For people with respiratory difficulties it can be deadly. Sulfur dioxide kills from 1,000 to 2,000 New

Yorkers per year, according to a Ralph Nader study. Sulfur dioxide may cling to tiny particles sucked into the lungs, eluding normal respiratory defenses, and be slowly released into the fragile air sacs of the alveoli. Laboratory studies link it to heart attacks, mutations, and cancer. We could be heading toward disaster, says Dr. Robert Shapiro, associate professor of chemistry at New York University:

> We know that sulfur dioxide causes certain chemical changes. If these same changes occur in human sperm cells, they could lead to mutations that would pose great threats to future generations. And if these changes occur in other cells, such as in the lung, then sulfur dioxide might be a cause of cancer.

Common belief holds that sulfur emissions can be eliminated by control devices. Unfortunately, this is not true. "Contrary to widely held belief, commercially proven technology for control of sulfur oxides from combustion processes does not exist," states a panel of experts from the National Research Council. The panel warns that the broad use of unproved processes exposes us to great risks, and even with concerted action there is little hope for doing more than preventing sulfur dioxide pollution from rising beyond present levels by the year 2000. "Without such action the amount is expected to triple in the next 30 years. Prospects for control now appear dim, with the growth of industry and increasing use of fossil fuels. Sulfur dioxide accounts for about 17 percent of all air pollution by weight.

Through the use of low-sulfur fuels Manhattan, New York City, has been able to reduce sulfur dioxide in the air by almost 50 percent. Despite this gain, concentrations are still often three times the established level of "safety" (0.04 ppm). In the city as a whole, sulfur dioxide levels average almost double the so-called safe limits.

CARBON, CARS, AND CONTROLS

Another well-studied substance is carbon—a constituent of coal, petroleum, limestone, and other matter. When combustion of carbon is complete it forms carbon dioxide. Normally this is not considered a pollutant in itself because it is an essential part of the life process. But it can also be treacherous. In the presence of moisture it converts to carbonic acid and can erode stone and is partially responsible for corroding magnesium and possibly other metals. It also may cause adverse atmospheric conditions by slowly heating the earth's atmosphere.

When carbon is not completely burned it produces lethal impurities. The most notorious is carbon monoxide. It comes from the combustion of any carbon-bearing substance (oil and its products, including gasoline, coal, natural gas, wood, tobacco, etc.). Most carbon monoxide comes from automobiles. In the United States there are some 110 million cars, the number rising each year. Each car burns an average of 700 gallons of fuel annually and discharges more than 1,600 pounds of carbon monoxide, 230 pounds of hydrocarbons, and 77 pounds of oxides of nitrogen. Automobiles produce some 60 percent of the nation's air pollution, up to 90 percent in some places. Their collective effect is staggering. In 1968 they dumped into the atmosphere 200,000 tons of lead, 500,000 tons of carbon monoxide, 314,000 tons of particulates, and 394,000 tons of sulfur dioxide. These are merely the better-known pollutants. Some investigators believe there may be as many as 150 to 200 individual contaminants from automobiles. The poisonous miasma from traffic hangs over the cities in a gray-blue blanket that gradually spreads over the countryside like a creeping deadly fog.

Carbon monoxide is a known killer. Acute carbon monoxide poisoning reportedly claims more than 1,000 lives in the United States each year. Two scientists from the California State Department of Public Health at Berkeley, Alfred C. Hexter and John R. Goldsmith, recently reported statistical evidence that carbon monoxide in the polluted air of communities like Los Angeles was "associated with excessive mortality." Breathed in a closed room, carbon monoxide can bring death in minutes.

Los Angeles has a daily output of more than 20 million pounds of carbon monoxide. Investigators report that at times there is enough carbon monoxide in the air to reduce the blood's oxygen-carrying capacity by 20 percent. The heart and brain are the organs most sensitive to lack of oxygen. The higher the levels of carbon monoxide, the harder the heart must work to supply oxygen, thus putting a strain on the entire circulatory system. There can also be adverse secondary effects. One researcher reports that carbon monoxide in the bloodstream of animals leads to cholesterol deposits three to five times above normal and results in numerous injuries to the arterial walls.

In less than fatal amounts, carbon monoxide can cause a

variety of symptoms of poisoning, such as dizziness and head-aches; and it can affect the brain. Three years ago Dean Myron Tribus of Dartmouth's School of Engineering warned: "We're on our way to a public catastrophe. . . . Carbon monoxide levels in New York City are approaching the lethal level." Aside from its lethal effects, carbon monoxide may be a cause of accidents; it lessens a driver's alertness and warps judgment of time and distance.

The government defines the levels where carbon monoxide intoxication sets in. In cities these levels are almost constantly exceeded. In midtown Manhattan carbon monoxide remains above "safe" margins all day every day. During daylight hours, when traffic is heaviest, the levels are often doubled. This is said to have an impact on the lungs equal to smoking two packs of cigarets a day. One investigator compares it to losing a half pint of blood a day—a loss that would weaken all but the very fit, and they are few indeed.

Where does this monoxide go? Does it convert to carbon dioxide? Is it breathed in by people who act as huge sponges? The answers are not known. Hemoglobin is known to absorb carbon monoxide much more readily than oxygen. It is said to take eight hours to expel what is breathed in during one hour. Maybe, says a whimsical investigator, the lunch hour monoxide from Manhattan is expelled during dinner in New Rochelle. And what effects have taken place meanwhile? The answers, unfortunately, are not known.

The lungs are also vexed with a variety of other car emissions. Nitrogen combines with oxygen to form oxides that can give the sky a bilious yellowish-brown cast. It also reacts with moisture in the air to produce nitric acid, which corrodes metal surfaces and damages plants. Relatively low doses of nitrogen oxides have been found by researchers at the University of California in Los Angeles to reduce the oxygen-carrying capacity of laboratory animals' blood by up to 38 percent and inflame respiratory tracts. Higher doses restrict breathing.

Nitrogen oxides are a key component of smog and a cause of illness. When concentrations of nitrogen oxide in the atmosphere go above 0.083 parts per million there is an average 18.8 percent increase in respiratory ailments, according to federal studies. In New York City nitrogen oxide readings now run about 0.09 parts per million—almost double the levels that federal authorities say are safe.

Smog, once a simple mixture of smoke and fog, is now a toxic broth cooked by the sun. Oxygen, nitrogen, hydrocarbons, and other airborne pollutants react and interact under the influence of sunlight. They form exotic and often unexpected mixtures. Almost invariably they are poisonous—irritating lungs, causing coughing and fatigue, and restricting breathing. They damage plants, crack rubber, fade colors, and deteriorate fabric.

Much has been written about lead. It is added to gasoline so Detroit can produce more powerful cars. Lead is reaching dangerous concentrations in the atmosphere. In some places, where traffic is heavy, it is double the amount where biological effects begin to show. The body is believed to store lead in increasing or decreasing amounts, depending upon exposure and intake. Lead poisoning can interfere with cell metabolism, affect the brain, blood, nerves and other organs. Dr. Clair C. Patterson, a geochemist, says we are all now on the threshold of chronic lead poisoning—and may have gone beyond it.

Cars also give off a variety of other known contaminants. From incomplete combustion come benzopyrene (a cancer-causing hydrocarbon) and PAN (peroxyacetyl nitrate), which makes eyes burn and tear, irritates the lungs, and damages plants. From tires come fine particles of rubber. From brakes comes asbestos dust (as well as from many manufacturing and construction processes, a proven cause of cancer in humans). From roads come dust and abrasive materials, fragments of metals, bits of plastic, phosphorous compounds, and multitudes of other substances, visible and invisible.

No one can predict the effects of these substances singly or in combination with other air pollutants. Automobile fumes are known to be especially damaging to the young. They inhibit learning and thought processes of children more than adults. Pregnant animals exposed to low levels of fumes over long periods of time tend to give birth to mentally deficient offspring. What does this indicate for the children growing up in congested areas?

The immediate prospects are not bright. No smog-free engine is now on the horizon. Detroit says it cannot meet the 1975 deadline that calls for an almost emission-free engine. Existing control devices are largely ineffective. A few hand-adjusted prototypes have reduced carbon monoxide and hydrocarbons by two thirds. But after a year or two only about 10 percent of the devices could meet emissions standards. NAPCA found in the summer of 1970 that cars certified as meeting government standards were actually emitting double the amount of carbon monoxide and hydrocarbons they were supposed to.[4]

A 1971 study in New York City revealed that new model cars, fitted with exhaust pollution controls, fared worse than older cars without controls. More than 60 percent of the 1971 and 1972 model cars tested in a voluntary curbside program emitted excessive carbon monoxide. Only 51 percent of all the 372 cars tested emitted excessive carbon monoxide, according to Commissioner Robert N. Rickles of the Department of Air Resources.

Most of the new vehicles were said to be taxis driven about 35,000 miles. Federal requirements for the new controls require that they remain effective up to 50,000 miles. Manufacturers have certified that they meet this requirement.

Earlier in the year 887 cars were tested at drive-in banks in the Bronx and on Staten Island. More than 43 percent of those cars were said to emit excessive carbon monoxide.

For a system to work efficiently it must be maintained; a single dead spark plug can raise emissions at least ten times over the legal limit. It is an unhappy fact that most cars are not well maintained. The majority are left much to themselves as long as they run, according to numerous studies. In one case a motorist, said a stunned mechanic, had not changed his oil for 20,000 miles. Even conscientious motorists have a devilishly hard time finding a mechanic who knows what he is doing and takes time to do it.

The problem is compounded by driving conditions. The more stop-and-go there is, the more inefficient cars become. The more cars there are, the slower they go. One study showed that cars in Manhattan spend 34 percent of their time idling. City driving produces four times more hydrocarbons and twenty times more benzopyrine than open road driving. In Manhattan a car operates at less than 10 percent efficiency, according to Brian Ketcham of the city's Bureau of Motor Vehicle Pollution Control. "This means that more than 90 percent of its fuel is thrown away as waste heat and air pollution. Of 85 billion gallons of gas that American cars burn annually, 75 billion gallons are, in effect, thrown away."

It is "unlikely" that the auto industry will be able to meet stringent federal emissions standards by 1975, Ketchum points out in a letter to *The New York Times* in February 1972. Further, he says, "strong evidence exists" that present vehicles tend to pollute much more than is reflected by federal emissions tests. "Data collected by New York City's Bureau of Motor Vehicle Pollution Control reveals, importantly, that accessories such as air conditioning and power steering can double and even triple vehicle emissions. . . . Unfortunately, the federal government has based its certifications standards on vehicles with no accessories operating. Its data hardly reflects the real world."

Here we have the lesson that "more makes less." Even if gains are made in emissions control, they are largely wiped out by the increasing number of cars and car use, unless emissions are sharply curbed. The total number of autos is expected to increase by half within slightly more than the next decade. Between 1964 and 1966 mileage traveled by automobile rose from 170 billion miles to 470 billion. It is still rising at a phenomenal rate.

Some recent developments are encouraging. New Jersey has announced a plan to start testing automobiles for emissions. All makes and models will be checked by a newly developed device during their annual inspection. The program was carried out with a federal grant. Jersey was picked as the demonstration state because it is the most densely populated state with the highest urban concentration in the nation and has the most densely traveled highways. Any car that cannot meet the state standards within a two-week grace period, after failing the emissions test, will be banned. The driver of a car used after the grace period will be arrested. At least a third of the state's 3.3 million cars are expected to fail the test. Under the testing program the state expects to remove about 20 percent of the 4.5 million tons of carbon monoxide gas and 32 percent of the 750,000 tons of smog-producing hydrocarbons emitted by cars during the first year of the program. California has an automobile emissions inspection program, but it is a roadside spotcheck operation that applies only to models produced in 1966 and since then.

A second encouragement is the number of cities that are starting to restrict automobiles. New York City has carried on an experimental program banning traffic on Madison Avenue, one of the city's major thoroughfares, on certain days, and automobile traffic is also banned from Central Park on weekends. Highland Park, Michigan, adjoining Detroit, is using a more indirect method to reduce traffic. It is turning major one-way streets into local two-way streets with plenty of stop signs to discourage motorists from using their cars for commuting. City officials said they took the action because the automobile is ruining the quality of life in their community.

THE PERVASIVE CARCINOGENS

All of this bad news does not even take into account the known cancer-causing chemicals in the air, except for a couple named. "We are surrounded by a sea of carcinogens," according to Dr. W. C. Hueper, former head of the Environmental Cancer Division of the National Cancer Institute. The air around us swirls with arsenic, mustard gas, benzol, tar, asphalt, creosote, oil, carbon black, paraffin oil, chromium, and many others running into the hundreds, possibly thousands. Since research has been limited, little is really understood about the cancer-causing mechanism. We are proceeding with some kind of blind faith that we are not on a disastrous and irreversible path. Cancers can take up to thirty years to develop after a single period of exposure. It is known that certain air pollutants can cause cancers in animals, and others (such as asbestos) can cause cancers in humans. Animals have developed cancers when certain air pollutants were applied to their skin. Mice developed lung cancers when exposed to tars taken from the air of American cities and from asphalt road sweepings and soot.

Of all the air pollutants, the carcinogens are the least understood and the most treacherous. Many years passed and many people died wretchedly before it was discovered that a contaminant resulting from the aniline dye and benzine industries was causing bladder cancer. Workers in the plants and people living in the immediate vicinity were dying of the disease. The offending pollutant was especially hard to trace because its target organ, the bladder, is so far removed from the lungs. The problem might have gone undetected indefinitely had not the number of bladder cancers been sufficiently high to attract attention.

These warnings neither intimidate nor deter. We fearlessly breathe in the carcinogens, deadly poisons, exotic gases, compounds known and unknown: lethal military nerve gases that have killed sheep grazing in fields miles away; fluorides from fertilizer plants that have ruined beautiful groves of citrus and other trees and shrubbery, settling on pastures and poisoning cattle or making them so sick they had to be slaughtered. PVC (polyvinyl chloride) is one of the new type plastics (3 billion pounds produced in 1968 alone); when burned in incinerators it can irritate the skin and upper respiratory tract. Certain plastics can produce one-half their weight of hydrochloric acid gas when incinerated; this makes gases increasingly corrosive as more plastics are used. These contaminants are in the life's breath of the old, the young, and newborn babies.

What happens to this mysterious collection of guck and poison so recklessly hurled into the air? Where does it go? What is its ultimate disposal? Very little is probably lost. Very little is believed to escape into space. Most of it stays right here, recirculated in a variety of forms through nature's own system of recycling. There have been many ingenious attempts to improve this system—men are forever trying to instruct nature in her duties and morality—but most have not proved successful. Nature is set in her ways. One creative New Yorker wanted to erect, high along the Hudson River, a battery of enormous fans to blow back New Jersey's migrating fumes and flavors, along with a few originating in New York for good measure.

The idea, although mischievous, illustrates a valid point. Virtually all pollution control is based on the concept implied by the fan. Even the most modern technology is often a refinement of that basic approach. It is expressed in the cliche: The solution to pollution is dilution. This means, in effect, sending your mess elsewhere in a less concentrated form. This is the principle behind the smokestack, the effluent pipe, and dumping wastes in inland waters and the ocean. By this logic and its offspring techniques we reduce the severe local problems to a milder worldwide problem. If we have not shared our great wealth with the less fortunate nations, we have not been stingy in sharing our wastes.

Air pollution is a gypsy. It does not stay where it is generated. The heaviest particles may land near their source, to vex the offender and his neighbors, but the others take off. The higher the smokestack, the more they are propelled in their flight. The particles hitch a ride on passing air currents. They go downwind, much as raw sewage and industrial wastes dumped into a river become the problems of the communities downstream.

The lighter particles remain aloft, transporting their toxic load, the cancer-causing substances that cling to them, exotic gases that envelop them. They may circle the globe many times. They may stay aloft hours or years. Eventually they sift down or are washed out of the sky by local precipitation. By then they may or may not be the same compounds that left the home stack or bonfire. They have been exposed to chemical changes due to wind, sun, and water vapor. They have assumed new shapes and forms as they weather and age. Some remain or become harmless. Others turn increasingly savage.

Like their chemical fate, their journeys and destinations are unpredictable. They flake down on fields, streams, streets, and buildings. They are sucked into unsuspecting lungs, retained or liberated, and possibly take flight again. They tinge and kill vegetation. They etch an indelible mark on materials, buildings, and monuments. Ultimately most of these transient substances percolate into groundwaters or wash into streams and reach the sea, where they travel up the food chain. The more indestructible may be evaporated into the air to resume their endless journey like the water molecule that bears them.

The process is infinitely complex, the possibilities endless. Air and water become one. There is no air or water pollution. There is only pollution. We use only 10 percent of the water supply, says Dr. Raymond L. Nace of the U.S. Geological Survey. "Ninety percent is a conveyor belt to carry wastes to the sea." The air over the North Atlantic Ocean and over the Indian Ocean off Southeast Asia is twice as dirty as it was in the early 1900s, according to a recent government report. "We know we are changing the atmosphere," says meteorologist William E. Cobb. The question is how and at what rate.

The dispersal of pollutants depends largely on weather patterns and atmospheric conditions. Most aerial garbage is discharged into the lower part of the atmosphere called the troposphere—a band about 12 miles thick around the earth's crust, containing some 95 percent of the total air mass. The remaining 5 percent is in an upper level that extends 250 to 900 miles away from the earth; there it gradually merges into airless, cold, empty space. To date it has been found to be lifeless and alien to man and his aspirations to escape his earth prison.

The sun powers our solar system, heating the earth unevenly as it spins on its oblique axis. This makes the air of the troposphere churn constantly. It gives impetus to the winds that disperse pollutants. These winds move vertically and laterally, diffusing, moving the air, carrying out the water cycle that cools the earth, refreshing it, preventing the sun from turning it into a hard-baked bare desert. Rains distribute the water. The balance of moisture in the air is altered by the sun's heat. This affects the fate of many pollutants.

Nature almost invariably works by subtle indirection. Her method of heating the earth is not to have the sun directly warm the air; our atmosphere is incapable of absorbing much of the sun's energy directly. Instead, the earth absorbs the waves of light energy radiating from the sun. Then the air molecules that touch the warm earth are themselves warmed in a process of "conduction." These heated air molecules, in turn, touch other air molecules and transfer their warmth to them. As the air heats, it becomes progressively lighter; it expands and rises. The cooler air above sinks and replaces the warm air. When this cool air is heated it too rises, setting in motion a continual flow of air like a fountain rising and falling. This is how the air in a closed room is heated and circulated.

Under certain conditions the system of air updraft and downdraft is stalled. At night, when there is no sunlight, the air chills and ceases to rise and churn. It is stable, much as the thin air in the stratosphere is stable. When pollutants are discharged during the day they tend to rise on the heated upflowing air until they are blown away by passing winds. If released at night they tend to sink nearby because of the lack of vertical movement of air. The prudent polluter rarely releases his discharges at night when they will rain down on him.

As long as the surface air is cooler than the air above, it cannot rise; it is said to be stable. In the morning, as the sun warms the earth, the heated air touching the earth starts to rise, dispelling the cooling air above that has formed during the night. But if the cool surface air becomes trapped by a freakish layer of warmer air above, or if it is pressed down by an invading high-pressure mass, it cannot rise or escape. The warm overlying layer of air acts as a lid. Then we have one of the dreaded inversions.

During an inversion pollutants build up in the stalled atmosphere. The fixed amount of air soon exceeds its ability to dilute the impurities released into it. The longer the inversion lasts, and the more stable the air, the greater the concentration of pollutants. The mix thickens and ferments. It may be compounded by moisture in the air. New and progressively complex chemical reactions result and profound changes follow. These may include the ferocious attacks by sulfuric acid and various oxides cooked up by the sun. Inversions have been responsible for the great pollution killer episodes in London, Donora, and New York.

No one has yet devised a way to get rid of inversions, short of waiting for nature to disperse them in her own time. One innovator suggested hanging a giant chimney from a balloon so the cool air below could pierce the warm air above, releasing the trapped pollutants. But how to get such a marvelous chimney aloft? And once in place, how to stoke the monstrous updraft it would require without causing even greater pollution?

The standard technique for combating inversions in most highly industrialized areas is to keep monitoring the air. As pollution builds up to dangerous levels a series of alerts are sounded. In effect the patient is warned that if he does not act quickly he will die. If the condition warrants, steps are taken to reduce emissions. This has generally proved successful in holding down concentrations of contaminants in local areas; at least large numbers of people have not been dropping dead in the streets. But at the same time the more subtle worldwide threat is becoming progressively greater.

Air pollution is now believed to be altering weather conditions and climate throughout the world. It does this in several ways. One is through the increasing amounts of carbon in the air. About 6 billion tons of carbon are mixed with the atmosphere annually. In the past century, with industrialization and the burning of fossil fuels, more than 400 billion tons of carbon have been artificially introduced into the atmosphere. The concentration in the air we breathe has been increased by approximately 10 percent, according to Lord Ritchie-Calder, United Nations science adviser. If all the known reserves of coal and oil were burned at once the concentration would be 10 times greater. This is something more than a public health problem, more than a question of what goes into the lungs of an individual, more than a question of smog, he warns. "The carbon cycle in nature is a self-adjusting mechanism. Carbon dioxide is, of course, indispensable for plants and is, therefore, a source of life, but there is a balance which is maintained by excess carbon dioxide being absorbed by the seas. The excess is now taxing this absorption, and it can seriously disturb the heat balance of the earth because of what is known as the 'greenhouse effect'."

The greenhouse concept is now familiar to most people. A greenhouse lets in the sun's rays but retains the heat. Carbon dioxide turns the entire earth into a huge greenhouse. The sun's ultraviolet rays easily penetrate the atmosphere. Upon striking the earth they are converted to infrared radiations that are trapped by the surrounding envelope of carbon dioxide, much in the way a greenhouse holds its captured heat. This heat warms the surface of the earth and is believed to be modifying climate.

By the year 2000 the extra heat from fuel-produced carbon dioxide in the air might be sufficient to melt the Antarctic ice cap, according to the President's Science Advisory Committee; other sources put the disaster further into the future. "The melting of the Antarctic ice cap would raise sea level by 400 feet," states the report. "If 1,000 years were required to melt the ice cap, the sea level would rise about 4 feet every 10 years, 40 feet per century." This would mean that much of the world's inhabited land and many of its major cities would be lost under the seas.

At the present rate of increase, the mean annual temperature all over the world might rise 3.6 degrees centigrade in the next 40 to 50 years, according to estimates cited by Ritchie-Calder.

The experts may argue about the time factor and even about the effects, but certain things are apparent, not only in the industrialized Northern Hemisphere but in the Southern Hemisphere also:

The seas, with their blanket of carbon dioxide, are changing their temperature, with the result that marine plant life is increasing and is transpiring more carbon dioxide. As a result of the combination, fish are migrating, changing even their latitudes. On land the snow line is retreating and glaciers are melting. In Scandinavia, land which was perennially under snow and ice is thawing, and arrowheads of more than 1000 years ago when the black soils were last exposed, have been found. The melting of sea ice will not affect the sea level, because the volume of floating ice is the same as the water it displaces, but the melting of ice caps or glaciers, in which the water is locked up, will introduce additional water to the sea and raise the level. Rivers originating in glaciers and permanent snow fields will increase their flow; and if ice dams, such as those in the Himalayas, break, the results in flooding may be catastrophic. In this process the patterns of rainfall will change, with increased precipitation in some areas and the possibility of aridity in now fertile regions. One would be well advised not to take 99-year leases on properties at present sea level.

It is not quite this easy to write off the world. Even doom is not so predictable. The gradual heating of the earth has been confounded by a sudden and mysterious interruption. For the past twenty years the temperature has been dropping approximately 0.2 percent per year. Why? Scientists do not know. Is it a short-term phenomenon that will reverse itself, or the beginning of a long-range change that could end in a new ice age? These are the imponderables no one can answer. But it is believed that the odd changes are due to man's activities.

It is known that particulates in the air keep sunlight from reaching the earth and exert a cooling influence. Recent government studies show that air pollution has reduced the amount of sunlight reaching Washington, D.C., by 16 percent in the past fifty years; most of the sharp reduction is believed to have occurred fairly recently. This can have health effects. In the German city of Gelsenkirchen, an industrial center on the Rhine, children were found to have more rickets because of less sunlight due to air pollution. The increasing number of particles in the air also provides nuclei for water vapor to cling to, increasing the amount of cloudiness and the likelihood of rain. One study found that LaPorte, Indiana, twenty miles southeast of Chicago, had rainfall about 30 percent in excess of nearby areas. At least 15 percent of the rainfall was due to Chicago's effects on the air, according to an investigator.

The influence cities have on weather has long been known. Only recently has it been related to air pollution. The pavement of a city tends to draw heat. Unlike the country, where soil is warmed to a depth of only a few inches, heat penetrates deeply into a city's concrete and metal; it can sink in 30 inches to 30 feet or more. In the country the stored heat is released quickly at night and the air cools. But the city surrenders its heat slowly and reluctantly. It escapes gradually into the air to give people restless and uncomfortable summer nights. Even the escaping heat is not dissipated quickly; instead it plays against the soaring vertical surfaces of tall buildings, bouncing back and forth like an endless game of catch. Meteorologists call this the "heat island effect." New York City averages 10–15 degrees hotter than its suburbs in Westchester county.

This concentration of heat sets up a whole chain of circumstances. Warm air tends to concentrate in the city's center, probably because of the mass of tall buildings, as explained by the National Tuberculosis and Respiratory Disease Association's *Pol-*

lution Primer. This warm air rises, bearing its burden of pollution. Then it expands, flows outward over the edges of the city, cools and sinks. Cooler air from the edge of the city flows into the center to replace the rising air. This is followed by the now-cooled dirty city air. A self-contained circulatory system has been set up; it can be altered only by a strong wind. It would seem that this circulatory system would assure ventilation of a city. But it does not, the *Primer* observes:

> City air contains large amounts of aerosols, tiny particles small enough to remain suspended in the air and to be moved about as the gases of the atmosphere are. As the dirty air moves up over the city, spreads out, cools, and sinks at the city's periphery it forms a quite distinctive, recognizable ceiling known as a *dust dome* or *haze hood.*
> The aerosols making up this ceiling can reflect heat back into space just as the earth does on a much larger scale. But the radiation takes place *above* the city. Meanwhile, the pollution formed ceiling prevents the sun from efficiently heating the surface air. And under this ceiling, each windless day the city's air grows thicker and grimier.

Natural atmospheric conditions often intensify the effects of pollution. As a rule, wind blows most lightly during the early morning and hardest in the early evening. At the time that pollution is greatest the winds are resting. In the early morning hours the pollution load is heaviest due to rush-hour traffic and freshly fired furnaces; this smog is likely to linger for hours until the winds begin to stir.

Location also has a profound effect on the degree of air pollution. Cities ringed by mountains or sunk in valleys, where breezes are stalled, tend to accumulate more pollutants. Open places, on the other hand, have much better natural ventilation. Coastal cities are usually in the best position of all. It is said that if New York were where Los Angeles is, backed by a protective barrier of mountains, everyone would be dead. Both cities are probably saved from disaster by their coastal positions. The adjoining land and water cause a regular exchange of air. The air over the land gets hotter during the day than air over the water. The layer of air touching the land warms quickly and rises, and the cooler air from the sea rushes onto the land. At night the process is reversed; the land cools faster than the water and the land air moves out over the sea under the warmer sea air. This ebb and flow of air gives pollutants less chance to build up to lethal concentrations.

On a local scale, pollution depends on several factors: what is put into the air, when, where, and the prevailing atmospheric conditions. It is also important how rapidly pollutants are stoked into the air; this affects the degree of combustion and nearby fallout, whether the city gets back its own wastes or fans them out via the winds to the surrounding suburbs and the globe. On an international scale the most important factors are quantity and quality, not when, where, or at what speed wastes are ejected. As pollution spreads throughout the world, the air tends to deteriorate almost uniformly beyond the centers of contamination. The increasing concentration of people and industry into small land areas exaggerates the problem. Gradually the fixed amount of air is being overwhelmed by the crud dumped into it. It no longer can cleanse itself through dilution and dispersion. The background pollution is starting to build up. The quality of our limited supply of air progressively deteriorates.

Despite all of the clamor and complaints about air pollution, progress in cleaning it has been spotty at best. In most cases it is nonexistent. As late as 1968, out of 7,300 communities reporting air-pollution problems, only 130 had control programs of any kind, according to government figures. A year later that dismal figure had not risen appreciably, according to Commissioner Middleton.

The first legislative act to correct air pollution was posted in 1955. It was followed by the Clean Air Act of 1963 and the Air Quality Act of 1967. Failure was written into these laws. They were weak and riddled with loopholes, practically dictated by industry. They were not enforced. They were ridiculously underfunded. Essential research was—and remains—almost nonexistent. Programs have been beset with red tape, muddled authority, conflicting jurisdiction, indifference, resentment, and opposition by states and localities to federal interference.

Industry pressures have successfully thwarted the laws, such as they were. On the state and local levels there was widespread refusal to alienate or drive away tax-paying industries; polluters were courted with variances, exemptions, delays. Some communities have actively aided polluters. Many entice them by offering enforcement-free havens or token wrist-slapping fines—in effect, a cheap license to keep on polluting. Air-pollution control in the United States has been largely a farcical charade. Conscientious officials are frequently subjected to abuse and unbelievable pressures. Senator Abraham Ribicoff told a congressional committee about the leverage exerted against him when, as governor of Connecticut, he tried to abate pollution:

> The mayor of an affected city pleaded for leniency toward polluters on the grounds they might leave town taking valuable jobs with them, if anti-pollution laws were passed. Industry and labor leaders backed the mayor . . . it takes a lot of courage to move into this field.

The 1967 Air Quality Act, for example, required the establishment of air quality regions, the publishing of criteria by the federal government from which states would develop air quality standards, and implementation plans requiring federal approval. In 1971, just prior to passage of a new air law, William D. Ruckelshaus, head of the Environmental Protection Agency, said:

> We are just now about ready to approve the implementation plans for some of the first regions designated. This means that now, in 1971, not one grain of dust, not one liter of gaseous pollution has yet been removed from the atmosphere of this nation as a direct result of the 1967 air legislation.

At the same time it was noted that in three years the government had designated only 34 of 109 projected control regions. And only one of them (Philadelphia) had set standards.

The regional approach to air quality made it possible for different regions to have different standards. Local officials often adjusted the standard to an adverse local condition rather than upgrading it to a tough uniform national standard. It accommodated the polluter, especially if he sat on a local board that determined the standard. Degraded air tended to become the norm. The temptation was to set the lowest possible goal because it was the cheapest and easiest to meet.

Pollution laws have been further weakened by their underlying philosophy. This is the basis for the so-called threshold concept—the mechanism used to legislate air and water as official sewers to accommodate industry. Industry, under the concept, is permitted to degrade the environment for its profit and the public is expected to clean it up from taxes. The law is designed not to protect the public as much as the sewer; it prevents the sewer from being overloaded to the point where it clogs up so badly that it no longer can function as a cheap and effective carrier of wastes. The critical point of breakdown for an air or water sewer is its "assimilative capacity."

In setting health standards, public health officials translate the engineers' "assimilative capacity" into "threshold levels." This, theoretically, is the point of corruption where water or air becomes dangerous to human health. To pollution experts, this is "air quality management" and "water quality management." These systems are inviting in their simplicity, says Environmental Protection Commissioner Richard Sullivan of New Jersey. "They are logically intact and they appeal intellectually."

The standards concept is based on the idea that experts have examined the literature on the subject to assess how dangerous the various contaminants are at various levels in air and water. Based on their findings they draw a line. This marks the threshold of harm. "This is *the line*," says Sullivan, and he underlines it. The line has almost the quality of magic:

> It is officially pollution if it is dirtier than this; it is officially unpolluted if it is cleaner. Then one devises regulations to bring us from where we are now, determined by measurement, to where we have decided we would like to be. This is the essence of the Air Quality Act of 1967; it is the essence of the Clean Water Act.

The approach is useful, observes Sullivan. "It gives dimensions of control. It allows us to circumscribe the issues at least to some extent, and to get a handle on what we need to be." But is it valid? Does it protect public health? Sullivan says the present standards concept has two basic fallacies. The experts not only do not know what they are doing; in some cases they are even encouraging more pollution. The first fallacy, he says,

> —and I think this even pervades a good portion of the scientific community—is that it is based on the unsound assumption that a sufficient degree of sophistication now exists to allow us to draw these lines—and it doesn't.
>
> We don't know where to draw the lines. We are drawing them, because we are going through this process. We have looked at the literature, and we say: "Well, the threshold seems to be about here, and now we will put in a stupidity factor and make it there, just to protect the public." This line is solemnly promulgated as marking the official definition of

pollution. In fact, it is a mistake to rely so completely on this line when the information that backs it up is so lacking in precision and reliability.

The second fallacy, he explains, is that

. . . the mere drawing of the lines suggests to the community, as a matter of philosophy of pollution control, that it is OK to pollute, so long as you don't go over this holy line. Thus, if the stream where you locate is clean, or if the air is clean, presumably you can shoot stuff right out the stack or the outfall until pollution is up to the official dirt level.

At that point we will try to push it back down to where we should have kept it in the first place. I think that is wrong. Yet on this issue we are being sued by the du Pont Company* which is challenging the application of our regulation to Salem County, New Jersey. The company contends that it is already nice and clean down there, so why should it have to spend extra funds to prevent the entry into the atmosphere of sulfur dioxide. That case is pending in the courts.

This means more pollution under the law than without it. Almost invariably the standard represents a compromise. Generally it has little to do with human welfare. It is not dictated by health but by economics. It does not prevent damage but merely minimizes it, at best. Local standards are often either set or approved by industry. The standards are fixed at the negotiated point where, as some corporate officials say, "we can live with them." That means the standard can be met with little or no expense. In reality it generally has little bearing on clean air or water; an arbitrary and often meaningless goal has been achieved. More often, even this minimal objective is not met. But at least the low standard does not make the community look quite so bad as if it were measured against a realistic level of purity and safety.

It was long obvious that the varying regional air standards were a deception. Either air is safe or it is not. You cannot say it is safe at one level in one air shed and not safe at that level elsewhere. Biology does not respect "economic feasibility."

In 1970 Congress recognized the weakness and travesty of the existing law and passed the Clean Air Act of 1970. This was a tremendous improvement on its predecessors but not without its

---

* The suit was dropped by du Pont.

weaknesses and ironies. The old law, proposing minimal standards that had never been met, was now replaced by a tough law that provided for tough standards. It was largely based on interpretation. Officials could be as tough, or lax, as they liked. The law would be as good as the men who administered it.

In at least one respect the law was historic. For the first time Congress deleted that ignoble phrase "economic feasibility." The only criterion Congress considered was the effect of polluted air on human health and welfare.

The new act provided for national ambient air quality standards and a six-year deadline for the automobile industry to develop an engine that eliminated 90 percent of the emissions permitted in 1971 models. The act broke down standards into two categories: primary and secondary. Primary standards were those affecting health. Secondary standards applied to human welfare, such as environmental effects, plants, damage to buildings.

Under the law the states had to submit schedules for achieving standards set by EPA by January 1, 1972. EPA then had four months to approve or reject the schedules. If EPA rejected a state plan it had until July 1 to impose its own plan on the state. The states had until July 1, 1975, to meet the primary standards for health. Secondary standards were supposed to be met within a "reasonable time."

The strength and weakness of the law were virtually the same. It was primarily subject to human interpretation, as most laws are. It could be as good or bad as those who administered it wanted it to be. This left open to EPA the level of standards and determination of compliance. What was to happen if a state did not achieve compliance within the stated time? And who would decide what compliance was? Would carbon monoxide be measured at the curb or 50 feet away; in the center of a city or its outskirts? Would sulfur emissions be measured at the smokestack or after they dispersed?

Obviously, no one was optimistic that the big cities—the biggest polluters—could meet the 1975 deadline. Ruckelshaus, the head of EPA, tried to whip industry along to meet the new goals without waiting for the deadline. He rejected the common argument that it was only good sense for industry to wait for improved technology. The technology for eliminating smoke and soot is "available now," he said, and any industry that "does not apply it is clinging to the nineteenth-century environmental ethic. No matter what the current status of state, local or federal laws affecting their actions, not to clean up smoke is inexcusable."

He said that if present technology were applied by industry, apartment houses, and commercial buildings, the 17.5 million tons of smoke and soot (particulates) now emitted annually from controllable sources would be reduced to 700,000 tons, a 95 percent reduction. And power plant emissions would be cut from 5.6 million tons a year to less than 80,000 tons; iron and steel manufacturing would cut emissions from 1.5 million tons to 30,000 tons a year.

A few months after passage of the 1970 act, EPA issued national standards for the six principal pollutants: sulfur oxides, particulates, carbon monoxide, hydrocarbons, nitrogen oxides, and photochemical oxidents. The standards were relatively good but left little margin between "safe" levels and the point of known adverse effects at present limited knowledge (as more is known about pollutants, the tolerance levels are almost invariably reduced as subtle effects are recognized). Industry was not happy about the new standards. A major corporation almost at once challenged the secondary standards in court as unreasonably strict.

Soon after the new air law was passed, the Administration began to challenge the intent of Congress by injecting concern for its economic impact. Sydney Howe, president of the Conservation Foundation, expressed concern over reports that the regulations proposed by EPA "are being seriously altered within the executive office of the President."

His concern proved well founded. Once more the Nixon Administration placed economics ahead of human welfare. Under intense political pressure from the White House and its advisers, EPA issued federal regulations seriously weaker than the agency originally planned to issue. Several proposed improvements in standards and enforcement were, as *Environmental Action* observed, "badly mauled." It stated:

"Nearly all of the sample regulations dealing with permit systems, emissions monitoring, air pollution emergencies and compliance schedules vanished. . . . Six sections, each suggesting feasible emission limitations for a major pollutant, no longer appear as regulations but merely as a listing of possible control levels. With all regulatory language removed, the suggesting power . . . was reduced substantially. . . ." EPA meekly went along with the changes in its proposals and even defended them.

Ralph Nader was less tolerant of political expediency and commercial accommodation. He charged that "the Administration has decided that the lives and health of American citizens must take a back seat to corporate desires to minimize production costs. The changes were made at the request of key government officials who represent the views of industry, rather than the public."

It would be incorrect to say or imply that no companies and no states or municipalities are trying to clean up the environment. Some are. But they are exceptions. New Jersey has passed some of the toughest antipollution laws in the country and is trying to enforce them. Minnesota has set up a network of some 400 air monitoring stations in an effort to find out what is going into the air and where. Los Angeles is the only metropolitan area in the country that has virtually cleared its skies of pollution from stationary sources, and helped push the federal government into passing legislation demanding "pollution-free" cars. Chicago, Texas, and Kentucky have expanded their air-pollution departments. Several other cities are tardily formulating programs and appropriating funds. There are now about 200 state, regional, and local air-pollution agencies, whereas a decade ago there were none. There would be more progress if the federal government gave more leadership and showed more courage and integrity in cleaning up the environment.

Where is this leadership to come from when leadership is so much of the problem? The root trouble is that present air control laws are geared to political expediency rather than human welfare. The present strategy is still reliance on ambient standards and the "air quality management" concept which does not require that everything that can be done for control actually be done. This is explained by Virginia Brodine, writing in *Environment* (January-February, 1972):

> Essentially, ambient standards allow the air to get as dirty as we can stand it; the minimum tends to become the maximum, clean air tends to get dirtier, at least to the level of the secondary standards. The principle of ambient standards based on health effects accepts the notion that anyone has the right to use the air for waste disposal as long as he does not endanger the health of others. As one air pollution control official recently put it:

". . . the so-called 'air quality management' philosophy was developed originally as a means of justifying control standards in the face of otherwise insurmountable political opposition."

In the same article Miss Brodine points out that "what the public has yet to learn is that the present approach to air pollution control of stationary sources, in the long run, will *not reduce the total burden of pollutants* [emphasis hers]—from stationary sources—industry, power generation, agriculture, space heating, and waste disposal. Gains in controlling emissions from each single source will be offset by the increase in the size and number of sources. Moreover, it is still quite uncertain just what the control devices on automobiles will achieve and in any case these devices will have their maximum effect on pollution reduction about 1990 and emissions from this source will then begin to climb again if the number of automobiles on the road continues to increase. Either we will fail to meet the goal of improving air quality in the cities over the long haul, or we will fail to meet the goal of protecting air outside the cities from deterioration, or both."

She says that even if present control methods being developed are applied and work, there will continue to be an overall rise in total sulfur emissions, nitrogen oxides, radioactive discharges, and in particulates, especially the dangerous fine substances below one micron (0.000039 inch in diameter). "New pollutants can be expected to appear in the atmosphere from the manufacture, use, and disposal of new products, altering the composition of air pollution. New pollutants may be discovered, investigated, and finally controlled, whereupon they may be replaced by some other contaminant. Past experience and present standard-setting procedures suggest that control of a new or previously uncontrolled pollutant will lag years behind its introduction into the atmosphere."

THE UNHAPPY LESSON OF NEW YORK

Often the failure to make greater progress is due not to lack of good intentions as much as the magnitude of the problem. New York, as the mecca of pollution in the United States, has made heroic efforts to wash its skies with only limited success. The city's air is classified "unhealthy" nearly 25 percent of the time. It is "unsatisfactory" most of the rest of the time. The city has had to face almost every obstacle. It is assaulted by virtually every known chemical. Its skies are charred and discolored by smoke, cinders, sulfur you can taste, car emissions, factory discharges, a medley of strange and noxious odors that come from the city itself and neighboring Long Island and New Jersey. It is impossible to lie on an apartment roof in Manhattan without soot and grime smudging bare sweaty flesh. To dine outside requires, first, a thorough scrubbing, and then an umbrella to deflect the fallout. The tenacious soot piles up on window ledges, begrimes faces, and dirties shirts. It falls unremittingly, like black snow. A Manhattan woman forgot to close her kitchen window when she went away for a few days. Upon her return it looked like "the inside of a coal bin."

The city is assaulted annually by 69,000 tons of particulates, 400,000 tons of sulfur dioxide, 1,423,000 tons of carbon monoxide, and 303,000 tons of hydrocarbons. Some of this comes from Con Ed's 11 plants. A smaller amount comes from industry. Though industry contributes less than 10 percent of the city's overall air pollution, it creates severe local problems. For example, residents of Maspeth, Queens, have picketed the Phelps Dodge Corporation's copper refinery which annually discharges into the air 100 tons of solid copper, copper oxide, and fly ash, according to city officials. (The company insists that "there is no particular pollution at all" from the refinery.)

More than half the city's pollution comes from automobiles. New York has 2 million cars. Half are squeezed into tiny Manhattan, which could reasonably accommodate half that number. The fumes are trapped in the granite canyons and glass walls, unrelieved by sea breezes that are becalmed by the great outpouring of heat and soaring buildings. Much of the pollution comes from stationary sources—municipal and apartment incinerators and oil burners, from office and public buildings, public and private housing.

In 1970 the New York City Air Resources Administration cited a building at 40 Fifth Avenue, just north of Greenwich Village, as a typical example of how apartment buildings contribute to pollution. By the city's standards it is a medium-small building: 16 stories tall, containing 78 apartments with 400 rooms. The building's oil burner pours into the air annually 18 tons of sulfur dioxide and 140 pounds of particulates. The incinerator, which burns 575 pounds of refuse daily, discharges another 400 pounds of dirt particles a year. The city has a total of 15,000 incinerators and 50,000 oil burners. Together they dump a total of 667,840 tons of noxious pollutants into the air annually, according to Robert J. Kafin, President of Citizens for Clean Air, Inc.

Faced with the problem of controlling this blast of pollution is the New York City Department of Air Resources. The agency has 300 employees and a budget of $4.1 million ($914,000 from the federal government). It is in a building owned by The Cooper Union, an engineering school east of Greenwich Village. Its laboratories are a maze of complex monitoring devices, testing devices, and the center of an electronic network that links 10 automatic sampling stations in a complex of 38 stations throughout the city.

In charge of public affairs is Mrs. Carolyn S. Konheim, a chance convert to the cause of fighting air pollution. She had taken her baby to the park several years ago; when she lifted him from the carriage his outline was etched in soot. The experience shook her. She began worrying about what the dirty air would do to her child over his lifetime. Then she began to brood about the effect on everyone. She organized a citizens' group that was to become Citizens for Clean Air. Later she joined the city agency. Unlike many antipollution professionals she still has an emotional involvement in her work. She speaks with deep feeling about the complexity of the problem of trying to learn what is in the air in trace amounts, the effects these pollutants have on people, the difficulty of being limited to crude instruments that often can pick up contaminants only at high levels when even low levels over long exposure can have significant effects. There is also the problem of the changing concentrations and character of the pollutants themselves. They change with the time of day, often from hour to hour and block to block. They vary with the season, with atmospheric conditions, in the concentration they appear in, individual makeup, and combinations. Some chemicals are diluted or altered by the water that traps them. Some burn off as gases. Some interact with others in the presence of sunlight to form more toxic end products for which there are no really satisfactory measuring techniques. Research is invariably hampered by lack of funds.

"We now know the volume of dirt, but we're just starting to try to learn what's in it," she says. "There are a number of different types of organic materials—some are likely carcinogenic. How many there are in the air, nobody knows. The monitoring system is picking up volumes of metals, many of them highly poisonous—lead, nickel, copper, iron, manganese, cadmium, chrome, zinc, cobalt, vanadium, beryllium, mercury, silver, tin —almost anything of industrial production and combustion in high-temperature incinerators." She extended her arm in a sweeping motion that took in the laboratory where she paused during our tour. "We are always uncovering new problems, like concentrations of lead and asbestos. This leads us to solutions, such as the banning of asbestos for fireproofing and lead in gasoline, under the city's new air-pollution code. Of course, then we have to be on guard for new problems resulting from the use of substitute materials."

New York's major advance in air-pollution control has been to reduce the amount of sulfur permitted in fuels. This has cut sulfur emissions about 50 percent and flying dirt by 25 percent. The city is, properly, proud of these gains. But sulfur dioxide is still well above federal limits. Carbon monoxide, as previously noted, is three times the "safe" amount; on occasion it has been five times higher. Air pollution in New York City kills from 1,000 to 2,000 people per year, Ralph Nader estimates. He cites sulfur dioxide as the chief culprit.

Analysis of the breath of 500 residents of the Bronx revealed that 17 percent had symptoms of serious respiratory diseases that the victims themselves were unaware of. The most prevalent ailments were asthma, emphysema, and chronic bronchitis; if not caused by air pollution they are at least exaggerated by it.

The enforcement of air-pollution control regulations in New York City is "far behind schedule," according to Kenneth L. Johnson, regional director of the Federal Pollution Air Control Administration. This raises questions of the city's continued eligibility for federal aid. City Council Minority Leader Eldon Clingan, who led the fight for the city's tough new air law, says the city is in a "crisis situation."

The city came in for an unpleasant surprise in February, 1972, when it learned that despite all its efforts at cleaning up the air, it was, in fact, getting progressively dirty. The news must have been especially disquieting to Mayor John Lindsay who was, at the time, in Florida, campaigning for the Presidential nomination, boasting how much progress New York City had made in improving its air. *The New York Times* reported that "City officials conceded for the first time that one of the key air pollutants here—suspended particulate matter, or dirt—had increased 7 percent over the last three years despite stepped-up control efforts." It explained that the city's daily air pollution index, which classifies the air on a scale ranging from "good" to "unhealthy," does not include the readings on suspended particulates. "The index, therefore, has indicated that the air was getting better and better over the last few years. Yesterday, for example, air quality was reported as 'good.'" The index, officials said, would have to be revised.

The lesson is implicit. New York is often a proving ground for the rest of the nation. The quality of its air and the city's inability to improve it holds "frightening" implications for all of us. *The Wall Street Journal* observed: "New York's air is not much different from the air of many other U.S. cities—just somewhat dirtier because the city is bigger."

The lesson of New York is discouraging in many respects. The city's major push to control air pollution was the famous Local Law 14. It met the fate of almost all pollution legislation. Local Law 14 required the upgrading of incinerators and oil burners. Installation would cost about $8,000 per incinerator and $5,000 per oil burner. Since the law was passed some five years ago, only a little more than 15 percent of the city's 15,000 incinerators have been brought into compliance.

The law has been fought every step of the way. Apartment owners long foiled the law by a series of court challenges. In July, 1970, the Court of Appeals, the state's highest court, temporarily restrained the city from enforcing the regulation. But in November, the court unanimously upheld the law and the U.S. Supreme Court ruled against further challenges. Since January the city has been cracking down on landlords, forcing compliance schedules through the threat of sealing equipment. All incinerators are expected to be converted by the end of 1972 and all oil burners by the end of 1973.

The magnitude of the problem makes it almost insurmountable. It is impossible to police every smokestack. Cagy building superintendents surreptitiously burn trash in defective incinerators under cover of darkness. Others have openly defied the law by burning during daylight hours, gambling that they won't get caught. Some unscrupulous building owners have found it cheaper to pay bribes than buy scrubbers and other antipollution devices: recently 13 inspectors—15 percent of the city's air-inspection staff—were suspended on charges of accepting bribes; the city charged that they had overlooked violations and falsified reports on the sulfur content of fuel oil.

The landlord has polluter logic on his side. He feels that he is being discriminated against. Why should he be picked out to clean up the air when others go on dirtying it? He points to those bizarre red-white-and-blue Con Ed stacks brazenly spitting out their black poison. He can point to the city's own 47 incinerators that are among the worst violators. He can mention 1,100 incinerators in municipal housing that are just slightly ahead of the private sector in compliance. The landlord can point to the city's 2 million cars spewing their toxic load into the air. He can talk about the thousands of jets daily adding to the city's pollution burden as they land and take off at LaGuardia and Kennedy airports; it is estimated that one four-engine jet at full takeoff throttle produces the per-minute equivalent pollution of 6,000 autos (the jets are supposed to be provided with antipollution devices by 1973). He can point to the aerial ferments from industry. "Why pick on me?" he says. "I'm one of the smallest offenders. . ." (which is patently untrue, since he and his kind contribute more than half the sulfur dioxide and particulates, the two most hazardous pollutants).

Even official opinion supports him. Jerome Kretchmer, the city's Environmental Protection Administrator, recently charged that "industry, with a few honorable exceptions," was "fighting tooth and nail every step of the way" against the city's efforts to end health and environmental dangers.

As one example, Kretchmer said that restrictions on asbestos spraying (see Appendix) were being violated by "most" contractors throughout the city. He complained that the real-estate industry has delayed upgrading incinerators and oil burners by more than four years of court fights, and twelve major business organizations campaigned against an air code that had been pending for seven months. He said a "major newspaper whose yearly output costs us and neighboring sanitation departments an estimated $15 million to collect and dispose of" asked him to drop his appeals to "consider some recycling of paper" because there were capital gains and business-deduction tax advantages in the use of virgin newsprint.

Kretchmer, who spoke at an American Bar Association panel session, told the lawyers that the packaging industry had beaten back Mayor Lindsay's proposed tax on nonfood packaging to reduce solid wastes; brewers and other major users of bottles and cans had "threatened political retaliation," if tough pollution controls went into effect; he had been personally disparaged by industry spokesmen; and a brewing firm that sponsored free concerts had threatened to discontinue the concerts. The environmental chief also accused Con Ed of refusing to spend "any money at all on stack-sulfur removal research until forced to by the mayor last year." The report said Kretchmer also called on Con Ed to "reexamine the relationship between its fat advertising budget," including sponsoring of telecasts of Yankee baseball games, and "lack of funding for antipollution research." Finally, Kretchmer assailed the motor vehicle industry. He said the city had gotten a federal grant to test vehicles causing low air pollution—"we are not getting them from Detroit."

THE RELUCTANT ENFORCERS

The plaints and problems of New York are heard throughout the nation in various degrees. Most cities are even less equipped to deal with their pollution problems than New York—or at least they are often more reluctant. Boston was cited by the government as the seventh most smog-afflicted city in the country in 1967; it only got around to considering control regulations in 1969 with a three-member staff and a $60,000 budget, according to *The New York Times*. Detroit issued 4,100 citations but only 85 offenders were brought to court. Louisville issued 769 citations in 1969, but rarely were first offenders penalized. Massachusetts authorities brought a mere 22 court actions against polluters during a recent five-year period. Michigan has issued just 22 cease-and-desist orders since 1965. An Illinois legislative committee criticized its State Air Pollution Board as being remiss in protecting public health and enforcing air-pollution rules "in virtual disregard of the intent of the law."

The *Times* pinpointed Las Vegas as typical of many of the nation's cities in its position on environmental problems: "Most notably it exemplifies the wide gulf between President Nixon's sweeping proposals for controlling air pollution, and conditions as they actually exist." Following the Federal Clean Air Act of 1967, Nevada adopted an air-pollution control law "for which smog officials' kindest description was 'industry-oriented.'" Instead of giving polluters a reasonable time to install fume-control equipment or stop operating (following the successful Los Angeles formula), the law imposed only "after-the-fact" sanctions: "Officials have to wait until a facility actually contaminates the air; then begin tortuous abatement proceedings that can end in no more than a citation for a misdemeanor."

Plants were said to have obsolete control equipment, if any. The misdemeanor law invokes token fines. Regulations provide for only the most elementary of industrial discharges—soot. Variances are readily granted. The county (Clark) has only four inspectors "to ride herd" on its 8,000 square miles of sky. Since 1967, more than 1,000 "notices of violations" of even the mild existing regulations have been issued. Of these only 35 cases have gone to court, and among these, there have been only three convictions. The highest fine imposed was $75:

> All of this is taking place in a county with a progressive control setup compared to many places. Its control agency has an annual budget of $180,000, half of which comes from federal grants. The total amounts to more than 50 cents a person for the county's 300,000 population—a rate of expenditure considerably higher than the national average.

The *Times'* report concludes on a somewhat less than hopeful note: "The same pattern, federal air pollution experts say, might be found in hundreds of localities across the country."

3

The

Pervasive

Poisons

It is not illegal to poison your neighbors as long as you don't do it all at once.

—Rafferty's Rules*

THE UBIQUITOUS DDT

DDT is probably the most publicized of all environmental pollutants, and for that reason alone it deserves special attention. But it is important for another reason. It is representative of our attitude and approach toward poisons. It tells much about the nature of the problem of environmental pollution with man-made substances. It is not as a single substance that I focus on it—for it eventually may pass out of use—but what it stands for in our way of life.

DDT's status is unique. For twenty-five years it has been praised and honored while poisoning the earth and its creatures. "There is no animal, no water, no soil on this earth which at present is not contaminated with DDT," says the prominent French cancer investigator Dr. Lorenzo Tomatis.

There is a common belief that DDT has been banned and is no longer used. This is a false notion. DDT is still in common use and will be with us for a long time to come. If it were actually outlawed today or tomorrow it would remain in the environment for decades and possibly generations.

DDT is one of a group of compounds known as chlorinated

* Apocryphal "folk" wisdom.

hydrocarbons because of their chemical structure. Their most noteworthy characteristic is their persistence. Curiously, for all the attention focused on it, DDT is one of the weakest of the chlorinated hydrocarbons. If it is eliminated and the others remain we have achieved little. But if DDT goes, then others would probably follow—the so-called "domino theory." That is why the big chemical interests and their blood brothers in government have fought so fiercely to keep DDT in use, insisting, against ever-increasing evidence to the contrary, that it is safe.

DDT is one of the so-called pesticides. This is a catchall name that includes a wide range of toxins used to kill plants, insects, and animals. There are 900 basic pesticide chemical formulas in the United States marketed under 45,000 brand names. All have one common characteristic: they kill. The name "pesticides" suggests that they kill only pests. In truth they kill man's enemies and friends indiscriminately. A spray can has neither judgment nor ethics, and both qualities are often lacking in the person who uses it. Simply by walking into a neighborhood supermarket or garden-supply outlet, any person (sane or not) can buy some of the deadliest products known to man. A single drop of parathion (derived from war nerve gases) on a man's skin can cause death in minutes. Thousands of people have died from the careless handling of these things. Many were young children.

Pesticides have a rare distinction. Unlike other pollutants they are not loosed upon the environment accidentally as a result of some other operation. They are unleashed with the express purpose of killing or destroying. There is foreknowledge of many of their disastrous effects. It is known that they cannot be confined or controlled. Virtually nothing is known about their more subtle effects on the environment or the creatures that ingest them. Practically nothing is known about their chemical fate when and if they do break down. Some of their metabolites do become less poisonous. Some become more poisonous. Some appear to be harmless. Some disappear. One thing is certain: they are not altogether destroyed.

The chlorinated hydrocarbons are remarkably long-lasting—the secret of their effectiveness and popularity. They resist water, sunshine, soil and microorganisms. DDT is said to have a half-life (the time it takes for half of it to disappear) of 16 to 31 years depending on physical and biological factors. This represents more optimism than proven fact. DDT has not yet been in use for 31 years.* In one test 31 percent of the DDT applied to soil still remained after 17 years.

DDT represents a triumph of propaganda. Most Americans are convinced it is safe despite overwhelming proof to the contrary. They have been indoctrinated with an irrational fear of germs and a worship of cleanliness. They insist upon unblemished fruits and vegetables. They are more terrified of a worm inside the apple than the poison on the outside. Few homes are without their spray can. Few people read the labels, which generally are inadequate and often misleading. We spray dutifully and routinely against the known threat and the unknown menace that we are continually warned might appear in devastating numbers to overwhelm all life. We are ever vigilant to creeping, crawling things, the subliminal beating of invisible wings that may haunt the memory of living man from his prehistoric past when great winged creatures filled him with terror. This psychic echo may still live in our unconscious and explain our urge to kill. "The fundamental practice has always been to kill an animal because it was there," says a biologist, ". . . a horrendously erroneous idea." Others say we have a passion to kill, that we are a death-oriented society. The spray can symbolizes our national phobia for sanitation. It is our united front in the unrelenting war we have declared on nature, and we refuse to recognize that nature has struck back with ferocious yet subtle intensity.

Basically DDT is a nerve poison. It affects the entire nervous system. The exact way it works is not known, but it appears to bind itself to the nerve fibers and cause overstimulation. This produces the violent, uncontrolled twitching associated with DDT poisoning. Ultimately the stricken insect dies from muscular paralysis. In sufficient doses DDT can kill at once. In sublethal doses it can cause a variety of effects: nervousness, digestive upsets, heart palpitations and other irregularities, liver disorders, loss of coordination, and behavioral abnormalities. The creatures dispatched by pesticides with such cavalier indifference die wretched deaths.†

* DDT was introduced for wide-scale use in 1945.

† The effects on mice and birds from a rodenticide (thallium sulfate) were recorded by D. Heinrich Mendelssohn of Tel Aviv University and reported in *Pollution by Pesticide*, published by The Conservation Foundation:

"It works slowly, causing paralysis, and causes the death of the mice after several hours or up to two days. As the poisoned mice move slowly on the surface of the ground, and have difficulty in reaching their bur-

The more subtle effects of DDT have been demonstrated through many animal experiments. DDT's defenders like to say that man is not a rat or a mouse or whatever. But concerned scientists warn repeatedly that such findings cannot be disregarded. The *Yearbook of the American Journal of Public Health* underscores the warning. It observes that while animals and humans differ in their response to poisons, "contrary to previous beliefs, it now seems likely that a substance which is poisonous to one form of life is very apt to be found to some degree toxic for other animals, including man." Effects may vary from animal to animal, species to species, but we ignore these warnings at our peril. DDT is now known to cause cancer in a wide variety of animals and is suspected of causing cancer in humans. It has caused mutations in animals. It works as a sex hormone, interfering with reproduction. It changes human responses to various drugs. It masquerades as other diseases.

DDT's various known effects were noted by the government's highly respected Mrak Commission that studied the pesticide menace. The committee was headed by Dr. E. M. Mrak, chancellor emeritus of the University of California at Davis. The panel was especially concerned about DDT's carcinogenic effects in animals and possibly humans. The report observed:

> A remarkable degree of concurrence has been found to exist between chemical carcinogenesis in animals and that in man where it has been studied closely . . . the observations of human experience have not been sufficient to eliminate the possibility that continued chronic exposure [to DDT] may slowly induce a low level of cancer in man . . . the evidence of the carcinogenicity of DDT in experimental animals is impressive.

Many studies heighten the suspicion that DDT can cause cancer in humans. Many people who died of the disease, as well as

---

rows, they are easy victims to birds of prey. Beginning from the fifth day after the distribution of thallium bait, paralyzed and dead birds of prey were found in the fields.

". . . First the flight of the birds is labored and unsteady, then they are unable to fly but are still able to stand. Later they are unable to keep their wings in the normal posture and the wings droop, then the leg muscles become paralyzed, the bird is unable to stand, it squats . . . leaning on the drooping wings and the tail. Soon it is unable to lift its head and eventually it lies prostrate on the ground and death soon follows . . . [This] may take from three to ten days . . . Even partly paralyzed birds are unable to feed, [and] unable to adopt the proper posture in case it rains; they become soaked, are unable to keep up thermoregulation and die of exposure."

of cirrhosis of the liver and hypertension, were found to have considerably higher levels of DDT in their tissues than victims of accidental death, according to one study. Several investigators, including Dr. W. C. Hueper, former official of the National Cancer Institute, warn that there is no assurance we are not risking a catastrophic epidemic of cancers in ten to thirty years from the substances now being dumped into the environment. Few of these compounds have been tested for their ability to cause cancer. An alarming percentage of those tested were found to be carcinogenic. The whole business of environmental carcinogens is fraught with dangers because so little is known about them or the mechanisms with which they work. Dr. G. Burroughs Mider, of the National Cancer Institute, notes that "no one at this time can tell how much or how little of a carcinogen would be required to produce cancer in any human being, or how long it would take the cancer to develop." It can take up to thirty or forty years. By then the victim would have long forgotten the insult that brought on his undoing. The weak carcinogens like DDT may also be the most treacherous.* Say that a substance caused one death per 10,000 population. In a population of 200 million persons this would account for many victims. Yet in laboratory tests it could be easily missed among fifty or so animals. Yet DDT and many other known carcinogens are regular components of our food.

Our daily intake of DDT is small yet never insignificant. This is due to one of DDT's many remarkable characteristics: its ability to accumulate in the fatty tissue of anything that eats it. Each creature that eats another adds its victim's toxic load to its own. With each meal the amount stored gets larger—a process known as "biological magnification." An animal can store many times the amount of DDT that could kill in a single dose; concentrations of residues may be a million times greater than appear in the environment. Such poison-bearing creatures are said

* The unpredictability of some food chemicals adds to their treachery. For years diethylpyrocarbonate (DEP) was considered the "ideal" food preservative, and was widely used in wines, draft beer and noncarbonated fruit drinks. Its virtue is that it does its job of preserving and then decomposes quickly; within 24 hours it virtually disappears from the foods to which it is added. For that reason the FDA does not even require that it be listed on the label of foods in which it is used. But in 1971 Swedish researchers found that before DEP disappears, it can react with ammonia in the beverages to produce the chemical urethan, a well-known cancer-causing substance. Once formed in the beverage, urethan stays there, although the DEP is long gone. (*New York Times*, December 21, 1971.)

to be "biological time bombs." They are potentially deadly for anything that eats them. Death may occur at once, or when a fatal dose is accumulated. Meanwhile each animal is a threat to anything that eats it.

The process of magnification may start with spraying for Dutch elm disease. Some of the spray drifts to the ground. It is absorbed by soil organisms and worms. Robins and other birds eat the worms and organisms. They, in turn, are eaten by higher species. In the sea the buildup begins in the water. The tiny phytoplankton absorb and store DDT. Grazing animals eat the plants. Larger fish eat smaller fish. Birds (and men) eat the larger fish. In one study plankton from the water contained 5 ppm of DDT residues. The plankton concentrated the pesticide 250 times. Small fish eating the plankton contained up to 300 ppm. Large fish eating the small fish had up to five or six times that total. Grebes (a kind of swimming bird) that ate the larger fish accumulated 2,000 ppm—more than 100,000 times the original amount in the water. What happens to the creatures that live in the water—and the people who drink the water and eat its contaminated creatures?

The average American is said to have about 10 to 12 ppm in his body; collectively we carry some 20 tons of DDT in our cells. The average Englishman is said to have about 5 ppm. The average person in India allegedly has 25 ppm.

Some DDT apologists insist that concentrations in a person's body level off at a certain point. What is that point? Again, there is no answer. Averages are meaningless—an excellent way to bury individual differences. The Mrak report observes that there is no convincing evidence available to clearly indicate whether tissue storage of DDT and its metabolites is increasing, decreasing, or remaining constant.

What happens to this stored DDT? Its defenders claim that it is inert, much like butter in a icebox. This would go against almost everything known about human biology. The *Journal of the American Medical Association* points out that this is most unlikely. The cell has a constant turnover, it observes. There is a rich blood supply and the use of the fat is regulated by endocrine, enzyme, and nerve influence. The *JAMA* says it appears to be a reasonable assumption that fatty tissue which has "these many important functions can be influenced by the presence of cumulative poisons such as the chlorinated hydrocarbons." Important gaps remain in knowledge about the storage, metabolism, and significance of DDT in human tissues, according to the Mrak report. This after we have been eating and storing DDT in our bodies for more than twenty-five years.

In nature the subtle effects of the long-lasting DDT compounds are known to be devastating. They can destroy the entire structure of life. Often the disaster takes place so gradually that it passes unnoticed until too late. Dr. P. A. Butler, a research consultant with the U.S. Bureau of Fisheries, points out that many animals die slowly and over long periods from pesticide poisoning and that even experts would be unlikely to notice this mortality rate: "They would only know, eventually, that the fish had gone somewhere else." Does this explain the mysterious disappearance of the vulnerable game fish along the Atlantic Coast?

It is not the occasional large amount from some accident that poses the major problem—rather, the low-level sublethal amount ingested day by day. Estuaries and lakes are often the hardest hit because the pollutants tend to concentrate in them. The pesticides may sift to the bottom and remain for years, gradually being stirred up and eaten. Any interference with the flushing action of the water hastens the ultimate disaster. In one experiment three researchers (Odum, Woodwell, and Wurster) took several fiddler crabs from a marsh on the north shore of Long Island and fed half of them decayed organic matter from a nearby marsh. Other crabs of the same species had disappeared from the marsh ten years earlier following a DDT spraying. The detritus contained only 10 ppm of DDT residues. By the fifth day:

> . . . all experimental crabs were uncoordinated. Instead of scurrying away when threatened by a hand, as usual with the control crabs, they moved a few centimeters, lost coordination, and rolled over once or twice before regaining equilibrium. . . . Such awkward and sluggish behavior is unusual and would certainly affect survival under natural conditions . . . the organic detritus appeared to be a reservoir of DDT residues in the environment, small particles sometimes containing residues thousands of times greater than the concentration occurring in water.

Every creature has its own toxic limit. It passes out of the picture when its limit is reached and the environment is too hostile for it to survive. Shellfish are among the most susceptible. The tiniest amount (50 parts per trillion) will kill a newly hatched fiddler crab. Baby fish are also highly vulnerable, and this is bringing about great changes in fish populations. DDT virtually eliminated baby trout in Lake George, New York. *Outdoor Life* tells how:

> Little fish come into the world with a bulging yolk sac from the egg attached to their bellies, and this serves as their only source of food during the first few days of their lives. As the trout fry from Lake George started to absorb this built-in food supply, they began to die. Within a month the entire lot was unknown.

Some creatures that prove more resistant may store huge amounts in their bodies and wipe out their predators. Oysters are especially lethal. An oyster pumps up to 40 quarts of water through its filter in an hour. In a month it can accumulate 70,000 times the amount of DDT appearing in the surrounding water. Many of the marine time bombs are eaten by birds. Twenty years passed before it was discovered how DDT was killing them—by affecting reproduction. It interfered with the hormone mechanisms that enable a bird to produce calcium for shells. The shells became so thin that baby birds could no longer develop. The ultimate came when a membrane was found with no protective shell. Several species have been cut down by DDT. The peregrine falcon is virtually gone. The bald eagle is almost gone. The osprey is on its way out. Others are threatened. The remarkable thing is that whole species can be wiped out without proof that DDT ever killed one single bird.

DDT worked with the same sly indirection in killing fish in New Brunswick. A mysterious spring die-off of salmon occurred after an earlier spraying with DDT, followed by a sudden rain that lowered water temperatures about 5 degrees. This should not have hurt the fish, but tests disclosed that even brief exposures to very low levels of DDT altered the fishes' temperature sensitivity, making a normal 5-degree drop fatal. The episode was recalled by Dr. Butler, who said he does not believe there is such a thing as a sublethal toxic effect—a sentiment echoed by many scientists as they probe more deeply and learn more about the awesome complexity of life and its fragile links. "I'm sure any toxic effect will, in the long run, lead to a lethal effect," said Dr. Butler. He said this concern applied not only to fish but also to humans.

Many investigations support this view. Two California investigators (Pottenger and Krone) found that rats fed chlorinated hydrocarbons had a higher mortality rate than controls, "yet the tissues of the animals showed no change on autopsy, demonstrating how easily such toxicity might escape notice. In short, one could not prove the diagnosis of insecticide toxicity in these animals by clinical or laboratory methods. But their lives were shortened by insecticides. We shall never know how many humans suffer similarly."

The evidence keeps mounting. DDT has now soaked land and sea. It has only tardily been found to interfere with the ability of certain soil organisms to fix nitrogen in crops. More recently it was learned that in some places in the ocean DDT has reached levels in phytoplankton high enough to reduce photosynthesis. The investigator who made the original experiments, Dr. Charles Wurster of New York University at Stony Brook, says this could have "worldwide implications." Others say it "may herald the death of a seacoast and an ocean."

At first the mechanism that led to universal contamination by DDT was not understood. How did it get into the bodies of Eskimos, caribou, and seals in an area where spraying had never taken place? At first it was believed it came from the water. But that left questions. DDT is not readily water-soluble, and only limited amounts of land applications are washed off into waterways. Most of the amount washed off the land finds its way into estuaries, concentrating in the mud and plants, wreaking havoc there. The mystery was compounded because as much as 50 percent of the pesticides sprayed in agricultural areas never reach the plants they are intended to protect. Where did the rest go? If it found its way into the ocean, how did it move so quickly? The fastest ocean currents travel only about 100 miles a day. Most laze along at 25 to 50 miles per day, giving solids a chance to settle into the abyss.

Gradually the answer emerged. DDT is transported in the atmosphere. It is not there in small fleeting amounts that quickly wash out and disappear; it is moved in massive amounts on the winds like other pollutants. It circles the globe, going from one hemisphere to another, often riding on jet streams that wing along at 250 miles an hour. It is estimated that in the atmosphere at any given time there is more than one billion pounds of DDT and its metabolites, according to studies.* Dr. Justin Frost

---

* See Woodwell, G. M. "Toxic Substances and Ecological Cycles," *Scientific American*, Vol. 216, No. 3, 1967.

at Southern Illinois University, writing in *Environment,* says it is reasonable to assume that of the 126,000 tons of chlorinated hydrocarbon pesticides sold annually, more than half enters the atmosphere.

This is the real price of our folly. Once DDT is in the atmosphere there is no escaping it. It becomes part of our environment. It enters the atmosphere as small particles or as vapor that may cling to the surface of dust and tiny particulates from smokestacks and tailpipes, adding its treachery to theirs. It travels thousands of miles, sifting down to contaminate the lands and waters of the earth. It is in the fish eaten by bears, seals, and birds. It is taken up by the delicate lichen that is the primary food source for the caribou which, in turn, are eaten by Eskimos. It becomes part of the texture of meat, vegetables, milk, water.

It is incredible how far and fast it travels. The man spraying his roses in Kansas City today can contaminate a polar bear in the Arctic tomorrow. Dr. Frost gives us an example of how far and fast it can travel. In 1965 dust was picked up in a storm in western Texas, where pesticide use is heavy. The following day a light noon rain deposited five or six tons of dust per square mile on Cincinnati, Ohio. The dust was found to contain "appreciable amounts of DDT, DDE and chlordane, and trace amounts of heptachlor epoxide and dieldrin, totaling 1.3 ppm of chlorinated hydrocarbons." In Cincinnati total dustfall averages 15 tons per square mile per month and can contain large quantities of pesticides, according to the report. "Samples of Cincinnati dust all in June and July of 1965 showed DDT in concentrations ranging from 3 to 90 ppm along with unidentified pesticides."

Pesticides volatilize from the soil when the earth is worked and mix with the air. Contaminated soil is whisked away as dust. The portions reaching the water as soil runoff or atmospheric fallout may vaporize as the water evaporates to be returned to the air. They resume their restless journey along with all the other contaminants in transit, mixing and reacting in unknown ways. Studies of British rainwater indicated that in the United Kingdom one inch of rain would deposit a ton of pesticides. With an average rainfall of more than 40 inches it is cal-

---

Also: Paul P. Craig, Horton A. Johnson, Maynard E. Smith and George M. Woodwell, of the Environmental Sciences Committee, Brookhaven National Laboratory, Upton, N.Y. "DDT in the Biosphere: Where Does It Go?" Work performed under auspices of U.S. Atomic Energy Commission.

culated that some 40 tons per year would come down from the atmosphere.

The pesticides settle into the great abyss of the oceans. Then they are subjected to the leisurely pace of eons. The particles may remain buried at sea, locked up by time for decades, possibly for centuries, before churned to the surface by upwelling water. Those that find their way to mudflats are slowly released to feed plants. They go up in the food chain. They are excreted and continue the cycle. After an unknown number of years or decades, they are broken down into other compounds. These may be more or less toxic. They may do undetermined mischief. The same molecule that was joined in our great human experiment comes back to plague us again and again, possibly in amended forms. The great experiment goes on toward its unknown conclusion. The accumulating DDT is part of the buildup of background pollution that could ultimately destroy or irreparably disrupt life in the marshes, bring death to the sea, and push our frail life-support systems beyond endurance.

The great irony is that DDT has failed in its original mission. It was to save us from all disease-bearing insects. It was to wipe out every pest that cuts down crop production. It was to give man dominance over nature. Instead we have created biological havoc. Some 224 insects have developed resistance to DDT by manufacturing an enzyme that converts DDT to DDE—a substance that is less toxic to them. Many of the predators that once kept insect populations in check have been wiped out. Many useful insects have been reduced or altogether eliminated (in Japan so many pollinating insects have been killed that fruit trees now must be pollinated by hand). Species of insects that were once innocuous are now pests. Often where there was one pest before, now there are a half dozen. Each spraying makes more spraying necessary. As pests develop immunity to one poison, a substitute must be developed. Each adds to the environmental damage. In California, two species of mosquito have acquired complete immunity to all man-made pesticides, according to health officials there. The resistant insects, the first of their kind in the state, have been breeding unchecked for about a year—a situation that is said to be potentially dangerous, as one of the immune mosquitoes has the special capacity to transmit certain deadly encephalitic viruses to humans and animals.

It was wild optimism to think we could defeat the bugs so easily. They have experience and number on their side. There

are billions of varieties, but only about 800,000 have been classified. Less than one percent are harmful to man. Most insects
reproduce quickly and in fantastic numbers. If only 2 percent
survive a spraying and become resistant they can soon return to
their former numbers—possibly more if their predators have
been killed. A single pair of flies is said to be potentially capable
of producing 191,010,000,000,000,000,000 offspring in four
months—enough to cover the earth to a depth of 47 feet.

Fortunately the laws governing nature never permit a single
species, plant or animal, to dominate any environment completely. Weather and enemies cut down the proliferating insects.
Some birds eat their weight in bugs every day; no insect ever
becomes resistant to a bird. Insects are also attacked by a variety
of diseases, viruses, fungi, and bacteria that help control their
numbers. These natural checks do their work without threatening man. Chemicals could not do the job alone. Usually they
disrupt and interfere with the interacting forces necessary to
carry on the subtle business of life, including the lowly microbes
and insects.

Insects are marvelously clever. They not only are resilient but
profit from experience. It is not easy to admit that a cockroach
may be smarter than a man. But cockroaches have been around
much longer; time is on their side. Other insects are excellent
strategists. After a field is basted with poison, the victims are
soon replaced by newcomers from adjoining fields, much as re-
inforcements pouring back to a blood-drenched battlefield to resume the attack. Soon the field must be basted again. The now-
poison-resistant pests explode in numbers. Their allies join the
fray. The farmer desperately turns to new and more powerful
poisons. More imbalances result. More poison residues are in the
crops for people to eat. Food prices rise to pay for the new poisons. Some investigators claim that the futility is even greater
than it appears to be. They say that even after all the spraying,
losses due to pests are about the same as they were 50 years
ago—about 10 percent. But now we are locked into a chemically
dependent situation. Remove the chemicals and there would be a
disastrous loss from the insects. We have destroyed the old and
the new does not work. We set out to poison the bugs so we could
feed ourselves. We wind up feeding the bugs and poisoning ourselves.

Many of our best agricultural lands have become virtual disaster areas. One is California's Imperial Valley. It was recently the

scene of a vigorous effort to control an outbreak of the pink boll-worm with chemicals. Dr. Robert van den Bosch, professor of Entomology at the University of California at Berkeley, explains what happened:

> . . . insecticides used against the pink bollworm have in-duced a general outbreak of the cottonleaf perforator in the treated fields, and apparently have led to a devastating out-break of the beet armyworm in adjacent crops. As a result, there has been a massive use of insecticides in the Valley, which has been costly to the grower and to the environ-ment.

Dr. van den Bosch finds an irony in all this mass spraying. He says we have set ourselves an "ecologically and genetically impossible goal." We set out to assure pure, high-quality food, un-blemished by insects or their remains, and we wind up contami-nating the food and causing widespread environmental pollu-tion. We are involved in a biological arms race that we cannot win. Each escalation is a new defeat; van den Bosch says, "We are at the brink of economic and ecological chaos in pest con-trol: The insects are beating us in the competition game, and have forced us into an environmentally damaging strategy. We cannot continue on our present course; it is a one-way street to ecological disaster."

We talk about getting out of our bind by switching to "biologi-cal warfare" against the insects. This would include breeding predators in the laboratory and unleashing them in overwhelm-ing amounts. From the laboratory would also come legions of sterile males to copulate with the females and outwit nature's reproductive imperatives. There would be traps baited with sex enticements, as well as viruses, bacteria, and various disease-bearing organisms aimed at specific target pests. But this re-mains impossible while we continue to unleash a blanket of poi-sons every year that wipe out the released predators along with the pests. And a blanket it is. We now produce approximately 900 million pounds of pesticides a year, with a price tag of $1.7 billion. Since World War II more than 1 billion pounds have been manufactured and pumped into the environment. Much of it is still with us and will remain. The amount of DDT in the biosphere is steadily increasing, because the annual input ex-ceeds the rate of breakdown, as van den Bosch explains: "The material has penetrated into literally every 'corner' of the earth, its waters and its atmosphere. It is in the air, the soil, lakes and streams, the sea, the polar wastes and on the crests of mountains. We can find it almost anywhere we care to look."

It is unmanageable in the environment, he says. There is no way to prevent its movement out of the area of use. The continued "essentially unilateral reliance on synthetic organic insecticides in pest control programs will only result in worsening economic and ecological problems."

Many still cling to the illusion that it is safe to douse the earth with this poison. They accept the commercial propaganda as fact. They ignore or are ignorant of experimental evidence of the harm it causes animals. They accept without alarm the presence of DDT in mother's milk at levels that exceed those ruled damaging in cow's milk (mother's milk in any other container would not be permitted to cross state lines, says Dr. Wurster). They ignore many warnings: DDT intake by infants is now in the same range where animals begin to show biochemical change. Studies show that sex hormones in rats are affected by enzymes activated by DDT; the same hormones are found in man, whose DDT residue is now "within a range to produce the same effects." DDT causes cancer in animals. It is unpredictable, uncontrollable, and potentially catastrophic.

Experimental evidence against DDT has not deterred its champions. They remain passionate in its defense. They assail environmentalists who warn against DDT as hysterical. In the same breath they proclaim that DDT is necessary to prevent mass starvation, and they insist it is safe because it is consumed in small amounts. These arguments were all cited by Dr. Norman E. Borlaug, who won the Nobel Prize in 1970 for his contributions to the Green Revolution, at the 16th governing conference of the United Nations Food and Agriculture Organization in Rome during December, 1971. He called those who oppose DDT "hysterical environmentalists."

Replying to him, at the same conference, was Dr. Sicco L. Mansholt, vice president of the Common Market Commission and its expert in agriculture. He said he was "one of those hysterical environmentalists Dr. Borlaug talked about. I am very concerned." He focused on the "accumulation" of DDT and called for a study on the effects of pesticides.

"I am not attacking DDT," he said, "but when Dr. Borlaug says that DDT is only used in a quantity of 400,000 tons a year, I would point out that this means 110 grams per man in the world a year. Ten years' use of DDT means one kilo per man in the world, and it stays in the world; it is in the soil, the plants, the seas, the plankton, the fish, the man." A kilogram is equivalent to 2.2 pounds, the news report explained; a gram is a thousandth part of a kilogram.

DDT's myth of safety began with its use during World War II. It was then acclaimed the miracle pesticide: deadly for bugs and disease-bearing organisms and safe for man. Following the war it was used on crops with spectacular results. Soon it had killed off the natural controls and made itself indispensable. Industry and government alike shouted its praises and committed themselves to its spread. It was forgotten that disease vectors had been controlled in the past with mild transient pesticides and simple sanitation measures; in this way malaria was controlled in Panama long before the advent of DDT. By a curious twist of logic, DDT's use in agriculture is often justified because of its heroic war-time role—as if that obligated us to eat it at every meal throughout our lives. DDT is also celebrated and justified because of the lives it has saved in the so-called undeveloped countries. It is overlooked that people are ill served if they are spared from death by disease-bearing insects only to multiply in such numbers that they must die of starvation.

DDT's alleged harmlessness was not based on thorough experimental evidence as much as a lack of data.* To this day it has not been properly tested on women, children, the weak and sick. This is odd because they—and especially children—are the most susceptible to poisons. Our alleged safety is based primarily on a poorly conducted experiment by the USPHS on a handful of federal prisoners. Only young, strong individuals were used, the group most resistant to the effects of poisons. There was no adequate followup to find out how the men fared. The study failed to recognize the tremendous differences in the way individuals react to poisons, burying individual findings in group averages, according to Morton Biskind, a research physician and one of the earliest critics of DDT. Comparable medical data on the subjects at the beginning and end of the experiments were lacking;

* N. Bruce Haynes, associate professor in the New York State Veterinary College at Cornell University, testified before a New York State hearing on the use of pesticides in 1972: "As far as toxicity to animals is concerned, to my knowledge there have been no confirmed cases of DDT or other pesticide poisoning in New York State. There have been suspected cases but no confirmation because no public laboratory in the state is available for analysis of pesticide residues in animal tissue in cases of suspected poisoning. The same is true for heavy metal poisoning."

complaints of the subjects which could not be confirmed by laboratory tests were dismissed or called psychological in origin. An investigator, Dr. Wayland Hayes, concluded that after about a year DDT storage in the body reaches a maximum "and thereafter [people] store no more of the material despite continued intake." His own charts do not seem to bear out this conclusion. One subject showed a rise of almost 30 ppm DDT storage after only four months' participation. Other men not in the test group (whose normal prison fare also contained DDT) showed substantial gains. Fifty-four prisoners began the experiment; at the end, after eighteen months, there were only four. There were no real controls because all the men in the study were ingesting DDT in varying amounts.

Dr. Hayes revealed his own attitude toward DDT in testimony before a congressional committee. He conceded that DDT causes definite changes in the livers of rats fed the substance. But he would not characterize these changes as damage. Just changes. The "interpretation is going to require further scientific study," he stated. Most biologists regard any change in tissue structure as harmful. Dr. Hayes, curiously, did not test the livers of the prisoners in his study, although the liver is a primary target of poisons.

The experiment proved only one thing: that a small group of men, who had already demonstrated their ability to withstand repeated small doses of DDT in their regular diet, could withstand substantially larger amounts without immediate disastrous consequences. That is the basic rationale for some 200 million Americans eating small amounts of DDT and other poisons in virtually every bite of food they take every day of their lives. One critic compared it to studying a dozen smokers; if none developed lung cancer, "you conclude that smoking is, therefore, safe for everyone."

Our margin of safety is precarious at best. It is based on the risky proposition that we all conform to standards set for the average, healthy person. The fact is that most people are neither average nor healthy. In this toxic world we have created, it is not only unsafe to be physiologically different; it is positively and increasingly dangerous. Chronic illness and death may be the penalty for the eccentric who does not have the good sense to be average. How many biological nonconformists accumulate astronomical amounts of DDT in their cells because of some bodily deficiency or biological perversity? How many have already died, written off as victims of some other ailment not immediately associated with DDT?

The Hayes test, such as it was, was not conducted until a dozen years after DDT was in common use. By then there was already considerable evidence that DDT was harmful. Private researchers had shown that it was damaging to animals and caused adverse nerve and heart reactions in humans under experimental conditions. As early as 1947 the government's own tests had demonstrated that it caused cancer in animals. But nothing was done to eliminate the poison. It was too late to find DDT wanting. The profits were too high to back off. People were relentlessly and ruthlessly pushed into what Dr. Wurster calls "a biological experiment of truly colossal proportions, using the entire world's biota as an experimental organism."

The Mrak report refers to "the absurdity of the situation in which 200 million Americans are undergoing lifelong exposure [to pesticides], yet our knowledge of what is happening to them is at best fragmentary and for the most part indirect and inferential."

The report outlines at great length the multiple risks we are taking by using these things. Then it points out that their production is expected to grow at an annual rate of 15 percent per year. This is translated by Dr. George Woodwell, chief ecologist at Brookhaven National Laboratory, to mean that by 1974 we will be using twice as many pesticides. At the same time the hazard will double. Not only the hazards that science now recognizes—he says, but in addition those not yet uncovered because researchers have been busy elsewhere.

We are curiously complacent about this threat. For years Dr. Biskind was an almost lone and unheard voice as he warned against DDT and related pesticides. He was drowned out by official and industrial propaganda when he denounced the Hayes report as "an intricate, carefully contrived and ingeniously composed document, perhaps a classic of its kind." He said, almost wistfully, "A new principle has, it seems, become entrenched in the literature. No matter how lethal a poison may be for all other forms of animal life, if it doesn't kill human beings instantly it is safe."

This belief came about not altogether by chance.

THE PHILOSOPHY OF POISON

American life has become progressively wedded to the use of poisons. It is based on a philosophy that took shape around the turn of the century which, in the beginning, revolved around the use of chemicals in foods. The original storm center was Dr. Harvey Wiley, who headed the Bureau of Chemistry in the Department of Agriculture. The bureau was the forerunner of the present Food and Drug Administration that administers the nation's food laws. Dr. Wiley was the father of the original Pure Food Law. Arrayed against him were the big-moneyed interests that saw huge profits from the use of chemicals in foods. Dr. Wiley fought to retain purity in food and the integrity of the law. He would not compromise with his principles. For this he paid the price demanded of most purists. Dr. Wiley told the story of what happened, in a small volume which he published himself back in 1929, with the melancholy title *The History of a Crime Against the Food Law*.

Today the crime is the law.

The food adulterers had a novel theory. They held that when harmful substances are injected into food in small amounts they cease to be harmful. Dr. Wiley rejected this view. He held that poison is poison in any amount, that it is harmful in any amount: "The character of the offense is not so much the amount of the material used as its nature." Any harmful substance ingested produces some harmful effect; the greater the amount ingested the greater the harm. He used calculus to prove that damage begins as soon as a harmful substance is introduced, although it cannot be seen or measured.

He explained that the human body is required to do a certain amount of normal work to remain healthy. This is regulated by normal food eaten in a normal way. If this normal exercise of the organs is reduced there is atrophy. If increased, the organs wear out prematurely. He said the continued bombardment by small doses of toxic substances deadens the sensibilities of the excretory organs. Gradually, they cease to function, bringing on early old age, sickness, and finally death.

Dr. Wiley had to combat the same arguments that are still being used today. The adulterers said his attitude was unscientific. They testified to the harmlessness of eating small amounts of poison. They gave as proof that it is possible for a person to cram enough salt down his throat to kill himself. Dr. Wiley patiently pointed out that this compared unrelated substances. Salt was necessary to life in small amounts and most adulterants contributed nothing to the life process. Further, this logic claimed that because a little salt is innocent and a lot is harmful, it follows that all other substances that are harmful in large amounts are safe in small amounts.

The adulterers also observed that certain poisons appeared in foods in their natural state. Therefore, they contended, it was safe to add not only more of those particular poisons but also a host of other toxic substances. This too was submitted as "scientific" proof. Again, Dr. Wiley patiently pointed out that poison is poison, whether produced by nature or propounded by man. It imposes a burden on the excretory organs. Because traces of certain toxins do appear in a few foods, it should not be interpreted as a warrant to add more, he said. On the contrary it should be taken as "a highly accentuated warning to avoid any additional burden."

The battle was joined over benzoic acid. Industry wanted to use it as a food preservative. Dr. Wiley argued that benzoic acid took no part in the formation of human tissue and its degradation product was hippuric acid, "a most violent poison." In the end, he lost, and it was through an odd set of circumstances that had little to do with the merits or demerits of benzoic acid.

A review board was appointed to study the substance. It was headed by Dr. Ira Remsen, who had been given a medal by the Chicago Chemical Society as the discoverer of saccharin. Dr. Wiley had offended Dr. Remsen by ruling that saccharin was injurious to health—a finding that is still being argued among food chemists.* Dr. Wiley's ruling had an unexpected backlash. President Theodore Roosevelt had been taking saccharin. Dr. Wiley, trying to be helpful, told the President that saccharin was

* In January 1972, the FDA removed saccharin from its GRAS (Generally Recognized As Safe) list and put an interim limit on its use—one gram per day for an adult. FDA said that bladder tumors had been found in rats fed large quantities of the artificial sweetener in their daily diet. It has not been determined whether the tumors are cancerous. Saccharin, it noted, has been in use for more than 80 years. Dr. George T. Bryan, the first scientist to link cyclamates and cancer, says a similar danger may exist with saccharin and warns consumers not to use any artificial sweetener. "There is as much evidence for banning saccharin as there was for banning cyclamates," he says. A National Academy of Sciences–National Research Council panel reported that it found no evidence that saccharin was harmful when used as a food sweetener but suggested further research.

not safe. The President exploded into one of his violent rages. He roared at Dr. Wiley, "Anybody who says saccharin is injurious to health is an idiot."

Had Dr. Wiley understood human nature better this would not have surprised him. People are never grateful to those who deprive them of their comforts and consolations. They are likely to be much angrier at the one who tells them they are being poisoned than at their executioner. Every honest public health official who closes a polluted public beach trembles in his boots, knowing that he becomes "an enemy of the people."

Roosevelt's fury was the turning point in Dr. Wiley's struggle to keep poisons out of food, and it was a turning point in American life. After the incident the President no longer supported Dr. Wiley and the administration of the Pure Food Law. The Remsen board, Dr. Wiley charged, became a vehicle to thwart the law. The Bureau of Chemistry under Wiley had resolved every case of doubt in favor of the consumer. The board overruled almost every decision of the bureau in favor of industry. If industry lost a decision it could appeal to the Remsen board. For the consumer there was no recourse. The Department of Agriculture supported the Remsen board in its policy of adulteration, Dr. Wiley said. Adverse reports on food chemicals were never published and were buried. Years later, in sadness and lingering anger, he wrote:

> All of these publications are in the morgue. They were objected to by parties using preservatives and coloring matters and articles adulterated with arsenic, and these protests against publication were approved and put in force by the Secretary of Agriculture. . . . The whole power of the Department of Agriculture was enlisted in the service of adulteration which tended to destroy the health of the American consumer. On the appointment of the Remsen Board further investigations [of food chemicals] by the Bureau were ordered to be suspended.

This is the background of what has been called "the philosophy of poison"—a philosophy that now largely dominates our lives. The substances at issue in Dr. Wiley's time were innocent compared to the violent poisons in common use today. The use of small amounts of poison in food provided a mechanism for a rationalization that ultimately was to shape our view of the entire environment. It marked the beginning of our journey down the chemical primrose path: We ate the poisoned apple and found it tasteless but unblemished and large; therefore it was said to be good. The same principle that permitted food adulteration became the basis for our modern "pollution-control" practices. Standards are set to regulate the amount of toxic substances that can go not only into foods but also into water and air—the permissible cumulative poisons such as DDT, lead, mercury, arsenic, radioactive substances that we carry in our blood, tissues, and bones. We have traveled far since Dr. Wiley's day.

The theory of "safe" amounts of poisons is based on nature's benevolence—all her creatures are provided with some mechanism for self-cleansing and protection against toxic insult. This is our margin for error. The air is purified by wind, sun, and rain; water by movement, dilution, and bacterial action. The soil is alive with microorganisms that break down organic matter and degradable substances. Animals have their own built-in protective devices that ensure survival. Insects and fish—the so-called lower orders—reproduce so rapidly and in such vast numbers that they are preserved from being wiped out by catastrophe, or they develop resistance to harmful substances. Men have organs of detoxification to protect them from the harmful substances they eat, and there is the filterlike mechanism in the lungs to prevent and repair damage to the respiratory passage.

The self-purification process of the body is complex. Biologists point out that every poison ingested, if it cannot be excreted rapidly in its original state, must be detoxified. This places a heavy and continuing burden on the liver and various other organs. In the process of detoxification certain vitamins are used up, primarily the vitamins B and C. The detoxifying organs, in order to get the extra nutrients necessary to do the extra work demanded of them, take vitamins from other parts of the body. This can cause a general vitamin deficiency. Eventually it can set off a chain reaction. Certain organs may be damaged or break down from the additional burden and become unable to do their job. This, in turn, can lead to serious ailments—the widespread degenerative diseases, including cancer, and even death. Yet the original toxins that caused the mischief could never be pinpointed or directly implicated.

Instead of respecting this human defense we have used it for economic exploitation. We have polluted our bodies, the water we drink, the air we breathe, the food we eat. For the essential

life-support systems—soil, air, and water—we have set not
standards of purity but lower limits of abuse beyond which we
dare not pollute. Then we do not meet even these minimal
standards.

For our protection from immediate disaster we have standards
called "tolerances." It is assumed that our bodies can tolerate
certain amounts of harmful substances. There is no limit to the
number of toxic substances that can be used in foods. People
whose physical limitations cannot meet these standards are expendable.

The tolerance concept is based on the premise that a harmful
substance can be reduced in amount until it is no longer harmful. In other words, poison in small doses ceases to be poison.
This assumes that for almost every harmful substance there is a
"safe" threshold dose where damage does not occur. The concept
takes a quantitative rather than a qualitative view of the matter.
It assumes that the bulk of a man is assailed by a minuscule
amount of harmful matter rendered harmless by diminution. It
makes no allowance for the minute size and sensitivity of a human cell and its vulnerability to foreign invaders. However, it is
now recognized that the cell is vastly more complex than once
believed; systems, such as hereditary instructions, once thought
to be complete in themselves, are now known to be but part of
other systems. These mechanisms are easily damaged or disrupted. Pathologists know that the body can suffer great damage
without that damage being visible; by the time it can be measured it is extensive. The tolerance concept does not respect this
vulnerability of invisible life processes, or the fact that cancer,
irreversible degradation of flesh, and death begin with injury to
a single cell.

The threshold for food is much the same as the magic line
that Commissioner Sullivan spoke of for air pollution. Everything above the line is safe; everything below is harmful: there
is this abrupt break-off point with no graduations in between.
Only the most primitive logic can support such a view. The line
for food tolerances, as in air and water pollution, is arbitrary. It
is drawn on the basis of animal tests. The lethal dose (L.D.) is
determined by finding the amount necessary to kill half of the
creatures. Then, in most cases, the margin is increased some
100 times as a safety factor. This is based less on fact than on a
projection of optimism. An animal is a poor silent suffering creature. It can only obligingly develop unmistakable symptoms of

poisoning and offer definitive proof of harm by dying. An animal cannot report on its state of being. It cannot tell that it has spots before its eyes, dizziness, backache, upset stomach, and similar complaints that cannot be picked up by routine procedures. Nor can humans who suffer such indispositions trace them precisely to any of the thousands of chemicals in their diet and in the environment.

Each poison permitted to remain in food has a tolerance level fixed by the FDA. Few substances are considered so harmful that some level of use is not found. DDT, for example, has a tolerance ranging from 1 to 7 parts per million. The higher tolerance is on some fruits, and in meat where the DDT concentrates primarily in the fat. The naive might be perplexed because more than 1 ppm of DDT is safe on one food and not on another. This has nothing to do with human safety. The higher levels accommodate industry; they are set because the residues cannot be reduced. The law can be very practical where commercial interests are involved. Despite the government's zeal for correct labeling, there is no provision to list on products the amount of contamination present. This tends to follow a law of human nature —we generally ignore that which we can't control.

For many years there was no tolerance for DDT in milk. In effect this meant zero. Milk is a basic food of babies, the sick, and many old people. But at the same time that DDT was not safe in milk these vulnerable people were getting it in vegetables, meat, eggs, and other articles of diet. In 1968 the government finally recognized the futility of trying to keep DDT out of milk. It set a tolerance of 0.15 ppm for whole milk and 1.25 for milk fat. Overnight DDT became safe in small amounts in milk. With the stroke of a pen the harmful becomes harmless.

Tolerances now exist for hundreds of pesticides. Some crops can have residues of a score or more of pesticides. Apples alone are often sprayed more than a dozen times. In addition to 45,000 pesticide formulations there are an estimated 3,000 other additives and contaminants that appear in the American diet. Most of these are toxic and require tolerances. The list of chemicals that are intentionally injected into food and that migrate in accidentally is almost endless. As I pointed out in *The Poisons in Your Food* (Simon and Schuster, 1960):

> Virtually every bite of food you eat has been treated with some chemical somewhere along the line: dyes, bleaches, emulsifiers, antioxidants, preservatives, flavors, flavor en-

hancers, buffers, noxious sprays, acidifiers, alkalizers, deodorants, moisteners, drying agents, gases, extenders, thick-
eners, disinfectants, defoliants, fungicides, neutralizers,
artificial sweeteners, anticaking and antifoaming agents,
conditioners, curers, hydrolyzers, hydrogenators, maturers,
fortifiers, and many others.

The amounts used are staggering. Present use is estimated at
more than 850 million pounds annually—approximately three
pounds of additives per person per year according to estimates
by Arthur D. Little, Inc., and reported in *Chemical and Engineering News*. Each person reportedly gets an additional annual
dose of four grams of coloring matter. By 1975 the use of additives is expected to soar to 1.3 billion pounds a year. Yet it is a
rare chemical that finds its way into the food stream to meet
nutritional needs. Their primary function (and usually sole
function) is to make more profit for the manufacturer. Cake
mixes alone can contain more than fifteen chemicals.

Many additives now appearing in foods are known to cause
allergies. For hypersensitive people there is no protection. Some
may have died from using these compounds. "When antibiotics
and dyes have been shown to be allergens, what other sensitizing
perils ply this vast and uncharted food sea?" asks Dr. William C.
Crater, a professor of medicine at Southwestern Medical School
in Dallas. He challenges the FDA and American Medical Association Council of Food and Nutrition stand that the present use
of chemicals in foods is safe. "How does this fit in with the presence of pesticides in meat, penicillin in milk, and cottonseed
meal in pastries?" he asks.

He challenges an FDA statement that additives are permitted
only after evidence of safety is provided by expert scientific
opinion. The FDA has made this statement so many times that
even it might believe it by now. The Federal Register lists several
thousand chemical additives to food, but only a few appear on
labels.These additives are supposed to be used at levels of "safe
use." Is the acceptable level the same for the hypersensitive person as the normal person, asks Dr. Crater. "More than 16 percent of the population is hypersensitive. How can these people
avoid their sensitizing chemicals?"

The problem was dramatized by a four-year study of 8,000
patients at Tufts University. Doctors were shocked to learn that
in one case out of 20 a prescribed drug causes an adverse reaction. In one case out of 100 the reaction is so severe as to

threaten the patient's life. These are prescription drugs supposedly thoroughly tested and administered under medical supervision. The human body is perverse and unpredictable. What is "safe" one time may not be safe another. Many of our assumptions are based on unfounded presumptions.

There are many instances in which a material is believed to be safe and later is found to be unsafe, Dr. Crater points out. "Cobalt, once added to beer as an antifoaming agent, is an example. Of the thousands of chemical additives to foods, how many could stand strict scrutiny?"

Under the "philosophy of poison," there are legal tolerances for chemicals that cause mutagenic effects, birth defects, and cancer in animals. Some unknown in the FDA's bureaucratic maze has decided to accommodate industry by permitting them. He has decided that the economic benefits outweigh the risk. Frequently these decisions prove wrong. Many substances have been in use for years, even decades, before being found to be harmful. Most have never been adequately tested. This has ominous overtones when it is considered that of 1,329 chemicals tested by the National Cancer Institute, about 25 percent turned out to be carcinogenic. Most of these never appeared in food. But no one knows how many of the substances now appearing commonly in our daily diet will turn out to be carcinogens, just as did the cyclamates, the artificial sweetening agents that were included in the GRAS list (Generally Recognized As Safe). After the furor over the cyclamates, the FDA announced that it would review the manner in which the GRAS list was drawn up, as well as the materials on it.

Permitting carcinogens in foods is folly, warn cancer experts. The International Union Against Cancer, made up of specialists from all over the world, passed a resolution warning that any substance that causes cancer in animals at any level of use or under any conditions should not be permitted in foods. This warning (and many others like it) is ignored by the FDA. The FDA repeatedly violates the law (the Delaney amendment to the food law) that states that no carcinogens are to be permitted in foods. It does not give the FDA the right to decide what is a carcinogen or under what conditions. The law says that any substance that causes cancer under any conditions is illegal. This includes DDT, which the FDA has permitted in foods with a "legal" tolerance for more than twenty years. Some bureaucrat, satisfying some commercial interest, or aiming at job security for himself after he leaves government service, can sign a paper and the public must gamble with its lives. These decisions are made in secret. The FDA refuses to make public information on the toxicity and carcinogenicity of food substances tested. The present FDA policy of strict secrecy in these matters is definitely against the public interest, says Dr. Hueper: "It prevents an examination of the submitted evidence, as well as independent appraisal of reasons for any approval by competent investigators not connected with interested industries or with governmental agencies involved in these decisions."

Tolerances are set on the basis that each dose is small. Each purveyor assures us that his small amount is safe. Individually the amounts are small. In the aggregate they are enormous: three pounds of food additives alone per person per year. This makes no allowance for all the complex interrelationships of chemicals or for cumulative effects. Indeed, it is now believed that for many substances, particularly the carcinogens, the cells have a "memory." The damage accumulates bit by bit. Each dose pushes the injured cell a little farther toward ultimate disaster. Each substance may affect only a limited number of people, but these small segments together may make up a majority of our population. The conglomerate of chemicals—each labeled "safe" by itself—stalks us as an army, seeking out their specific targets. Together they exert a continuing, relentless pressure. Slowly the body loses its resiliency—its "adaptive energy," as it is called by a group of Canadian researchers.

Crisis follows crisis. Each new assault leaves an indelible scar. Each time we become a little weaker, a little more vulnerable. But the damage is infinitesimal, so low on the curve that it cannot be measured by the most sensitive instrument. Who is able to prove that a man is older or weaker one day than he was the day before? Yet this process goes on inexorably, aided by environmental insults; only after enough time has passed can the cumulative aging be demonstrated by the cumulative result. "The organism pays for its survival after each stressful situation by becoming a little older," as Dr. Wiley noted. Finally the reserves are exhausted. The body falls victim in the next battle and there is no single assailant to blame.

The combined assault from the ingestion of the great variety of synthetic substances was described by an international team of eight leading scientists as "internal pollution." They called for a program to spare mankind from its long-term effects. The panel, which included several Nobel Prize winners in medicine and biochemistry, said that the populations of developed countries are eating more and more synthetic foods or food additives whose subtle effects on the body have not been adequately explored. The panel warned that man's technological advances are "endangering his very survival." It said chemicals made by man "are the primary offenders, because they are ubiquitous and being exploited to the limits of their possibilities."

What makes up this legion of poisons and compounds so wantonly taken into our bodies and dumped into the environment? Some have been mentioned or touched upon. Many are released accidentally. Most are by-products of some other process. The number of fish killed is a good barometer of this toxic assault. Between 1960 and 1968 a total of 103,380,000 fish were reported killed in 2,830 incidents. This represents only the number reported; it is considered a conservative estimate. Death comes from industrial wastes, municipal sewers, and a galaxy of careless or stupid acts—a farmer washing out his spray tank in a stream, a bulldozer breaking a buried fuel pipeline that poisons a stream, a muskrat digging a hole through a dike that holds back settling toxic wastes from a factory, crop dusting and pesticide runoff. Small tragedies are lost in impersonal statistics. We focus on one impersonal event: carbine herbicide was applied to a field beside a brook and X number of fish died, states the report. But the observer of the incident sees something else: "The eyes of the fish were white prior to death. The fish gasped and struggled at the surface before succumbing."

We are products of this environment, the expression of the "philosophy of poison." The typical American today is a toxic creature himself—a walking experiment. We carry asbestos in our lungs, DDT in our fat, lead in our bones, radioactive iodine-131 in our thyroid, strontium-90 in the marrow of our bones, as well as other cumulative poisons and compounds of a more transient nature.

The list of poisons we are routinely exposed to is formidable and virtually endless. Some of the more lethal substances, such as mercury (now a universal contaminant, its presence discovered only recently and then by chance), have been publicized. Others are virtually unknown. It is a major disadvantage in a book of this type that each substance must be treated separately. Our attention is focused on only the compound being written about. For the moment we forget that the others exist. But this is a laboratory artifice. In the real world these substances never exist in isolation. They are forever mixing and interacting, exerting their various effects on the body in exotic combinations. It is this total effect that must always remain in doubt. Our only clue to what is happening is in observing the ultimate ravages on our bodies from outraged nature.

We must begin with the assumption that good health is the normal state for men, as it is throughout the animal kingdom. Bad health is a departure from this norm. Our problem then is to find what environmental influences have brought about these changes. The substances named here barely scratch the surface. Volumes would be needed to cover all the substances we have unleashed against ourselves. Hundreds of those used on foods are listed in the National Agricultural Chemicals Association "Official FDA Tolerances." Several of the more common environmental poisons, including mercury and lead, are discussed in the Appendix.

One group of widely distributed environmental chemicals deserves special attention—those that cause genetic changes. The scope and severity of this hazard is only starting to be recognized. Genetic factors appear to be either primary or "very significant" in the illness of at least one child in five admitted to the pediatric service of a large hospital studied recently. The figure, cited by Dr. Barton Childs, professor of pediatrics at Johns Hopkins University, indicates the importance of genetic problems in human disease.

Few studies have been done on the mutagenic effects of drugs, said the doctor. A mutagenic effect is one that causes a mutation or other inheritable disturbance in the genetic material of cells or of whole creatures. Most mutations are harmful. They are known to cause diseases such as the blood disorder hemophilia and the form of mental retardation called mongolism. The way most abnormalities arise is unknown. "Sometimes, presumably, they represent mistakes made by individual cells in copying the genetic material that is passed on from one generation to another," according to a *New York Times* dispatch. Radiation can cause mutations in living cells. So can some chemicals. This is of special importance today because of the number of chemicals being injected into the environment and used in foods without proper testing. An especially vicious group of mutagens are the so-called herbicides used in food production in the United

States and dumped on Vietnam in millions of pounds. These are discussed in the Appendix.

The overall problem all of the environmental chemicals pose becomes increasingly threatening as they multiply. Equally important and even more puzzling is what to do with their toxic wastes. They have already contaminated air and water. Where to turn next? To the ground, of course. That's the only place left. Bury them and pretend they do not exist. Leave them as an unpleasant surprise for future generations. The newest and most popular disposal site is the so-called deep well. It combines the virtues of being cheap and, according to those who use them, "safe." So, into the earth go our deadly wastes—the cyanides, carbolic acid, radioactive materials, hydrochloric acid, chlorinated hydrocarbons, chromic acid, sulfuric acid, organic phosphates, "steel pickling" liquors, brine, and all the others that we do not know what to do with.

But the earth, like the sky and water, has been known to reject these toxic offerings. It has returned them with violent and bizarre effects. Earthquakes have been touched off. Cyanide has oozed from the ground to taint drinking water supplies. A well on the shore of Lake Erie blew out to spew 150,000 gallons of stinking poisonous paper-mill wastes into that already vexed lake every day for three weeks. Agitated saltwater has violated the earth's diminishing supply of sweetwater. Cellars have mysteriously flowed with used motor oil. Lawns have split open to vomit acids and salts that killed flowers, shrubs, and grass.

These deep waste wells have multiplied from only a couple in 1950 to more than 150 known today and others that are not known. If not curbed soon there may be hundreds of thousands within a few years, it is predicted. Their safety is based on risky hypotheses. In one survey 14 percent of 114 known wells had to be shut down or suffered failures. Others are of questionable safety. They corrode, leak, and taint groundwaters. They are frequently unpredictable and their contents often unstable.

Little is known about the buried world where we are now injecting our leftover poisons. It is a great network of rivers, lakes, crystalline springs, and connecting channels and fissures. Water percolates through the soil, seeking out the path of least resistance, finally finding its way back toward its distant home, the eternal ocean. We know as little about the subterranean black caverns of the earth, the complex deposits of minerals, gases, salts and acids, the timeless mountains of coal and ancient seas of oil, boiling waters that erupt in geysers, raging fires that can rip the earth open in volcanic fury to vomit flames and void ash. We know little about the faults that slip suddenly, unexpectedly, after billions of years of perilous balance, to bring on devastating earthquakes, the subtle relationships of an unseen world too fantastic for the understanding of the most profound mind or the liveliest imagination. Scientists say this subsurface world is the least understood part of the entire earth. But that does not deter us in our daring experiments.

In theory a deep well seems a clever way to get rid of unwanted poisons, if only nature would be more cooperative. Such a well is simplicity itself. You need only bore a small hole, five or six inches in diameter, that reaches into some buried reservoir 2,000 to 3,000 feet deep. This, theoretically, is below the shallow groundwaters that freshen the earth and provide our drinking water supplies. The hole is fitted with a steel pipe encased in cement to prevent corrosion and leakage of wastes into groundwaters along the way.

One technique is to dump wastes into buried saltwater deposits. This can have a disadvantage, forcing the saltwater up and contaminating groundwater supplies. A more sophisticated method is to drill into porous rock, such as sandstone, limestone, or dolomite; these can soak up fabulous amounts of wastes, like a huge sponge. Wells range from the small up to 20,000 square miles, holding billions of gallons of wastes. Ideally these underground sponges are sealed by surrounding layers of impermeable rock, such as clay or shale, to prevent leakage. But, unfortunately, such ideal conditions are almost nonexistent. Buried wastes also have an unpleasant habit of refusing to stay put. They seek out flows in the stone tombs and migrate into other underground regions, to subsurface streams, or they bubble out at some distant surface point. They can and do travel unseen and undetected for miles. In one case brine wastes emerging in Michigan were traced to a well in Canada. Even when toxic wastes do stay locked up in their subterranean prison they render the area forever useless. And there is always the threat that future generations will unwittingly tap this buried toxic reservoir.

Some wastes are injected under great pressure—a process known as fracturing. The purpose is to make new cracks in the reservoir to increase its holding capacity. But this can bring on unforeseen difficulties: unanticipated breaks permit wastes to seep out, or force the liquid to seek out previously undetected flaws. Even without the additional man-made disruption the waste wells are risky enough. One solemn warning comes from Dr. William T. Pecora, director of the U.S. Geological Survey. In a congressional hearing he testified that there is a constant interaction of the water locked up in the porous and permeable rocks of the earth—a constant interaction of this water supply with the streams:

> No one is properly prepared to recharge water or liquid wastes into complex underground systems with certainty of the results that such an operation may yield. Hopefully the time is approaching when large quantities of polluted material cannot be dumped into the ground without considerable thought being given to the consequences.

The consequences may be bizarre, spectacular, or tragic. They are nearly always unexpected, the result of some miscalculation, some unfounded optimism. Several are outlined in *Environment* by David M. Evans, a geologist, and his co-author Albert Bradford. At the Rocky Mountain Arsenal in Denver the army, after contaminating miles of groundwater with toxic discharges from the manufacture of nerve gases and pesticides, pumped 150 million gallons of poison into a deep well. This caused Denver's first earthquake in 80 years, and during the next five years it led to 1,500 quakes of varying intensity, correlated to the frequency and intensity of the pumping rates and pressures. The pressure of wastes has upset underground pressures, sending geysers of toxic wastes and gases shooting out of abandoned or forgotten wells—sending wastes down one hole and up another. Oil rigs have been toppled. Salt springs have mysteriously erupted in dry creek beds.

Chemicals in wells sometimes react, as mentioned, with odd and unexpected results. In Denver the army created a mix that spontaneously formed the weed killer 2,4-D. It emerged on a nearby farm, killing grass and crops; cows aborted and lost their calves; young pigs and registered calves died. The farmer was able to collect damages because he could prove the source of contamination. Many are not so lucky. A federal official said we have been lucky that no widespread catastrophe has taken place. The potential is there, he said.

Still industry insists the wells are safe. They are monitored, we are assured. But what happens if the company goes out of business, or if the well develops a leak? These are very real possibilities, according to Evans:

> We are not told what to do once we discover that a city's water supply has been poisoned. One possibility would be to keep pumping and abandon the city. These "checks" illustrate the point-of-no-return of disposal wells: when something goes wrong underground, it is impossible to retrieve the poisons that have broken loose.

Almost no restrictions exist on disposal wells at this time. Only a handful of states have passed regulations; often they are undermined, as it were, by some variation on the theme that disposal wells are not permitted "unless other methods are not economically feasible." More expensive methods are almost never economically feasible. Usually this means that some local official eventually decides what is, or more likely is not, "economically feasible." And local officials are notoriously subject to local pressures. They sometimes quietly issue permits without holding public hearings called for. Other times they quietly look away when wells are knowingly dug without permission.

The federal government has shown a similar disinterest in imposing restrictions, despite many warnings about the seriousness of the threat. The hazard becomes greater as populations grow and well and land use increases. Many geologists are very concerned because we are increasingly making unusable underground supplies of brackish water that could be cheaply and easily desalted as a future water source to replace dwindling supplies of freshwater. Time is always on the side of the polluter; the damage becomes progressively harder to recall or undo, and the cheap and shoddy become the norm. We continue to get many assurances that waste wells are safe, but there are no guarantees. Who will guarantee that they will not leak in the future and possibly cause mass poisoning of drinking supplies? Who will be responsible? Who is to pay for the damage and purify the water? Who can tell us with certainty that eventually these stored poisons won't migrate to the sea? We wait in vain for such assurances.

The unfortunate part is that these wells are not necessary. They are cheap and that is their only justification. They cost from $20,000 to around $200,000 and average about $100,000. A modest sum indeed compared to a waste treatment plant that might cost $500,000 or more. They are also cheap to operate. A textile company that buries flammable materials figures it saves $500,000 a year, according to the *Wall Street Journal*. One possible solution is new regional treatment plants that are being developed to serve many industries in the area. One of the first is in Bridgeport, New Jersey. It sends tank trucks to industrial plants, collects their wastes, and discharges them into incinerators and through chemical processing. We are not told how these poisons are decontaminated or what compounds they break down into with what effects. But at least the problem is being recognized and some attempt at correction taken. To permit the wells to proliferate and poison the earth is not a solution. It merely leads to new dilemmas and more deferred payments. The wells represent just another subsidy to industry and a further narrowing of our ever-dwindling margin of safety.

"PROTECTIONS" THAT DON'T PROTECT

The "philosophy of poison" rests on an article of faith: the assumption that the people are protected by government from avaricious commercial interests. Unfortunately, this is not the case. For more than two decades the FDA, in conjunction with the Department of Agriculture, has permitted DDT to remain in food despite devastating evidence against it. During that time the government itself has been the greatest offender in poisoning the earth. The Department of Agriculture has been, in effect, the major sales arm of the chemical industry. It has promoted and financed vast spray programs that have saturated the earth with multiple long-lasting poisons. It has used its official position and great authority to mislead the people about the safety of these programs. It has justified the unnecessary use of poisons by propaganda that whips up hysteria and a sense of crisis. It has withheld information about the effects of these spray programs, suppressed adverse results, and viciously attacked critics of its policies. In court cases it has supported the poisoners and adulterers. It has fought every effort to curb its power to poison on whim or at will, usually with success in the federal courts.

The pests the spray programs are aimed at are never "exterminated" according to promise. Each spray program leads to others. Even as DDT's threat is recognized and proof of the great ecological damage from sprays builds up, the Department of Agriculture plans new spray programs with substitute poisons. We are told these poisons are safe because they break down faster than DDT. One is called Sevin, a carbamate compound. It has been found to be potentially carcinogenic.* In animals, researchers say, it has caused cystic kidneys, skeletal deformations, cleft palates, facial malformations and other birth defects. It reportedly breaks down into a substance (alphanaphthol) that a group of investigators rated "very toxic." In 1969, there were 14,964,000 pounds of Sevin produced. It appears in milk, is sprayed on cattle, pigs, sheep and poultry, on barley, oats, pecans, beets, tomatoes, broccoli, rye, poultry houses, and sugarcane. It is deadly to honeybees (they perform 50 to 90 percent of all pollination) and is destructive to insects eaten by fish and birds. As of 1964, the last year for which data were available, it was not registered for use as an aerial spray over populated areas but was used extensively in that way.

Meanwhile the myth persists that DDT is now banned. It is not. It continues to be used in large quantities and apparently will continue to be used for a long time to come. This is due to another of the government's accommodations to industry. Under federal law every pesticide must be registered with the government before it can be sold. The approved uses and amounts are stated on the label.

If it turns out that a poison is more dangerous than originally stated, the government usually has the label rewritten and permits the sale to continue. In emergencies it has the authority to remove any substance from sale by revoking its registration. This can be done in two ways: one is called "cancellation," the other "suspension." Cancellation has a fine sound of finality about it. But in the strange world of USDA bureaucracy it is less final than suspension. Suspension means that the sale must be halted immediately. Cancellation means that the sale must be halted at some specific future date. If industry cares to challenge this order it can do so, first through prescribed administrative actions; then, if necessary, in the courts. Such challenges can be drawn out for years. Under the cancellation proceedings the sale can continue indefinitely. Such a ban is, in effect, meaningless.

The government never took any action against DDT until forced to do so by mounting public pressure. Several conservation organizations challenged the government's right to spray DDT in a number of court actions. The publicity alarmed many people who demanded a ban. The Mrak report also urged that

* M. B. Shimkin et al. *Cancer Research* 29:2184, December, 1969.

DDT be phased out of use in two years. Robert Finch, then Secretary of Health, Education and Welfare, was asked to set a zero tolerance for DDT on food. He held a press conference. With much fanfare he announced that DDT would be banned in two years and he was appointing a committee to see to it. But he denied the zero tolerance. Secretary Finch seemed to feel that DDT would be harmful after two years but was not so in the meantime, and people could go right on eating it.

An appeal to the Department of Agriculture was equally futile. Secretary Clifford Hardin was requested to *suspend and cancel* DDT's registration. The request was accompanied by a massive petition, assembled by Dr. Charles Wurster, chairman of the Environmental Defense Fund's Scientists Advisory Committee; it cited 88 scientific articles on the effects of DDT and was accompanied and supported by a bibliography of 268 articles. As usual no one was listening. Secretary Hardin's response was to *cancel* the registration of DDT for a few minor uses: on shade trees, on tobacco, in the household, and, except when "essential for control of disease vectors as determined by public health officials," around water. The action meant that DDT could still be used on cotton plants, fruit trees, berries, vegetables, and so on. Inevitably, industry challenged the order, as the secretary had to know they would. The "ban" was then lost in a maze of administrative proceedings, court actions, reviews, studies, reports, committees, evaluations—the whole baffling and unending series of bureaucratic delaying tactics that are the life's blood of polluters and their friends in government. The action amounted to a public relations triumph. The public believed a ban was in effect, while in fact the sale of DDT continued.

Later, in November, 1970, the Department of Agriculture quietly announced that the "ban" had been rescinded. It said it had issued the order only at the urging of the Interior Department. "But then it dawned on us that all these questions are under review anyway and will not be completed as soon as expected," said a spokesman. "We had to get off the hook." The department announced that it had sent manufacturers a notice, canceling the 1971 deadline and automatic bans. Once more DDT had triumphed.

In late 1970, control over pesticides passed from the USDA to the new Environmental Protection Agency headed by William D. Ruckelshaus. He promptly announced cancellation of DDT—in accordance with court orders—and said his agency would also start a review to determine whether all uses of DDT and the herbicide 2,4,5-T should be suspended pending the outcome of the long cancellation proceedings. A few months later, in March, 1971, Ruckelshaus announced that his agency would not suspend the sale of DDT or 2,4,5-T. This meant they could continue to be used for at least two more years.

The New York Times observed that conservationists who had been seeking suspension of DDT and 2,4,5-T were disappointed. They had believed Ruckelshaus had been inclined toward suspension.

> At the same time they recognized that the National Agricultural Chemical Association, farm organizations, and some members of Congress, including Representative Jamie L. Whitten, Democrat of Mississippi, had been putting pressure on the White House against immediate suspension of the pesticides. . . .
>
> Mr. Whitten is the author of a pro-pesticide book entitled *That We May Live*, which appeared in 1966 and attacks the anti-pesticide arguments of the late Rachel Carson's book, *Silent Spring*.
>
> He is chairman of the House appropriations subcommittee that handles appropriations for the New Environmental Protection Agency. In an article last Sunday, the *Washington Post* said that three pesticide manufacturers had subsidized "sales" of the Whitten book. Asked about this, the *Post* reported that Mr. Whitten said, "That's our business."

Ruckelshaus was asked if he had been under any pressure from Whitten; the *Times* continued: "Mr. Ruckelshaus said, 'I haven't talked to him.'"

This adds up to one thing. There is no ban on DDT. But the government now exults that the sale of DDT has fallen off following the "ban." This is a hollow victory at best. The "ban" underscores many ironies. DDT is neither more nor less dangerous now than it was twenty-five years ago when the government insisted it was safe. The only difference now is that people are aware of some of the risks that were long withheld from them. When they learned what these risks were, they became alarmed. Many stopped using DDT. But any damage that has been done cannot be undone. We cannot even take comfort in the thought that more damage will not continue. DDT's use may have fallen off in the United States, but it goes on spiraling throughout the world. We still manufacture large amounts of DDT, but now about 70 percent of it is being exported. At the same time new DDT plants are springing up throughout the world, especially in underdeveloped countries—India, Turkey, and perhaps elsewhere. Much of this DDT will enter the global distribution system. Ultimately it will be shared by the world community; what Americans previously got directly they will now get indirectly.

Much of the environmental contamination that we now must live or die with is due to the breakdown in enforcement of the limited protection afforded under the 1947 pesticides law. Until registration of pesticides was put under control of the Environmental Protection Agency in 1970, it was administered by the Pesticide Registration Division (PRD) of the United States Department of Agriculture.

A 1967 congressional hearing detailed the operations of this government agency that virtually worked in the service of some of the companies it was supposed to regulate. The report documents a failure to enforce the law, favoritism toward some companies over others, an almost complete indifference to public welfare. There was an "almost incredible failure" on the part of PRD to enforce the law, reported the committee chairman, Representative L. H. Fountain.

For more than five years PRD had specific authority to cancel registrations—to begin proceedings to halt the sale of dangerous products. But in that time it did not once secure the cancellation of a registration in a contested case. As late as 1969, the committee found, PRD "had no criteria for determining whether a particular product was hazardous enough to warrant suspension of its registration so that it could be removed from the market immediately nor did it have established procedures for implementing such a determination." Only one suspension action was taken (despite the alleged lack of procedural rules), "but a product containing an identical amount of the same active ingredient was allowed to remain on the market without even bearing a required warning notice on its label." The report left no doubt that the one suspension taken was to accommodate a rival firm.

According to a 1969 report of the House Committee on Government Operations, PRD had as consultants who helped determine registration criteria three employees of the Shell Chemical Company, a division of Shell Oil Company, one of the nation's largest pesticide manufacturers. Shell repeatedly received favored treatment. It virtually dictated the labeling on one of its products that had been contested by a PRD employee. Hazardous chemicals were permitted for use around food despite warnings that they were unsafe. PRD approved for sale undetermined numbers of pesticides over unresolved safety objections from HEW. PRD kept no record of the number of times it rejected HEW safety protests or even refused to inform HEW of the rejections. Of 1,600 unresolved objections required to be referred to the Secretary of Agriculture by interdepartmental agreement, not one was referred. In 1969 PRD certified 252 pesticides over objections raised by the Public Health Service. PRD approved products for sale despite inadequate and even contradictory labeling, and "in some instances . . . knowingly left illegal and potentially hazardous products on the market."

One such self-contradictory label was the one approved by PRD for *Super Hy Kill*, a concentrated insecticide spray carrying this warning:

> Use in well ventilated rooms or areas only. Always spray away from you. Do not stay in a room that has been heavily treated. Avoid inhalation.

On the other side of the container the directions advised:

> Close all doors, windows and transoms. Spray with a fine mist sprayer freely upwards in all directions so the room is filled with vapor. If insects have not dropped to the floor in 2 minutes repeat spraying, as quantity sprayed was insufficient. After 10 minutes doors and windows may be opened.

PRD's relationship with Shell is worth examining in greater detail. The one time that PRD took cancellation action against a product was when Shell complained about a competitor to its *No-Pest Strip*. Shell's *No-Pest Strip* was registered in 1963 after John S. Leary, Jr., then PRD's chief staff officer for pharmacology, overruled an objection by one of his subordinates who had recommended that the label include the word "Poison" and a skull and crossbones. In 1966 the Public Health Service recommended that the *No-Pest Strip* be denied registration "because the devices used deliberately subject human beings to continued exposure to a pesticide." Five months later the same John Leary who had been instrumental in registering the product resigned from PRD to go to work for Shell. Before leaving PRD he filed a memorandum stating that the PHS report "serves no useful purpose" and did not justify changing the *No-Pest Strip* label.

The odd adventures of *No-Pest Strip* include its testing by a Shell medical consultant, Dr. Mitchell R. Zavon. He found that food samples exposed to the strips were free of pesticide residue. Other tests by different people found the opposite to be true. For six years, from 1963 to 1969, Zavon served as consultant to Shell and PRD and was involved in discussions of *No-Pest Strip,* one of Shell's biggest moneymakers. Apparently PRD saw nothing extraordinary about a Shell representative helping shape government policy on a product whose safety was being contested and that his company manufactured, although federal law specifically bars consultants from serving in official matters in which they have a financial interest. "The U.S. Department of Justice has as yet shown no great interest in Dr. Zavon's dual roles," observes *Consumer Reports* in November, 1970.

Zavon, representing himself only as a "health officer" (he was an assistant health commissioner for Cincinnati), endorsed pest strips in a letter to the Public Health Service and urged that the entire subject of pesticide vaporizers "be re-reviewed." At least in part through Zavon's zealous efforts, the difference between the PHS and PRD concerning the safety of *No-Pest Strips* was resolved in 1967 with a somewhat stronger label warning. PRD, however, did not require Shell to recall and relabel its inventory of *No-Pest Strips.* That would have cost the company money. Instead the public was forced to take the risk of continued exposure to the pesticide with no warning of the danger. Nor did PRD or its parent agency, the USDA, issue a press release announcing the new warning, notes *Consumer Reports.* It is not easy to understand a Zavon, but a rare insight into his mind is afforded by a fatalistic quote by him that appears in Frank Graham's *Since Silent Spring:* "We are natural creatures, and who is to judge whether or not our destructiveness, however we may deplore it, is not an ordained path in nature's road of terrestrial development."

PRD was not alone in its derelictions. The USDA's Agricultural Research Service was similarly taken to task by the Fountain Committee. Since 1947 ARS had been empowered by law to seek the recall of unsafe products, but the first recall did not take place until twenty years later—exactly two days before the Fountain Committee began its hearings. ARS was similarly rebuked by the government watchdog agency, the General Accounting Office. It found ARS had not reported a violator of the pesticide law for prosecution in thirteen years despite "repeated major violations of the law." In that period several shippers were cited for from four to twenty major violations each. The shippers cited "did not take satisfactory action to correct violations . . . and ignored ARS notification that prosecution was contemplated." In 1960, ARS tested and reviewed 2,751 product samples and found 750 to be in violation (20 percent in major violation). Yet there were only 106 seizures.

The indictment continues—a devastating documentation of a government agency not responsive to public needs or desires and indifferent to public welfare. Many of the people I talked to, particularly the scientists, are dismayed by the response of government and industry to the pesticide problem. In place of any real effort to resolve it, they have encountered only new tactics of delay and denial. Dr. Wurster told a congressional committee that he was "greatly discouraged and disappointed by the attitudes and actions" of certain manufacturers of pesticides and their trade association, the Manufacturing Chemicals Association:

> Instead of helping to solve the problem, they deny its existence. Rather than recognizing incontrovertible scientific evidence, they attack those scientists who do the research. They treat serious pesticide problems as no more than problems in public relations, and assign public relations men to disseminate nonsensical propaganda that confuses the public.

Wurster and many other scientists have been bruised by their encounters with industry and government. Such encounters usually take place in the courtroom. For most of these men it is a new experience, and often unsettling, as they see fact distorted by clever corporate lawyers and bought scientists. But as Wurster said to me during our conversation, "The battle must continue. There is no alternative, or the world as we have known it will cease to be inhabitable."

4

Too Many People
and
Their Wastes

The heart of the problem as of all human problems is one of meaning, it being altogether a meaningless and bankrupt enterprise to try to make of the human enterprise a simple exercise in production, trying to make of it an attempt to see how many people one could keep alive on the surface of the earth. Our problem isn't one of numbers but of the quality of human life.

What we need to do is to produce that size of population in which human beings can most fulfill their potentialities, and from that point of view, in my opinion we are already overpopulated, not just in places like India and China or Puerto Rico, but here and in Western Europe, and with that overpopulation in our western world there has gone, I think, a signal deterioration in our culture all through the last century.
—Professor George Wald (Harvard University biologist)

BOOMING POPULATIONS

The population problem is usually defined in terms of statistics: so many people are born every year, so many die, leaving us with a net increase of X number of mouths to feed and bodies to care for. This seems a reasonable enough way to approach the problem. But the solution remains more elusive. It becomes less statistical than political and is complicated because procreation involves more than an equation in logic and logistics. It is rooted in every psychic fear, prejudice and subliminal impulse known to human beings. We are not ruled by our minds as much as we are driven by the deep and mysterious forces that ferment in the unconscious.

The multiplication of populations is patently irrational, destructive, and potentially cataclysmic. It defies the simplest mathematical truth—that the world is finite and will feed and support only a given number of people. As the numbers increase, the pressures on the earth are intensified. At a given point the earth no longer can support the number of inhabitants. Then, of necessity, there must be starvation and death.

No essay on population would be complete without statistics. I cite a few of the more meaningful figures, in passing; but it is their effects that are of primary interest here. A few years ago these statistics were dramatized in an exhibit at the New York World's Fair. Each time another baby was born, a "humanometer" reflected the new world population. Fair visitors stared at the whirling figures, fascinated, as if trying to translate the abstract numbers into reality.

Dr. Paul Ehrlich has called this "the tyranny of arithmetic." The tyranny is expressed like this: 235 babies are born in the world and 96 persons die, for a net population increase of 139, per minute, 8,309 per hour, 199,450 per day, and 72.6 million a year—a rate that threatens to double the earth's present 3.5 billion inhabitants in about 30 years. Every night the sun sets on some 200,000 additional human beings.

The problem is compounded by what demographers call doubling. As numbers swell, the time it takes the total to double becomes increasingly shorter. In 1850 there was a total of one billion people. Eighty years were required to achieve two billion (1930). But only 45 years must pass for us to reach four billion (1975). By 2000, if we are still on this suffering old planet, we will be seven billion—approximately double the present population. Ultimately there will be standing room only on earth. Then we will be balancing on one another's shoulders. But experts console us by saying it will never come to that. By then we will have resolved the problem rationally, irrationally or accidentally —by planned population control or some epic misadventure such as pollution, famines, thermonuclear wars, plagues, and viruses attacking hunger-weakened nations to cause mass death on a scale unknown even in the Middle Ages. Through one disaster or another the problem will be solved.

But the immediate question is what to do with all these people? They must be housed, fed, educated, jailed, taken care of when ill and old and, eventually, their corpses disposed of. Between birth and death they will demand some of the amenities of life, but these amenities become increasingly scarce. As the mass grows, comforts decline and even necessities diminish. Wealth must be spread thinner. The worth and weight of the individual are diluted. As we become collectively greater in number we become individually of less consequence. First we struggle for recognition, for some sense of self, and finally we settle for survival. Personal worth must diminish in the hand-to-mouth struggle for subsistence. We become wretched beggars, asking no more than a slab of concrete to sleep on and a crust of bread. We do not live. We survive.

Those are extremes, but already we can visualize the new world that must emerge. A forest of steel and glass skyscrapers, with each man his own jailer. One prophet sees a new city of 250,000 population being built every 40 days from now to the end of the century. Another visionary predicts "a new Chicago every 18 days."

Secretary of Commerce Maurice H. Stans described this evolving world as an "anthill society"—an enormous megalopolis: *

> BosWash, an unbroken stretch of people, homes, factories, highways, railroads and power lines from Boston to Washington; ChiPitts, a solid belt of heavy industry from Chicago to Pittsburgh; San San, from San Francisco to San Diego; JaMi, the fourth megalopolis, along Florida's east coast from Jacksonville to Miami.

Some seers go even farther. The Greek architect Doxiadis, as Lord Ritchie-Calder notes, "has been warning about such prospects. In his Ecumenopolis—World City—one urban area would ooze into the next, like confluent ulcers." He sees the East Side of World City having as its High Street the Eurasian Highway stretching from Glasgow to Bangkok, with the Channel Tunnel as its subway and a builtup area all the way. "On the West Side of World City, divided not by the tracks but by the Atlantic, the pattern is already emerging, or rather, merging." You don't need a crystal ball to predict it, he says. "We can already see it through smog-covered spectacles. A blind man can smell what is coming."

We do congregate in cities, as if seeking comfort in numbers from the distempers brought on by overcrowding. More than half

---

* Stans resigned as Secretary of Commerce on January 27, 1972, to become chief fund raiser for Nixon's reelection campaign.

of the United States' 200 million people are squeezed into about 10 percent of its land. Prospects are that by 1975 the number will have reached 235 million with three quarters pressed into small areas. Such crowding almost invariably leads to urban chaos—physical discomfort and psychological dislocations. Every human problem is multiplied. There is more crime, more insanity, more unhappiness, more alienation, more loneliness, more desperation—more everything. It's harder to find a place to live or simply to sit down and rest. Noise makes it hard to get a good night's sleep. Nerves are on edge. Annoyances become crises. Crises become disasters. Tempers flare. Violence erupts. Space diminishes. Buildings rise higher and rooms become smaller. Wastes and pollution soar. Welfare costs spiral. Services break down and the city starts to decay. The middle classes flee. The crumbling city is left to the very rich, who can afford to isolate themselves from its unpleasantness, and the poor who must adapt or perish. All live behind barricaded doors and windows in terror of their neighbors; we become our own jailers.

The problem does not end there. They flee to the suburbs, whites and the few blacks who can afford it, and soon the suburbs become small cities themselves with all the familiar problems of big cities: lack of services, racial tensions, soaring prices, competition for small comforts, rising costs. The suburb pushes out until, ultimately, it touches the decaying central city. They are now one. The nightmare vision becomes reality. Even today we are laying the foundations for tomorrow's phantom cities as we breathe life into the ghostly statistics that will become the people who occupy the terrifying World City of the future.

The problem of servicing the projected nightmare monolith boggles the mind. *Municipal News* says that based on present U.S. conditions, every 1,000 new people in metropolitan areas require

> 4.8 elementary school rooms, 3.6 high school rooms, 8.8 acres of land for schools, parks and play areas; an additional 100,000 gallons of water per day; 1.8 new policemen, 1.5 new firemen; 1 additional hospital bed; 1,000 library books; a fraction of a jail cell; sewerage and treatment for 170 pounds of organic water pollutants per day.

In addition, we are reminded, each person is responsible for 4 pounds of solid wastes to be disposed of every day, plus 1.9

pounds of air pollutants. Each throws away 250 cans and 135 bottles or jars a year. These usurpers will get in your way in the supermarket, queue up in front of you at the movie, litter parks, preempt the bench you intended to sit on, bump against you in the streets. Their cars will block and smash into yours on the freeway. You will have to reserve space in a national park months in advance, possibly years. Getting to the seashore will be almost impossible, even if the water is not too hopelessly polluted to bathe in.

Recreational facilities will become increasingly scarce. Already they are inadequate, part of the competition for breathing space. More and more shore communities are restricting their beaches to local residents. Only 5 percent of the ocean frontage is publicly owned. (The government has repeatedly failed to follow urgings to buy such lands for public use, and now most of the areas recommended for purchase have fallen to industry and private developers, beyond reclamation and lost forever. Only 3 percent of the entire United States today remains in the wild state.) For those limited recreational waters available, four hours of driving for a few hours in the sun is not uncommon in metropolitan areas. The need of people for relief from the pressures they impose on one another is pitiful. Representative Morris Udall of Arizona said he has seen fishing streams and beaches "where one had to stand in line or elbow somebody out of the way to get to the water . . . there's more privacy at Kennedy Airport than at a camp ground I visited on a North Carolina beach last summer."

The new world with its multiple tensions and frustrations will inevitably be one of violence as the aggressive or panicky individual attempts to assert himself above the mass with force and is met by counterforce, and large numbers are less able to accommodate themselves to fewer choices imposed by the necessity of political and social logistics. There will be little opportunity for solitude or refuges for self-communion. The press of numbers alone makes the city a place of external turmoil and inner disassociation. Those who survive must find a strategy to preserve their sanity. They may withdraw and block out the world or try to outshout it. Either can mean disaster. Withdrawal will often mean drugs, excess alcohol or psychological immobility. Survival must mean some sacrifice of sensitivity. The Super City must inevitably tend more toward housing robotlike people living artificial lives in vertical glass and steel boxes, completely isolated from nature, from one another, and from themselves.

How might these people act under such overcrowding? Animal studies give insights into the bizarre behavior that can take place when numbers become excessive. The classic illustration is the lemmings of Sweden. These rodentlike creatures multiply wildly, exploding into millions almost overnight. Mysteriously, as if at some given signal, they suddenly stampede. From their normal home in the heights they head for the lowlands. Popular belief insists that lemmings are motivated by some suicidal urge. Biologists think these death migrations are due to overcrowding. Thousands are crushed as they swarm through ravines and hurl themselves off cliffs. They trample one another and drown as they cross mountain rivers. Many are killed by cars. They are slaughtered by trains in such numbers that the tracks become too slippery for cars to move. They are preyed on by foxes, hawks, and owls. Those that survive march dauntlessly into the sea, some 250 miles from the starting point. They advance resolutely until they drown—the ultimate reward of a lemming.

It is not easy to identify with a lemming, but they may tell us more about the human condition than rests comfortably. In the animal world they have no monopoly on erratic behavior under the stress of overcrowding. There is the death dance of the marsh hare. The mother guppy devours her own young (balancing not only the number to the size of the tank but preserving a precise male-female ratio). Can we believe that we are so much different than the other species? That we are not subject to the same natural laws? We have eliminated all our natural controls and now are multiplying at an uncontrolled rate. When this happens in the animal world the population is invariably brought back to a reasonable balance by some catastrophe: disease, an invasion of outside enemies, starvation, or some internal regulating device such as the lemmings' death march or the guppies eating their young. By what divine right can we escape the tyranny of nature's balance? Indeed, our "regulator" may already be at work. We may be our own predators, suffocating and poisoning ourselves to death with our own wastes. We may do slowly and painfully for ourselves what nature did for us with merciful swiftness. Primitive man may have been more intuitively understanding of human limitations—with his rough abortion, ritual wars, geronticide and massive infanticide—than we are with our well-meaning charity. Is it more cruel to abort or slay an unborn baby or to preserve it to starve to death?

Many animal studies attest to the bizarre effects of overcrowding and the need for population control. The need for a certain amount of space may be such a control factor, according to anthropologist Dr. Edward T. Hall. His researches suggest that space is as essential to the maintenance of life as food or water. Each organism, no matter how simple or complex, has around it a "sacred bubble of space," he says. Wherever the organism goes it takes with it this invisible surrounding bubble. Few other organisms are allowed to penetrate the bubble, and then only for a short period of time.

The space bubble appears to vary in size depending on such factors as the emotional state, immediate activity, position in a social hierarchy, and cultural background of the individual. The bubble may vary between individuals and cultures. What is comfortable or necessary for one may be unbearable for another. The Englishman may despise crowding as much as the Japanese craves it. Tensions or external forces, such as the design of a public housing project, can force an occupant to feel that he has been, in Dr. Hall's words, "thrown into aggressive relationships with strangers or has been sealed off and removed from people. If one's bubble is crushed or dented, or pushed out of shape, he suffers virtually as much damage as though his body were crushed or dented or pushed out of shape. The only difference is that the effects take longer to make themselves evident."

These aggressive relationships lead to what Dr. Hall calls "behavioral sink"—a form of social chaos. In a study of Norwegian rats, in which the usual number of animals in a pen was doubled, there was a severe breakdown of the normal structure of rat society. Family groupings were abandoned, many males lost all sense of sexual discrimination, mounting aged and infant rats or other males. Sadism in the form of tail-biting became prevalent. Other rats lost all interest in sex. "Courtships and mating . . . customs were abandoned in favor of promiscuous affairs." Females stopped taking care of their young and let their nests become cluttered. Some animals went abroad only at night, while others slept. Certain aggressive aristocrats were able to protect their territory within the pen and continued to observe the rules of rat society. But the mass of the population, under the stress of crowding, became unruly and neurotic:

> The constant turmoil of the "sink" resulted in a sharp increase in the death rate, especially among females and the young. Infant mortality rose to 75 percent. Kidneys, livers,

and adrenals of dead animals showed signs of adrenal hyper-activity usually associated with extreme stress. Such experiments indicate that animals regulate their own density as a function of self-preservation.

What does this mean to humans? Fortunately, no positive parallel has been proved. But it raises fascinating speculations. Dr. Hall believes many of the symptoms of social disorganization in our cities might be traced to the effects of crowding—crimes of violence, sexual deviation, the breakdown of family ties, lapses in habits of cleanliness, and others. We still seem far from the mass-death condition of lemmings, he says, but this by implication is a possibility if nothing is done to alleviate the stress of urban life, especially among the poor. "I feel that the vital question raised by these studies is whether man can learn enough about the relationship between space and human behavior and put his knowledge to work with sufficient speed to save himself from disaster."

In observing the effects of overcrowding, there is a corollary health problem that should not be overlooked. It applies to urban dwellers everywhere today. Like the rats, people are under enormous stress from the sheer pressure of numbers. At the same time they are being bombarded by the multitude of contaminants in the environment. Thus, the human organism is under stress on all levels at once: socially, psychologically, and physiologically. Man's inner resources are being drained and exhausted at a time when they are most necessary to meet the mounting external assault. The cost of environmental disease was recently pegged at $17 billion a year. That is a guess. What is the real figure?—if it can be measured in dollars.

While the problem of overpopulation is worldwide, it varies in severity among countries. Population growth is usually figured on the number of live births for every 1,000 babies born, compared to the number of overall deaths in the population. Most governments hope to stabilize the growth rate at zero or at least reduce it to 1 percent. Zero theoretically would put births and deaths in balance. A 1 percent growth rate means the population will double in about 70 years; 2 percent in 35 years; 3 percent in 24 years; 4 percent in 17 years.

In 1955 world population was increasing at a rate of 1.8 percent annually. By 1985 it is expected to reach 3.0 percent. It is climbing fastest in the so-called underdeveloped countries. Latin America has the highest rate with 3.6 percent. Europe has 0.8

percent and that is expected to remain constant over the next couple of decades. The United States rate is slightly over 1 percent.

The mathematics of doubling has been projected to mischievous limits by Dr. Paul Ehrlich, a biologist and population expert. Nine hundred years from now there would be 1,000 people per square foot of earth's surface, he says; about 1,000 years after that there would be a weight of people equivalent to the weight of the earth; a couple of thousand years later the entire universe would be solid people, and the ball of people would be expanding at the speed of light. Just to keep populations within global limits it would be necessary to export people to other planets, assuming that were possible. Some 70 million a year would have to go in order to keep earth's population constant. Now, he says, suppose we built rockets immeasurably larger than any in existence today—capable of carrying 100 people and their baggage to another planet. Almost 200 such monster ships would have to leave each day.

> The effects of their exhausts on the atmosphere would be spectacular to say the least. And what if through miracles, we did manage to export all those people and maintain them elsewhere in the solar system? In a mere 250 years the entire system would be populated to the same density as the earth. Since population explosions could not be permitted on the star ships the passengers would have to be willing to practice strict birth control. In other words, the responsible people will have to be the ones to leave, with the irresponsible staying at home to breed.

Until recently it was forecast that the U.S. population would double to 400 million by the year 2000. Following the 1970 census the estimate was changed to a predicted range of 266 to 336 million. This is encouraging, say biologists at Yale University: "But we dare not become complacent about the situation until our average family size drops to a level that will permit long-term stability—not merely growth at a slightly reduced rate." They point out that another researcher, Thomas Frejke, calculates that "even if average family size in the United States were reduced immediately to only 2.1 children (the number necessary to ensure that each person is on average replaced by one descendant in the next generation), the U.S. population would not stop growing for about 70 years, and it would then be almost 40 percent higher than it is now."

But family size in the United States has not yet dropped that low. It is still well above 2.1, and probably over 2.5, according to the biologists. "The longer that this discrepancy persists, the larger our population will get before it stabilizes."

We are reminded of other menacing statistical truths by *New York Times* columnist James Reston. During the decade of the 1960s the proportionate increase in the American population was 13.3 percent, the second smallest rise in the nation's history. Even so, it raised the population in 10 years by almost 24 million—half the total population of Britain, France, or West Germany. The increase was proportionately smaller in the 1960s because it reflected smaller families born in the depression years of the 1930s coming of age. But now the famous baby boom of the 1940s is coming into its peak childbearing years, foreshadowing an even larger proportionate increase: "Even if the birth rate were to drop to replacement levels—two children to a family—the fact is that the *number* of women in the peak childbearing years—20 to 29—will increase by nearly 40 percent by the year 1980." A more immediate and menacing baby boom is seen by Dr. Philip M. Hauser, a University of Chicago sociologist, a population expert. He says the United States is now "at the beginning of a new tidal wave of babies . . . our second postwar baby boom—an echo of the first. Beginning with about October, 1968, the number of births each month has been greater than the same month in the preceding year."

There is a tendency to blame the poor and ignorant for our population problem, but this is not justified, many experts say. It's true that the very poor tend to have more children than anybody else. But the fact is that the much more numerous white middle-class families with their three and four children add most to the total. They are raised with a so-called higher standard of living.

The United States is not like the underdeveloped nations where large populations mean want. In America there is no lack of food and goods, as much as inequality of distribution. Numbers are not directly equated with the exploitation of the environment and pollution. We take and pollute out of all proportion to our population. We are not a threat just to one another but to the entire world. By now we have been reminded often enough that we use a disproportionate amount of the earth's wealth. We represent only 6 percent of its population and account for some 50 to 60 percent of the total resources used annually. To provide the minimum needs of a single American child from birth to age 18 costs $30,000, according to the Institute of Life Insurance. In many parts of the world a worker's annual income is less than $300 a year. Our affluence should be an embarrassment. Instead we boast of it. In America dogs and cats eat better than many people elsewhere in the world.

Our excessive use of goods and production of wastes imposes enormous strains on the environment—so great indeed that 200 million Americans are equal to some 5 billion people in India. The birth of an American baby in the middle class is at least 25 times and, by many standards, 50 times the "disaster" for the world as the birth of an Indian baby or a ghetto child, according to Ehrlich. Why? "Because, we, the affluent people in the United States, the Soviet Union, and Western Europe, are the super-polluters and super-consumers of the planet."

The figures are impressive. Every man, woman, and child in this country uses 45,000 pounds of raw material per year—food, building materials, cotton, wool, pulpwood, metals, nonrenewable resources such as chemicals, and 15,000 pounds of fuel. In a lifetime the average American personally pollutes 3 million gallons of water (6 gallons each time he flushes the toilet, 58 gallons when he takes a shower). He burns 21,000 gallons of leaded gasoline, consumes 28,000 pounds of milk and 10,000 pounds of meat.

This rapacious devouring of the world's resources has not made us beloved elsewhere. Many poor nations now realize that by taking from them their only real wealth we are reducing their prospects of future development. At the same time we take from the rest of the world we are heedlessly robbing future generations of Americans. If the entire world lived at the U.S. standard it could support fewer than one billion people—one third the present number, according to one estimate. If all the world's people suddenly began using the same amount of petroleum products used by each American, all known petroleum reserves would be gone in about six years, says Representative Morris Udall.

America's ever intensifying plunder of the world's resources is not limited to devastating environmental effects. It also has enormous moral and political implications. As the rich get richer and the poor get poorer, the threat of nuclear war increases, Dr. Ehrlich points out. Morally our position is even more indefensible. Dr. Ehrlich observes that with our great wealth we can

afford to raise beef for our own use in protein-starved Asia. We
can afford to take fish from protein-starved South America and
feed it to our chickens. We can afford to buy protein-rich peanuts from protein-starved Africans. The consequences could be
severe:

> Even if we are not engulfed in world-wide plague or war
> we will suffer mightily as the "other world" slips into fam-
> ine. We will suffer when they are no longer willing or able
> to supply our needs. It has been truly said that calling the
> population explosion a problem of undeveloped countries is
> like saying to a fellow passenger "your end of the boat is
> sinking."

We have been less aware of the poverty problems elsewhere
than our own striking economic successes at home. Between
1950 and the early 1960s, the number of two-car families in
America jumped from 4 to 18 percent. Motor travel increased
nearly 70 percent. The use of boats and private planes increased. We ignore the simple truth that for us to live so well
others must live poorly. And most of the world does live poorly.
It is basically a worldwide network of slums with a few islands
of affluence, according to Professor Georg Borgstrom of Michigan State University. Roughly only about 15 percent of the
people in the world live anywhere near the level of Americans.
Dr. Ehrlich points out that the average person among some 2
billion Asians has an annual income of $128, a life expectancy
at birth of only 50 years, and is illiterate. A third of a billion
Africans have an average life expectancy of only 43 years and
an average annual income of $123. Of Africans over 15 years of
age, 82 percent are illiterate.

Millions live in the most wretched of conditions, far worse
than the most miserable American slum. Home is often a tar-
paper or tin shack, a mud hovel, bare earth or a patch of cement.
Of the 3.6 billion people in the world today, between 1 and 2
billion have an inadequate diet—they do not get enough calories
to provide the animal heat necessary for proper bodily function,
or they lack the quality proteins essential to nourishment. Even
as the earth must support 70 million more mouths a year, an
estimated 10,000 people are dying from diseases brought on at
least in part by malnutrition. Most of these victims are children.
Their only heritage is suffering and early death. Hunger is usually an indirect killer. It weakens bodies so secondary disease

can finish them off. Governments are egoistic and this oblique slaughter is a bureaucratic convenience. Death solves many problems. There are fewer mouths left to feed. Officials are spared the embarrassment of admitting that their people died from starvation. It is more honorable for a government to attribute death to tuberculosis than hunger. Tuberculosis implies individual structural weakness, while hunger is the undeniable failure of government to provide an essential human need.

Since the mid-1960s we have been warned of the dangers of overpopulation in the underdeveloped nations—of proliferating disease, possible plagues and epidemics such as cholera that even now is beginning to strike around the world, small and large wars that sap the earth of its resources, vitality and the dwindling reservoir of hope, wars in which all lose in varying degree, victor, vanquished and nonparticipant; all emerge poorer as the irreplaceable whole is wasted and forever lost—for what? What war has ever really been won?

Expert after expert warns that massive famines are inevitable in various parts of the world. Food supplies cannot possibly keep pace with the burgeoning numbers. The experts differ only on the timetable. The "time of the famines" will begin about 1975, according to the authoritative Paddock brothers.* Many others predict the same catastrophe. Only timetables differ. Mass starvation is expected to cut down millions annually in Egypt, India, Pakistan, China, and Africa, and subsequently in Latin America. Dr. Garrett Hardin, a biologist and population expert, says it is a "probability that in 10 or 20 years there will be a dreadful catastrophe in the world, of people starving—50 or 100 million in a single year." Chester Bowles, former ambassador to India, sees an approaching world famine that will be "the most colossal catastrophe in history." The distinguished scientist-author C. P. Snow offers an even more unnerving preview, of millions in the poor countries starving to death before our eyes. "We shall see them doing so on our television sets."

The greatest population growth the world has ever encountered will take place in the second half of the 1970s, according to a recent U.N. study. This will coincide with the projected beginning of the great worldwide famines. To feed these multitudes, food supplies would have to be doubled by 1980 and tripled by the end of the century.

* Paddock, William, and Paul Paddock, 1967. *Famine — 1975!* Little, Brown & Co., Boston, Toronto.

Where are the new food sources to come from if disaster is to be averted? Already the ocean is rebelling against being over-harvested and poisoned. Soils are exhausted and poisoned by new agricultural tricks and techniques; these tactics produce immediate big yields but must be paid for ultimately by reduced yields. The earth is not so easily outwitted. It is, in fact, a precise and demanding bookkeeper, an unerring and uncharitable broker in repaying us for our follies.

Many countries even now are able to keep their people alive only through imports. Japan gets about half of its food from abroad. India's margin against mass starvation is some 27,000 tons of wheat a day from the United States; this grain is so laden with pesticide residues that Indians have referred to it as "the mother of disease." In parts of India there have been food riots; more are expected. Latin America has increased its total food production over the past five years, but with 25 million more people to feed the average individual has 7 percent less to eat. This figure is misleading, of course. The rich are not affected, so the whole reduction must be divided among the poor, making the actual percentage of food loss much higher than averages indicate. How blithely we dismiss the emptiness in a man's belly by giving it a number. Egypt built the massive Aswan dam to help feed her wildly multiplying millions by providing new sources of energy, but long before the dam was finished it afforded means to provide food for only one third of the new population, even at present poor nutrition levels. In addition the dam has caused vast social and environmental problems.

All these impersonal statistics of food shortages must be translated into human effects. Two thirds of the world's children live in developing countries; for most of them malnutrition is a fact of existence, says nutrition expert Dr. Nevin S. Scrimshaw. He points out that animal tests have "led to the stunning impli-cation that infants and young children whose physical growth is stunted by malnutrition may also be prevented from attaining their full mental capacity and social development." Unfortu-nately, improved diets later on cannot make up for this early deprivation; most humans are lifelong victims of their childhood deprivations.

The effects of diet are dramatically illustrated in Japan. Since World War II the Japanese have added more protein to their basic rice diet, Dr. Scrimshaw explains. They are now taller and huskier. "The stereotypical diminutive, long-waisted bandy-legged figure of the old Japanese woodcut print is becoming harder to find among the young."

The average Japanese today consumes 2,254 calories daily, less than 100 calories more than he did in the 1930s, but the big change has been to add animal protein to his diet. The Japanese diet now includes meat, milk, bread, cheese, and other high-protein foods that supplement the basic diet, of which rice is the mainstay. In 1968, according to the Health and Welfare Ministry, the average 20-year-old Japanese man was just under 5 feet 6 inches and weighed 130 pounds. Thirty years before, the average height of the 21-year-old was 5 feet 3½ inches and his weight was just over 120 pounds. Six-year-olds were three inches taller in 1968 than in 1938.

In former days lack of food tended to be a kind of population control, in the classical Malthusian sense. This was coupled with disease, often the result of lack of food or malnutrition. It was savagely inhumane but effective. Now populations are artificially maintained by food imports and medical aid from more advanced countries. The combination has cut down the old barbaric infant mortality, but in its place is a new humanistic horror. As more children live, the rate of doubling increases. This imposes an accelerating need for more food, schools, homes, services. Where there was starvation there is now malnutrition; swift death has been replaced by lingering illness. The young tend to marry earlier and reproduce sooner, especially in the undeveloped countries. This speeds up the doubling rate, along with all its related problems. Usually there is little or no money to care for these new hordes. Latin America, for example, has the fastest growing population in the world. Its per capita income is $237, less than one-tenth that of Americans but double that of many countries in Africa and Asia. All of this is comparatively recent, an ironic perversion in human lives of the original good intentions behind it. Ehrlich states:

> Death control did not reach Colombia until after World War II. Before then a woman could expect to have two or three children survive to reproductive age if she went through ten pregnancies. Now, in spite of malnutrition, medical techniques keep seven or eight alive. Each child adds to the impossible financial burden of the family. According to Dr. Summer M. Klaman, the average Colombian mother goes through a progression of attempts to limit forms of contraception and moves on to quack abortion,

infanticide, frigidity, and all too often to suicide. The average family in Colombia, after its last child is born, has to
spend 80 percent of its income on food. . . . That's the
kind of misery that's concealed behind the dry statistics of
a population doubling every 22 years.

Populations must be cut. But how to go about it? The basic
problem is that it takes more intelligence and effort to avoid having a baby than to have one. Birth control programs are voluntary and therefore they fail. People just will not stop having
babies despite all the expert advice, birth-control information
and devices available. The various programs go under the catchall banner of "family planning." Family planning includes expert advice, the use of birth-control pills, intrauterine devices
(IUD), condoms, instructions in rhythm and other "natural-
control" methods, sterilization, relaxed abortion laws, and sex
education in schools. In the United States tax laws have been
proposed to offset the present "entrenched system of incentives
to reproduce." There have been various bizarre schemes for state-
imposed control, including dumping sterilizing drugs in water or
food supplies and requiring a license to have a baby.

There have been proposals to cut down economic aid to undeveloped nations that fail to reduce birth levels. This is something
like threatening to cut off the funds for the poorhouse if the
inmates don't take themselves in hand and become self-
supporting. Harried governments often try desperately to stem
the flow of babies. They distribute the supply of devices provided
by the affluent-nervous-industrial-polluting nations. In India
they pay small bonuses to men who have themselves sterilized;
mobile vasectomy laboratories go around performing the procedure.

The various programs have been spectacularly unsuccessful.
The technology of birth control cannot compete with the natural
desire of people to reproduce. India has had birth-control programs for 17 years and the present growth rate is 2.5 percent a
year compared to 2.3 when it began. In Latin America control
efforts have been overwhelmed by the general level of a 3 percent increase per year; in some countries it is as high as 4 percent, which means doubling the population in only 17 years.
Some of the most intensive efforts have been made in Egypt with
its runaway populations and devastating poverty. Eighteen years
ago King Farouk was deposed by Colonel Nasser and the young
officers who set out to reform the country with such high expec-

tations. The new rulers boldly built dams and factories, reclaimed land from the desert, transformed swamps into farmland, irrigated, and called for restraint. Every gain was wiped out by the relentless outpour of babies.

Some 3,000 babies are born every day in Egypt. Population has soared to nearly 34 million and is expected to reach 45 million by 1980. By the end of the century 70 million people will be crammed into a country now considered one of the most overpopulated in the world. Egypt's problems illustrate what is happening elsewhere; from 1940 to 1967 the death rate dropped from 26.3 to 14.2 owing to improved health care, medicine, vaccines, and sanitary measures. Forty-two percent of the present population is under age 15 (compared with 27.6 percent under age 15 in the U.S.). In Egypt people traditionally have married early, bred fast, and died young. Tradition supports large families. Men wanted children to help them in the fields. They wanted sons to prove their fertility. Women wanted children as protection against divorce; a father was less likely to run away with another woman if he had to pay support for a dozen children or so.

The Egyptian government has threatened many restrictions to induce smaller families. It would raise the marriage age from the present 16 to 20. It would limit free education privileges to the first two children born, and prohibit the issuance of sugar, tea, and cooking oil ration cards for more than two children. But Egypt's birth-control chief, Dr. Dhalil Mazhar, protests these measures. He is a practical man, a humane man. He says the raising of the marriage age would only increase homosexuality. Cutting down on ration cards would not limit families but only make life harder. He advocates voluntary sterilization.

But still resistance continues, fanned by superstition, rumor, and the force of tradition. Some Egyptian women fear that the pill and the loop cause cancer. Some, according to *The New York Times*, abandon the pill because a child has died and they want another, or because the marriage has broken up, or because their husbands insist on it, or because they are afraid people will think they are sterile. Some women refuse to take the pill because it requires a medical examination. Many people fight it because they are afraid it will encourage illicit relations. That is why there is no sex education in the schools, said one worker. "That is why there are illegal abortions," she said. "But that is also why we avoid talking about legal abortion and sterilization."

There are other factors that affect populations. Many appear irrational to the Western mind, which is blind to its own irrational conduct. The Japanese, for example, are awesomely crowded, with 102 million people (half the population of the entire United States) crammed into a string of narrow little islands smaller than the state of Montana. The crowding is unbelievable, according to Philip M. Boffey, in *Science*. Every inch of the land is utilized, crops grow everywhere:

> Up the sides of steep hills, in the narrow alleys between adjacent railroad tracks, even at the front stoop, where one ordinarily expects to find a lawn. In the cities, and even in rural villages, tiny houses are jammed side by side, with little or no yard space and barely enough room to walk between. Living is so close that privacy is difficult. . . . Neighbors are squeezed so tight that they can overhear one another's conversations and smell their cooking.

Boffey recalls that a public health official complained that, in crowded Tokyo, "all you can find is a place to eat and a place to earn money. There is no green, no trees. I don't feel that people are living a very human life." Boffey himself was appalled by such overcrowding. But, curiously, he says there is some evidence that the Japanese have grown accustomed to their close living conditions and actually like them. Japanese travelers are said to be "overwhelmed" by the massive stone buildings of the West and they soon "begin to miss the light wooden structures and small landscape gardens to which they have so long been accustomed." A former Japanese diplomat is quoted by Boffey as saying, "For generations many of our people have been living under the same conditions, so they don't question whether it is wrong or right."

The Japanese are the only people who have been successful in their attempt to reduce their population. Following the disaster of World War II—a war that, ironically, was started to find breathing space to the west—the country found itself with too many mouths, a war-ravaged economy, and more squeezed in than ever. Propaganda reinforced a 1948 law that removed previous obstacles to birth control—abortion and sterilization. By 1957, Boffey reports, Japan's birth rate had tumbled to 17.2 births per 1,000 population (a growth rate slightly under 1 percent). This represented "a historically unprecedented drop of 50 percent in just 10 years."

Did this please the Japanese? Not at all. Prime Minister Eisaku Sato recently advocated increasing the birth rate. The reason was the same one that dominates most human affairs: economics. Japan's postwar economic miracle was threatened by a "severe labor shortage." Not a labor shortage as such, an economist explained. "What is deficient is young labor, which is very cheap." Boffey explains that Japan's decision to boost its birth rate may have an impact far beyond its own borders. It may throw a monkey wrench in worldwide efforts to curb population growth by somehow downgrading the importance of birth control. Others fear that Japan demonstrates that radical population control can never succeed, Boffey states, "for the minute a nation reaches the point where its population is apt to level off and then decline, various pressures—political, economic, and nationalistic—build up to reverse the trend."

What, then, is an ideal population size? The experts disagree. But experts are often limited by their professional viewpoint. Usually it is tied to economics and the carrying capacity of the land. Rarely is it determined by philosophical factors: what a nation's aspirations and beliefs are, what kind of people it wants to produce, what is ideal for a human being. We are more the victims of destiny than its masters. For different goals the ideal figures would be different. In the United States we represent a land density of only 56 persons per square mile, compared with 588 for England and 975 for Holland. But because of the concentration of two thirds of the people in metropolitan areas and our devastating impact on the environment, due to our standard of living, it has the effect of making us overpopulated. From a health viewpoint we are already beyond the optimum number, according to Dr. John H. Knowles, director of the Massachusetts General Hospital. He told a recent symposium of scientists, "We have exceeded it, gentlemen—we have already exceeded it."

Dr. Ehrlich insists that the number of people the United States can comfortably accommodate is 150 million—50 million less than the present population; the world figure should not exceed one billion, less than one third what it is now. The problem here and elsewhere is not just quantity of life but also the quality of life. The tragedy of overcrowding and its attendant ills is compounded because so many of the children born throughout the world are unwanted. This is true in the undeveloped countries where fornication is often a means of passing time and relieving boredom ("Get lights in the villages and keep the

coffeehouses open and you'll cut down the birthrate," a percep-
tive Egyptian commented).

In the United States, where lights are plentiful and all of us
are enlightened, we have our own forms of desperation. All
kinds of sexual perversions, psychological abnormalities, and
personal tragedies are revealed by population studies: young un-
married girls who unconsciously want to get pregnant, possibly
to punish someone else or themselves; women who want babies
and cannot have them; women who don't want babies and have
them; child abuse (much more prevalent than is commonly believed); bungled abortions; "welfare" babies so the mother will
get benefits; deserted children; a catalogue of adoption disasters.
Out of this tableau comes the vast human suffering and social
cost that must be added to the world's burden of misery, all of it
usually buried in statistics that hide terrible truths about the human condition.

The unwanted account for 35 to 40 percent of our population
growth in recent years, according to Dr. Charles F. Westhoff of
Princeton University's Office of Population Research. One study
disclosed that in 22 percent of the births, one or both parents did
not want the baby. Such children are often born to rejection and
unhappiness.

Large numbers of the unwanted are born to those least able to
take care of them—the poor. One study found that among low-
income families 54 percent of the children are unplanned and
unwanted. In 1966 alone, the 8.2 million poor and near-poor
women of reproductive age had 451,000 unwanted births, according to a Public Health Service study. Aid to Dependent Children in the United States costs $1.5 billion a year, and the case
load is increasing by more than 23,000 a month. This does not
take into account the misery visited upon child, parent, and society alike, a chain reaction that echoes from generation to generation. A vicious circle of poverty and fertility is at work, says the
National Advisory Commission on Rural Poverty. "Because they
[the poor] do not limit the size of their families, the expense of
raising unwanted children on inadequate incomes drives them
deeper into poverty. The results are families without hope and
children without future."

How do we break the circle? More education is the usual solution called for. And yet, education is often the first casualty of
excessive population growth. Throughout the world illiteracy is
increasing. A U.N. study of 90 countries concluded that the

number of illiterates has risen in this decade by almost 60 million, bringing the number in the world to about 800 million—approximately 35 percent of the total population. The irony is that those who need education and birth control the most to break their chains of misery are likely to be the ones who are deprived of it. Often they are the ones who resist birth control the most. In their ignorance they fail to see excessive reproduction as a cause of their plight.

In the United States, birth control is still vigorously opposed in many areas and on many grounds: religious, ethical, eugenic, nationalistic, social, cultural, economic, and emotional. The Roman Catholic Church sees it as a threat—although many Catholics, after having six or eight children, find themselves forced to practice controls, leaving them with both the problem of a large family and a compromised religion. Puerto Rican men in New York violently charge that the whites are sterilizing "our women," even when the women have voluntarily gone to clinics for help. Blacks charge it is used to keep them in a minority. Economic interests that depend on huge numbers for their markets sometimes carry on guerrilla warfare, citing pious reasons for resisting controls. Many opponents are sincere, their objections valid.

The clash of the opposing views can be explosive. Coalitions of birth-control advocates and their opponents often make strange bedfellows. They may fire at indirect social or political targets to achieve devious personal ends. Restraint of the reproductive urge strikes deeply at our cultural and psychic roots; it is a mysterious prompting tied to our unknown origins and destinations, a seething, dark morass of irrational fears, phobias, superstitions, prejudices, instinctive drives, religious inhibitions, egoistic thrusts, and economic alarms. For some it is a challenge to life continuity and personal immortality. Resistance flares at mention of sex education in the schools and even voluntary family planning. Hackles rise at the suggestion of compulsory restrictions.

Birth control is the most politically sensitive issue in American life, next to economics. Politicians recoil in terror before the force and pressures it unleashes, the incalculable risks that any stand entails. With some understatement, Senator Joseph D. Tydings observed that "strong resistance still exists in Congress to programs promoting voluntary family planning, much less compulsory population control." In 1969 President Nixon took a historical step by creating a commission to study population growth. He was the first president to take such a step, notes Dr. Donald Aitken, an astrophysicist and consultant on the space program. But the President was careful to defuse the political dynamite in his act. He negated it all at the end by promising that the governmental effort would not interfere with religious convictions, personal convictions, etc., complained Aitken. He said he would like to see Mr. Nixon stand up a few years from now and say: "Nothing has happened. Population must be controlled. We must set an example. So the Government has to step in and tamper with religion and personal convictions—and maybe even impose penalties for every child a family has beyond two."

As usual, Presidential rhetoric proved stronger than economic support. The new family planning legislation in the "landmark" population bill hailed by President Nixon authorized additional spending of $60 million for services and $50 million for research in the next fiscal year. The Administration, however, asked for increased appropriations of only $47.3 million and $10 million respectively in the 1972 budget. The request was seen as particularly disheartening because success of the program depends heavily on the development of new birth-control techniques and increased understanding of family planning. *The New York Times* saw "another hint of a weakened Administration attitude toward the population problem" in a speech by Conrad F. Taeuber, chief demographer of the Census Bureau. Taeuber implied that the present 1 percent rate of population growth in the United States is nothing to worry about, and other factors, such as per capita consumption, were of greater concern. This would seem to overlook the simple mathematical truth that each additional American adds new strain on the nation's dwindling resources.

Proponents of population control see it as a matter of survival. Opponents see it as a moral or tactical wrong. Many other groups and individuals waver between these extremes, trying to find a balance point between threat and relief. The questions raised are urgent, profound, and unanswered: Wherein lies equity? Is equity implicit in survival? What is society's role in protecting itself? When does the exercise of a basic individual right become a collective social wrong? Why should society subsidize the irrational and suicidal multiplication of the species when even present numbers cannot be adequately supported and the environment is being destroyed under its overload?

Opponents ask equally difficult questions: Who among us has the wisdom to regulate the power of life? Presumably controls would be administered by the state. This means the rise of a whole new bureaucracy geared to the most critical issue in life—the allocation of life. Competent bureaucrats are already in short supply, and governments already undertake a great deal of activity they are unfit to conduct, says Professor Joseph J. Spengler of Duke University. Would these bureaucrats wield the enormous power of deciding who should have children and how many? It poses a frightening delegation of authority in view of the failure of regulatory bodies in every other area of environmental responsibility.

The debate continues. The "humanometer" clicks fiendishly twice each second. Society and the individual gradually bend under its relentless imperatives. Stern abortion laws are relaxed. Religions reconsider and revise God's will or are weakened as they remain inflexible in its violation. More than 100,000 Americans annually seek sterilization, often with deep personal misgivings and guilt. Many scientists press for an end to voluntary birth control in favor of strict state regulations; they too have misgivings.

Compulsory methods are a dismal prospect except when viewed as an alternative to a cataclysm, says Dr. Ehrlich. To continue limiting population by voluntary means is insanity, says Dr. Hardin—"the result will be continued uncontrolled population growth." We cannot continue to place reliance for population growth on private conscience when the reproductive behavior of a considerable fraction of any population is little influenced by conscience, says Dr. Spengler—"herein lies the inadequacy of planned parenthood programs." People want too many children, says Professor Kingsley Davis—"family planning has not controlled any population to date, and by itself is not going to control any population." He includes Japan in his indictment, explaining that at its low point (when the people were said to be deterred more by adverse astrological signs than government propaganda and logic) the population would have doubled in 63 years.

Dr. Ehrlich says the decision for population control will be opposed by the growth-minded economists and businessmen, by nationalist statesmen, by zealous religious leaders, and by the myopic and well-fed of every description:

> It is therefore incumbent on all who sense limitations of technology and the fragility of the environmental balance to make themselves heard above the hollow, optimistic chorus—to convince society and its leaders that there is no alternative but the cessation of our irresponsible, all-demanding population.

We are told there is no more time to lose. Four years from now there will be 300 million more mouths to feed in the world, most of them hungry. We are told that 300,000 babies born every day are born to families that are hungry, poor, ignorant or ill, that runaway population is already responsible for cruelly crowded living conditions, famine, emotional stress, joblessness and loss of human dignity. Willard Wirtz writes autobiographically of a "Secretary of Labor who spends his years in office trying futilely to fight unemployment by creating more jobs; and only later, freed of the inhibitions of office and politics' restraints, faces the truth that there are too few jobs because there are too many people." We are told that the United States Government still spends on population control less than 4 percent of what it spends on space exploration and about one seventh of one percent of what it spends on the military—hundreds of millions for death control, "scarcely one percent of that amount for fertility control."

The warning is clear. But even in population control lurk other subtle dangers. We tend to think in terms of numbers rather than people. In trying to preserve the environment will we lose such humanity as still remains to us? Do we save our lives at the cost of our souls? The moralist asks what will be the ultimate effect of the relaxation of abortion laws. Will it further blunt our fading sensibilities, further brutalize us, make us even more antilife than we already are? Will it make it even more difficult to look upon ourselves with any sense of reality to avoid our own self-repugnance? Does life degrade into just so much flesh? Sex into a nonfeeling act of cooperative masturbation? Is the new abortion clinic, in reality, a new slaughterhouse for the yet unborn? Who is next? asks the moralist: the old, the sick, the unfit?

We can stand no further erosion of our humanity. One of our redeeming virtues has been our love of children. Any loss of this virtue would push us farther along the path of dehumanization. The threat was expressed in a touching letter to *The New York Times* from Dr. Charles U. Lowe, scientific director of the National Institute of Child Health and Human Development. He sees us laying our environmental ills at the feet of children: The public is misled into believing that the "amelioration of the environmental problem is directly dependent on controlling population growth."

This is not necessarily so, says Dr. Lowe. Population control and pollution are not a single problem with a direct causal relationship. Rather they are two separate problems that overlap. To resolve one does not necessarily resolve the other. "Family planning, which seeks to protect the health and welfare of mother and child and of society by improving the quality of life, is a humanistic goal. It should not be confused with the desirable but mechanistic goal of cleaning up the environment." Many population and environmental-protection programs are obliquely directed against children. If people become convinced that children are the cause of environmental contamination, "America would face a catastrophe because they would cease to support programs designed for the benefit of infants and children":

> More and more one hears the ecologist equating the occasion of an additional birth with an effect upon the welfare of the adults. Such arguments pander to the mechanistic and hedonistic mood of our society. . . . The shocking abuse of our environment is hardly due to the birth of babies. The onus for the unconscionable exploitation of the planet cannot be laid at the feet of children.

This strikes hard at the root of the population-environment dilemma. Dr. Lowe says the economic aspects of our resources have taken precedence over the control of waste and pollutants.

> It is fair now to place the emphasis on control of pollution at its source. We are in favor of family limitation as dictated by social and economic considerations. But we cannot accept the anti-child tones by some environmentalists and by polluters reluctant to modify their practices.

It is these practices and their effects that we are now concerned with. Not all lemmings must march to the sea to destroy themselves. It can be done quite effectively right at home.

MOUNTAINS OF WASTE

Dr. John S. Chapman, a poet-professor at the University of Texas, has a vision of New York City. It becomes "a dome, a mound of piled up jalopies, old tires, rotting banana peels, sucker sticks, and orange peels, all gently powdered over with fresh-fallen snow. And the last few remaining men are struggling through the whole thing to get to the ground level, which is the spire on top of the Empire State Building. . . ."

This apocalyptic vision threatens to become reality as we struggle with our wastes.

Solid wastes are our most pressing pollution problem. They are not like the gases and poisons that are invisible and insidious. They are not of the future. We can and do talk ourselves to death about other environmental threats, but solid wastes permit little conversational latitude. Something must be done about them now. They threaten to inundate us. Each day the problem becomes more acute. Each day there are more people, more services, more products. We produce more wastes and have less room in our crowded population centers to dispose of them. The dilemma drives municipal officials frantic. How to get rid of this daily avalanche of trash? What rug to sweep it under?

We produce more wastes than any other country in the world. We spend more on wastes than on any other form of pollution ($6.81 per person per year) because we cannot temporize with it. Many figures dramatize the magnitude of the problem. Each American generates about 7 pounds of solid waste a day—a combined total of 530,000 tons a day, more than 500 billion pounds a year. It's enough to fill the Panama Canal four times over. Enough to build a dike 6 feet wide and 22 feet high from coast to coast each year. It's said to increase at a steady 4 percent per year. The paper industry alone produces 30 million tons of scrap annually. Some 2,500 different telephone directories use 140,000 tons of paper annually. New York City alone distributes 13 million phone books weighing up to 5 pounds each and every year. We exacerbate the bulging rubbish heaps with an annual output of 48 billion metal cans, 26 billion bottles and jars, 65 billion metal caps and crowns, 4 million tons of plastics (it's supposed to triple in the next decade), 8 million abandoned cars. It's all there, measured in the millions of tons and billions of pounds, conscientiously assayed and broken down: 55 percent paper, 14 percent food scraps, 9 percent glass, 9 percent metallics, 5 percent yard refuse, 4 percent wood, 1 percent plastic, 3 percent miscellaneous.

Hidden in the impersonal inventory of our gross national waste are the artifacts that sketch our cultural portrait. There are unnumbered broken dressers, rusty bed springs, aerosol cans, cracked mirrors, refrigerators, toilet bowls, plastics, non-functioning labor-saving devices that worked so splendidly in the television ad. This trash that we treat so contemptuously contains much personal pain and heartbreak. But it also includes the "airmail" that rains down from the upper windows of tenements in ghettos, a message of social protest delivered by the angry black or Puerto Rican. Eventually most human endeavors find their way into the trash—a sobering thought.

The problem of waste disposal represents an evolutionary pattern. Our ancestors found it convenient to dump the profane and unwanted into streams, to the discomfort of their downstream neighbors. Others hurled it over cliffs or piled it in mounds; when the accumulation became too offensive they simply moved elsewhere to start over again. Those who lived near the shore dumped trash into the ocean. Raw garbage was regularly barged to sea until the practice was outlawed by a 1934 Supreme Court ruling. It was stopped not because of the marine damage but on aesthetic grounds; the guck washed back to vex fashionable· beachfront communities. Now we are more fastidious and only dump into the ocean sludge wastes from sewage plants, spoils from contaminated waterways, unwanted military poisons, and industrial wastes that include acids, phenols, and radioactive materials.

The most primitive form of disposal today is the open dump— a fascinating place for the young to explore and unearth abandoned treasure. Almost no town is without some kind of public dump, a ripe haven where solid waste is periodically burned to reduce bulk. The open dump pollutes air, ground, and water and creates health problems. Dumps are usually found in poor neighborhoods. The poor are notoriously more tolerant of nuisance than the rich, and traditionally more expendable.

Open dumping and burning has been replaced in some communities, usually in cities, by incineration and sanitary landfill. The latter is a refinement of open dumping; there is no burning and the trash is packed down and regularly covered with soil, like a multilayer sandwich. The ratio is usually about 10 or 12

feet of garbage and trash to a few inches of dirt. When the landfill is full, it is capped by about two feet of dirt and triumphantly
proclaimed as "reclaimed land." Many of our finest marshes and
inland wetlands have been "reclaimed" in this way. Despite the
obvious disadvantage of open dumping and burning, it remains
the most popular method of waste disposal. Its virtue is compelling: it's cheap. Dumping can be done for as little as 50 cents a
ton. Sanitary landfill costs range from $1.50 to $2.00 a ton. Incineration costs five or six times as much as sanitary landfill.
Once more pollution has its roots in economics.

Incineration has not been altogether successful. Real improvements have been slowed by lack of research. Most incinerators in use are inefficient and offensive. Some 75 percent are
inadequate, according to the Public Health Service; they fail to
burn enough of the garbage or they pollute the air, or both. Some
experts claim that practically none of the incinerators in operation are up to air-pollution standards. Half of the 175 communities with incinerators have stopped using them. Their basic
drawback is that they cannot produce the high temperatures
necessary for complete combustion. If they don't have expensive
antipollution devices, they emit heavy particles while reducing
bulk only about two thirds, leaving stinking, rotting, half-cooked
garbage-stew to be disposed of. New technology is expected to
produce incinerators capable of the heat necessary for almost
complete combustion (2,500 degrees compared with the present
1,800). At the same time incineration reduces bulk, it will emit
more invisible gas into the air.

Many states and localities have failed or refused to enforce
regulations against defective incinerators and backyard bonfires.
This saves the community the cost and inconvenience of getting
rid of the trash and relieves local officials of offending voters by
raising taxes. It also gets around the sticky political problem of
alienating and possibly driving away local industry with the attendant loss of jobs and income. Such communities resolve their
solid-waste problems at the cost of polluting the air. In effect
they make more law-abiding communities subsidize their own
negligence and economies.

It was not until after World War II that solid waste came into
its own. Overnight it erupted as a national problem. This reflected not only a spiraling of our numbers but a change in our
way of life. In the past we were a nation of tinkerers; we prided
ourselves on the homey virtue of making do ("waste not, want

not"). We had repaired. Now we replaced. Many new appliances were sealed so repairs were impossible. They were built to wear out fast. Rising labor costs made it cheaper to buy the new than fix the old. The new standard we embraced encouraged people to throw out the old while it was still serviceable and replace it with the new; often the replacement was inferior in design and function. We had entered an era that glorified the new and shiny.

At the same time merchandising changed. The package became more important than the product; this was true of goods and people as well—image rather than content. A simple paper wrapper or bag no longer sufficed. It had to be cardboard trussed with reinforced tape that broke the fingernails. Hardware items were embalmed in indestructible plastic bubbles or cellophane shrouds. In the supermarket, plastic thwarted the inquiring nose from sniffing out defective produce or the probing finger from detecting the telltale soft spot. Not only did it promote efficiency and get around rising labor costs, cutting down on clerks to weigh out goods; it also reflected our increasing horror of germs and dirt—a phobia fanned by the pesticide manufacturers.

Everything must be wrapped, bagged, or packaged. Even the package is packaged. A roll of brown paper is wrapped around a cardboard tube, covered with cellophane, labeled, and put into a bag with a sales slip. Everything bought at the market is lavishly bagged. The bag has a life of only a few minutes—the time taken to get from store to home, and then the bags go into the garbage. Trees are cut down for this. Compare this waste to the system that still exists in many cities and towns throughout Europe. Most people shop at least once daily for food. They go to the market and talk to one another. Merchants actually talk to the customers. The products are openly displayed, a marvelous aromatic medley of fruits, cheese, vegetables, and spices. Fruits and produce are put into the buyer's string bag that is used again and again. In Paris a bottle of wine requires a five-centime deposit that is returned when the bottle is returned. Loaves of bread are popped into the string bag unwrapped.

Plastic is now ubiquitous. The only thing that does not come in plastic is babies, and possibly that is being planned. My wife visited a private school in New Jersey where the entire lunch was served on plastic dishes, with plastic knives, forks, spoons, and cups. After the meal everything was thrown out. Tossing them out is cheaper than washing dishes and satisfies our neurosis for sanitation.

Plastic is one of the most serious waste disposal problems because of its popularity and insidious indestructibility. It has been shaped into 1,235,000 forms. The combinations are inexhaustible. A plastic bottle can remain in the ground for hundreds of centuries and probably emerge the same as the day it was buried. In an incinerator plastic may bubble, crack, melt, blister and change shape, but the volume remains practically unchanged. It produces such intense heat that it can burn out grates. It gives off particularly noxious gases. Usually it winds up as solid waste in dumps. "It's eerie," said a solid-waste expert. "You can go to the dump and find that same old plastic bag you discarded there five years earlier, blowing around in all that desolation as though it had just been thrown away the previous afternoon."

Aluminum cans are almost as vexing. They too resist time, refusing to rust and disintegrate. More lasting metals are being developed all the time. Ninety kinds of indestructible metal are now said to exist. They cover the earth and debase the seas.

A submersible naval craft is 50 miles off the Atlantic coast, 2,450 feet down. Admiral R. J. Galanson, chief of naval materials, expectantly approaches the porthole. He is going to view the wonders of the undersea world. The admiral peers out. Two feet away on the ocean floor is an empty beer can, an indestructible American artifact for the ages.

The automobile not only poisons us during its life but plagues us with its corpse. Each year the surplus of abandoned cars increases by at least one million over the number reused. Old cars formerly were used in making new steel. New techniques have reduced this form of reincarnation. The nation is decorated with hideous automobile graveyards, smashed rusting hulks piled atop one another in soaring mountains, metal monsters that don't even repent their past sins. Not only is the skeleton a nuisance, even the entrails are formidable. A $4,700 American-made sports car is said to be 88 percent plastic; 45 percent of many other cars are plastic—wiring, hoses, upholstery, and flashing. More than 1 million tires are scrapped annually. They form ugly foothills 70 and 80 feet high, mockingly durable.

Litter adds its bit to the national trash heap. A survey revealed that along a typical mile of highway are 2,665 discarded items, ranging from false teeth to refrigerators. Disposing of the cast-offs in that particular study cost about 32 cents per item. Philadelphia spends $2 million a year to pick up litter from its roads.

Washington, D.C., is close behind with an outlay of $1,500,000.

The power that underwrites our technological way of life produces not only air and water pollution but also solid waste. A typical coal-burning power plant builds a mound of 150,000 tons of ashes a year to be disposed of.

The problem of waste disposal becomes increasingly urgent as we rapidly run out of room. Most of the nation's 12,000 so-called sanitary landfills are full or almost full. The name itself has its ironies. Few are very sanitary by official standards. The Public Health Service has declared that 84 percent are unacceptable and that they represent a threat of potential disease and land blight. Into these burial grounds go not only the imperishables but also contaminants and rotting organic matter: vegetable wastes, decomposing meat scraps, dead animals, anatomical disease-ridden remains from hospitals, industrial wastes, unwanted stores of pesticides, sludge residues from sewage-treatment plants. Frequently they are on the site of former marshes, wetlands, or beside rivers.

In New Jersey the vast Hackensack Meadows, near the Lincoln Tunnel, have been largely converted to sanitary landfill. Once this tract was the habitat of wildlife. Hundreds of acres of marsh grass and reeds were washed by tides from the adjacent Hackensack River. The spongy earth slowly released refreshing moisture into the air, affecting local weather patterns. The meadows provided escape and pleasure for the urban dweller and passerby; the truck driver thundering past on the Jersey Turnpike was momentarily cheered by a pheasant rising from the brown cover.*

Now the meadows are largely covered by some 3 million tons of refuse. All day huge bulldozers and earthmovers crawl over

---

* The Hackensack Meadowlands Commission, an official state agency, has announced plans to build a huge sports stadium and arena on 750 acres of the 20,000-acre meadowlands. In addition, 80,000 new housing units will be built to house a population of 150,000 in what is already the nation's most heavily congested area. Conservation officials have bitterly attacked the plan, pointing out that it would destroy the only open space in densely populated North Jersey, increase the dangers of floods, heighten fears of water and power shortages, increase air pollution problems in an area that already fails to meet federal air quality standards, and destroy an ecologically important marshland. Others said it would add to already severe traffic problems, increase the populations of adjacent towns, and hasten the decline of Newark and Jersey City by "attracting what remains of the middle-class communities of both cities," and by usurping federal funds that could be used to help rejuvenate the older cities. Who, ultimately, will remember the gentle wetland that once graced this area?

the mass like monsters. A procession of trucks each day dump 13,000 tons of trash and garbage and 75,000 tons of toxic and volatile industrial wastes, the refuse of 15 million people in metropolitan New Jersey, Philadelphia, and New York, to corrode air and water; that statistical 7 pounds of waste per person per day becomes reality. All those no-return bottles, rust-proof cans, indestructible plastic, built-in obsolescence, no-repair, still-serviceable-but-unfashionable items, are ground down and sandwiched between crusts of dirt. Bedsprings and broken couches thrust up through the heap. The landfill spills down into the Hackensack River, now a dead gray stream that gradually absorbs the poisonous juices that leach out. The river carries them into Newark Bay and the ocean. If dumping were to stop today it would take at least two generations before the poisons percolated out of the adjacent land, some say. More likely it would take a century or longer.

New York also depends on sanitary landfills to get rid of its solid wastes. The city boasts that 17 percent of its glory was built on filled-in land, much of it "worthless marshes and swamps." The main fill area is now the southern tip of Staten Island. Within ten years, probably sooner, the city will run out of land to "reclaim." Officials are desperately looking for new sites.[5] One proposal is to fill in a large area of New York Harbor for a jetport. A more dismaying prospect is to fill in Jamaica Bay, "because there is nowhere else to go." This exquisite bay, a wildlife refuge, a human refuge, is one of the few estuaries left in New York.

Nothing is safe from garbage. Public officials know that garbage is political dynamite. Garbage dumps generally rise where political strength is weakest. That invariably means it becomes neighbor to the poor. Former Sanitation Commissioner Samuel J. Kearing, Jr., writing in *New York* magazine, says that as his education in garbage progressed he eventually came to understand that municipal incineration was, essentially, an attempt to solve a political problem with technology: "In places like New York City, incineration reduces the magnitude of the political problem. People are far more tolerant of ashes than they are of raw garbage being dumped nearby."

The dumping of garbage is both dirty business and big business. It attracts criminal elements just as it does vermin. Local officials with the power to license dumping sites control enormous power. Often these licenses are sold to operators fronting for organized crime. In return the politicians receive generous "campaign contributions." These bartered landfills are often unsightly and subject to lax code enforcement, says Kearing. "No matter how serious the violation, health officials cannot close the offending dump, because there is no other place to deposit the area's refuse."

As legitimate business, garbage is the fifth largest service industry in the United States; it is the third largest item in municipal budgets, trailing only schools and roads: about $1 in $10 goes to garbage. Collectively—for the entire United States—it costs us about $3 billion a year. Yet, to resolve this massive problem the government spends only $4 million a year for research. For the development of chemical and biological warfare we spend $300 million a year. The experts say we should be spending a minimum of $100 million a year on solid-waste research. Contrary to public belief there is still only a limited and inadequate technology available. Even the high-temperature incinerators are still in the development stage. Composting techniques have often proved expensive, and the product is rejected by the fastidious and miseducated public.

Only 5 percent of the nation's refuse now goes to sanitary landfills. The Citizens' Advisory Committee to the Council on Environmental Quality proposes that all open dumps be converted to sanitary landfills and that open burning near cities and towns be stopped. Realistically the committee notes that more money will have to be spent for research to develop an advanced technology for efficient recycling and disposal of the ever-growing mountains of solid waste. Short of recasting our way of life—producing less waste through reduced consumption of unnecessary goods—this seems the only hope of resolving the problem.

Various efforts are being made to resolve the solid-waste dilemma. They range from the conventional to the ingenious and bizarre. At the federal level we have the Resource Recovery Act of 1970. This provides for the study of recycling methods, as well as federal grants for actual construction of demonstration projects to recover and recycle used materials. The act was originally opposed by the Nixon Administration, which wanted only to authorize the study of recycling methods.

Several state legislatures are pondering bills to require deposits on disposable bottles. In 1971 Oregon became the first state to require a deposit on all soft-drink and beer containers. A simi-

lar bill was defeated in Washington State two years earlier, the
opposition funded by $171,000 from industry, according to Mrs.
Betty Roberts, chairman of the Oregon Senate Consumer Affairs
Committee.

Manufacturers such as the Reynolds Metal Company try to
avoid legislation by setting up redemption centers and offering
bounties for each can returned. Many communities have established recycling centers, often dispatching scouts as recycling
agents. Experiments are being made with grinding glass for use
in paving and construction materials, and even for use in children's sandboxes. There are industry surveys that seek to prove
that throwaway bottles reflect a new public temper and have
nothing to do with whether the product is disposable or not
(meeting public demand, they call it, as if they were forced to
create the product). The Glass Container Manufacturers Institute insists that in some litter surveys more deposit bottles were
found than no-deposit bottles; the average returnable beer bottle
used to make 31 round trips, now it makes only 19 and in some
cities only 4, says the institute. But Mrs. Virginia H. Knauer, the
President's Special Assistant for Consumer Affairs, did not buy
it. "Public relations flackery," she called it—part of the institute's $7.5 million campaign to promote the use of no-deposit
bottles.

The Japanese have developed a technique of compressing garbage into bales, reducing the volume to one tenth the original
size, for use as construction blocks. But critics object that the
blocks, if sealed, might ferment and cause a gas that would explode—an undeniable disadvantage in construction. At the same
time it overlooks the possibility of a new type of disposable
building that could overcome some of our current urban problems. There are suggestions, and even trial experiments, to fill
abandoned mines and quarries with waste from the cities, but
civic pride makes the receiving communities less than enthusi-
astic about the plan. The scheme was used successfully in Los
Angeles. When it closed down all the city's incinerators to stop
that source of air pollution, it disposed of the 4,500 extra tons of
daily refuse by filling a gaping quarry in nearby Palos Verdes.
Everyone is allegedly delighted with the result. Possibly there is
less civic pride in Los Angeles than in the Pennsylvania coal-
mine region.

Other proposals inevitably would use the ocean as the carpet
to hide wastes under. One would compress garbage in bales and

dump them in the deep. Another calls for having incinerator barges burning constantly at sea, dumping the ashes in the water and ignoring the consequent air pollution. Experiments have been carried out in dumping abandoned cars and stacks of weighted tires in the ocean as artificial reefs to encourage fish breeding. Certain fish are said to like this, but ultimately the man-made reefs decompose. The scrap metal in cars is forever lost, and the tires become a slow and indirect but nonetheless continuing form of marine pollution.

There is much brave-new-world talk about producing paper that mulches itself away when used, glass containers that dissolve when broken, and other self-disposable items. We are not told what chemicals would go into these products and where they would go. Other suggestions include a "fusion torch" that could break down solid wastes into their original mineral state; pneumatic tubes that carry community rubbish to a central incinerator that burns it and uses the heat for industrial and domestic purposes—a technique already employed in Europe and seen as a possibility here. A National Research Council study recommends banking our present wastes until such time as technology is developed to reclaim them.

The conservation-minded call for more recycling—not throwing away resources after a single use but reusing them again and again. This practice is called "looping" by Dr. Aaron Teller, former dean of engineering at Cooper Union in New York. He describes us as a "nomadic raw material culture," using up natural resources just as a nomadic tribe uses up land and then moves on. We rob not only ourselves but future generations as well, he says:

> We vent to the atmosphere approximately 12 million tons of sulfur per year worth a half a billion dollars while we extract 16 million tons a year of sulfur from the earth. By this act alone, we deprive future generations of sulfur at the rate of 12 million tons annually—and sulfur is one of the critical raw materials supporting a modern civilization. . . . We are now in short supply of this material on a world-wide basis, and the price has risen 60 percent in the last five years.

We waste and poison at the same time. We throw away resource at home and import the same resource from abroad. "The distraught fictional world of Alice in Wonderland seems almost rational compared with the man-made idiocy that we are engaged in," says Dr. Teller.

Recycling has run into many roadblocks. These obstacles are given a variety of names, but they usually add up to the same thing: it's cheaper to waste than to recycle. Companies that have gone into recycling often find themselves losing money as their products are rejected. The reuse of paper, for example, could save millions of trees annually. It takes 17 trees on the average to produce a ton of paper, according to the Association of Secondary Material Industries, a trade group. But of 58.5 million tons of paper used in the U.S. in 1969, only 11.5 million tons were recycled, meaning that 200 million trees did not have to be cut. If 50 percent of the paper had been recycled, the cutting of another 300 million trees could have been avoided. We get lost in claim and counterclaim about why paper is not recycled to a greater degree. The paper companies say used paper does not measure up to new paper and often bears the subtle aroma of the mother garbage. Victor Brown, head of a recycling firm, says the truth is that the paper companies have such a heavy investment in woodlands and in pulp-making equipment that "they simply aren't interested in recycling." Other sources, including some paper company executives, support this. Paper manufacturers are making too much money cutting down trees and selling new paper to invite competition from reclaimed paper. The situation is not likely to change while the government refuses to underwrite research that would develop improved technology, and even discourages it by discriminatory practices. The overall effect is to hold onto the status quo that encourages the exploitation of resources rather than their conservation through reuse.

Industry generally has shown little interest in conserving resources. A survey of the 100 largest manufacturers revealed little or no involvement in developing methods to alleviate the waste problem, even that generated by the use of their own products, according to the *National Review*. This amounts to some 182 billion tons of solid waste each year. Only rarely, it was stated, did industry acknowledge its responsibility for adding to our pollution and waste-disposal problems. This is not altogether surprising. Industry is not in business to conserve but to make money. Conservation becomes desirable only when it offers profit. The National Academy of Sciences, National Research Council, notes that "our whole economy is based on taking natural resources, converting them into things that are consumer products, selling them to the consumer, and then forgetting about them. The user does not consume [the product]—he just discards it."

Government alone could change the situation by instituting uniform regulations. No state or community can effectively cope with it. Local restrictions simply drive polluters into more lenient jurisdictions. The most enlightened corporation management could not affect the situation perceptibly and still operate at a profit under existing conditions. Kearing points out that as long as government permits a cheap, open-ended disposal system to continue, everyone will bury used-up products in the ground. Only regulation can force the investment necessary to recycle significant quantities of used materials and to render the residue innocuous so that the ground, air, and water will not be degraded. Says Kearing:

> But today the Federal Government permits the production of gaseous wastes that contaminate the atmosphere. It permits the use of streams and waterways to dilute and flush away waste products. It permits anything at all to be buried in the ground. This is the cheapest possible way to get rid of waste. It constitutes an enormous subsidy to industry and the consumer economy at the expense of the environment. It encourages, indeed makes inevitable, the practice of built-in obsolescence and disposability in everything from automobiles and furniture to clothing and containers. It eliminates craftsmanship while destroying the countryside.

The end of the road is implicit. Every city and village could become a mirror reflection of New York, spared only in degree. There awaits us Professor Chapman's vision of the spire atop the Empire State Building poking through a mountain of garbage.

The city has emerged as less a place of commerce or homes than a symptom of our national malaise, a projection of our collective ego turned sour. Dr. Teller says the city was not designed as an environment for people. It just happened—"the tyranny of small decisions." It reminds us that a high standard of living may prove to be the most lethal comfort ever yet devised. New York is our laboratory.

New York is the waste capital of the world, producing more than any other city on earth. Each day it grinds out 24,000 tons of solid waste, along with 600,000 tons of solid matter that boil into the skies to rain down on the city and foul its air. Disposal of this solid waste costs $30 per ton. The city has a sanitation department of 14,000 employees, more than the total population of many towns, and organized as precisely as an army. Disposal equipment is valued at about $20 million; the replacement cost of all facilities is fixed at $100 million. But it is a futility. On any given day, sanitation crews fail to clean 25 percent of their scheduled routes, according to a recent sanitation commissioner who was fired, victim to the job's impossibility. Even his estimate was understatement, according to John DeLury, president of the Uniformed Sanitationmen's Association, a harried labor chieftain who always keeps a bottle of aspirin on his desk. "I say they're 95 percent off schedule. I say the streets are filthy."

The city's triumph is that it is never quite overwhelmed, never quite inundated. It is like trying to clean the Augean stables. No matter how much garbage and trash is removed there is always more. If the removal system ever breaks down completely, the city goes, said a federal official. How narrow the margin of safety is was demonstrated in the nine-day sanitation strike of 1968. Before the end, garbage was piled a dozen feet high in many places. Rats came out of sewers and cellars to feast happily on sidewalks. An epidemic was feared. A few more days without collections and the city would have disappeared under its own wastes.

The city's residents get used to the trash; they die, flee, or stay and wheeze and await the mayor-messiah who will make good his campaign promises to deliver the people from the curse they live under.

As a tourist mecca, New York is shamelessly wanting. Baghdad on the subway! Where all is possible! It has become a huge polluted slum with patches of affluence where people work in vertical air-conditioned, sealed-window, wall-to-wall-carpeted cages and live behind barricaded doors, afraid to walk on their own streets at night, afraid to open their windows for fear of cutthroats, burglars, crazed drug addicts and poisoned soot-filled air. What a mockery of my distant, innocent youth: health rule one—sleep with open windows! It's not only unhealthful to open a window in New York. It's downright dangerous.

A visitor from California, a professor, complains to *The New York Times* of the dirt and filth. It erases all the good impressions of New York. He goes home and cannot remember the shows or the restaurants. "We can't remember anything good, only the mess. It's in your hair, your eyes, it's everywhere." The salesman from Westchester sees a downhill city. He hopes never to return. It's the dirt. "I see people living with dirt and disease like this and I know that their minds are deteriorating and they don't care about their surroundings any more."

Dog dirt on the sidewalks. It is hazardous to walk down those streets at night, even if you don't get mugged, raped, or have your skull bashed in. New York City alone has an estimated 500,000 dogs. Daily they leave behind 110,000 pounds of solid waste, plus 2,000 to 3,000 gallons of urine. In a year that adds up to 40 million pounds of shit and 600,000 to one million gallons of urine, all to be stepped in or washed out through municipal sewers to the sea as untreated sewage. Dog shit and all those disposable bottles await the unwary. In midtown there are the theater district, the skyscrapers, the affluence. On the perimeter is decay and then the ghettos with ripe garbage thrown into vacant lots. "How can you cope with the problem of rats without first coming to grips with the garbage?" asks the embattled city official. The Bronx janitor shrewdly builds himself a garbage shelter—a reinforced metal roof—as a sort of bomb shelter to avoid the airmail delivery. "Before, I'd get hit with something every time I went in or out," he cries triumphantly.

Every year about 60,000 cars are abandoned. Some 3,000 decorate the city's streets at any given time. Some are abandoned voluntarily, most involuntarily. Many a motorist has broken down and gone to find help, only to return to a steel skeleton. The strippers can pick a car clean as a turkey rack in 10 minutes. Why not just abandon the damned thing? The city provides a free service to haul away the unwanted hulks, but only one call was received in the first 10 months this year.

There are the millions of broken TV sets, ruptured overstuffed chairs and bed-wetted composting mattresses, deposed refrigerators, and no-down-payment-murderously-high-hidden-interest-charges furniture that collapses before the payments are completed and the garnishee worked out. All of it is hurled out windows, surreptitiously dumped in vacant lots, or conscientiously piled on the sidewalk in the spirit of good citizenship. All this garbage must go somewhere. Where? It's just moved from one place to another. About one third is burned in the city's incinerators that foul the skies. The rest is buried in municipal landfills that consume some 200 of the precious, dwindling acres a year. But wait! City Council President Sanford D. Garelik has an inspired solution: a 2,500-foot-high mountain of compacted and sanitized refuse on a swampland in Pelham Bay Park. A "recreation mountain and ski resort second to none." It dwarfs and embarrasses West Berlin's famous "Mount Junk," the artificial hill built of wartime bombing rubble. It is a vision more noble than Professor Chapman's; it is double the height of the Empire State Building. Mount Garelik boggles the mind of less visionary men —an artificial mountain covered with artificial snow. . . .

But the world is not ready for Mount Garelik. It still belongs to tomorrow. Today we remain earthbound to conventional procedures. In researching this book I took a garbage odyssey, following the trucks from street collection to final resting place. I talked to the men and inhaled the odors as trucks evacuated into bins and scoops, and conveyors fed enormous incinerators the fermenting mass. My host, the foreman, explained procedures for firing the incinerator—28 feet long, 25 feet high, 8 feet wide. The monster burns steadily six days a week. A big metal conveyor belt revolves to feed the fire. The mass is reduced by about three fourths. It takes 100 men divided into three shifts to tend the incinerator.

We walked through the immaculate building. My host pointed with the pride of vicarious ownership to the shops, the webs of pipes, and dials. "My machine shop . . . my grates . . . my scrubber that takes my gases and mixes them with water. The water washes out the heavy particulate matter. The gases go out that stack. It's the only stack I have. It's beautiful . . ." The pride of the polluter!

I walked along the East River to the marine transfer station at 91st Street that receives garbage collections from the area. It was early winter. The chill morning air was dirty and gray. The foreman in charge was expecting me. A transfer station is where street rubble is loaded onto barges for the trip to the landfill on Staten Island. It is little more than a corrugated-metal shed with a wet slippery cement ramp and floors. Trucks back up to the platform edge and dump directly onto the barges. There was a great stench and a few scroungy pigeons waddled around on the slimy pavement as if their feet hurt. "You get used to the smell," I was assured. That's the genius and downfall of humans. They get used to anything.

The residue from the municipal incinerators, along with other refuse, was brought to the station. The average truck carries five tons and most days there are around 275 truckloads. Inspectors are supposed to keep a sharp eye on the trucks. Wily drivers have been known to keep their load instead of dumping it; then they can knock off for the next few hours and bring back the same load.

A barge holds maybe 500 tons of garbage, but the newer ones hold 800 or 900 tons. The rectangular steel hulks—150 feet long, 12 feet deep, and 40 feet wide—are loaded with every kind of debris from the city. One in the station was on fire, its load smoldering and sending up a plume of choking smoke. Pigeons ignored the fire and rummaged in the trash on the barge. The foreman explained that the barges are moved by tugs at the convenience of the tides. They were then waiting for a tug to remove the smoldering garbage. "When they get this one with the fire out of my hair I'll be very happy."

Minutes later the tug appeared. "You're going to get to see a movement," said the foreman. He and the others from the station went to the side to watch the delicate maneuver as the loaded barge was pulled out of the snug berth and an empty one nudged into its place. It took delicate handling because of the tide and lack of clearance.

Minutes later a car from the sanitation department came to take me to the landfill at Pelham Bay, in the Bronx, near the proposed Mount Garelik. Rain had fallen the night before. There was an ocean of mud. Scores of sanitation trucks were lined up waiting to get to the landfill. They wallowed through the muck, up a steep, rough road that circled around the mountain of garbage. Against the horizon they spilled out their load, as if vomiting. Thousands of seagulls flew over the trucks. We were met by the site foreman who took us up to the top of the landfill in a pickup truck. The foreman, a spare man with an easy manner, said the landfill used to be a "worthless marsh." He explained that there were three separate lagoons covering about 70 acres. "Now there's no more marshland. It was below sea level." He pointed far below to the adjacent bay. "Now we're fifty-five feet high, and we plan to go to eighty or ninety. We're going to make a ski slope." He too spoke with pride, as if recalling the victory over some municipal affliction.

Each time another truck disgorged its load, the gulls rose in a frantic white cloud and descended on the fresh feast. They stood about with scraps of food—chicken, steak, and bones—in their bills, eating greedily, looking absurdly white against all the filth and garbage. Occasionally they would drink out of dark pools of water. Then they were again in motion, flying, fluttering, screaming and squawking, landing and taking off again, thousands and thousands of them, ignoring the men and the trucks, ever swarming over their heads in an agitated cloud of immaculately white feathers. The foreman said that he saw the Alfred Hitchcock movie *The Birds.* "The next day I was half afraid to come to work," he mused, and chuckled.

Several U-Haul trucks arrived with odd pieces of furniture and old appliances. The foreman explained that a section of the dump was reserved for private homeowners. We watched two men unload a truck. "That's the trouble with America today," said the foreman, as if uttering to himself a familiar thought. "People want to make America beautiful and they make slogans and wave the flag and at the same time they make America ugly." It was curious because I thought of how he had spoken of the "worthless marsh." We both wanted the same thing but had such a different vision of what beauty is. To him it was a ski slope made of garbage and to me it was the lost marsh.

The homeowners, oddly enough, suddenly ceased to unload the truck and drove away with most of the load. Unaccountable things happen in a dump. The foreman said the poor contribute more garbage than the rich, although you'd think it would be the other way around. But that's not the way it is, he said solemnly: "Low-income people are more concentrated than the rich, and they don't go on vacations. They're always there. And they buy cheap furniture that falls apart and has to be disposed of. If you buy good it lasts a lifetime, but the cheap stuff falls apart and you have to keep buying more."

For a long time we sat in the cab, watching the trucks arrive, insulated from the garbage but surrounded by it. It was like an old silent movie. There was the same feeling of unreality about it.

At the foot of the hill was the unending parade of garbage trucks awaiting their turn to inch up the hill and unload. At the edge of the hill was a patch of brown marsh grass not yet covered but destined to become a golf course and recreational field. The mountain of garbage keeps growing higher day by day as the trucks arrive. They come around the clock, 700 loads every 24 hours.

Adjacent to the landfill was Pelham Parkway. Cars hurried along, the drivers intent on the road and their own problems, eyes straight ahead, windows up against the chill winter day. Opposite the parkway was a new housing development being built for 10,000 more families. But now the new buildings were just huge cement skeletons that rose naked against the sky and their foundations scarred the earth. It was all there—the people, the buildings, the cars, the birds, the wastes, the dead marsh, the landfill that was poisoning the adjacent estuary. An inexorable drama was being played out, and I was watching the final act from a splendid perch atop an emerging ski slope.

5

Power

vs.

People

At its present rate of growth, the electric power industry will exceed the physical capacity of the environment to absorb its wastes in the not-too-distant future. Even if all of the problems of controlling sulfur dioxide, nitrogen oxides, and radioactive wastes from various kinds of power plants are solved, the waste heat from power plants now planned will eventually create substantial disruptions of the environment.

—*Environment*, March, 1970

BLACKOUT

Late in the afternoon of November 9, 1965, some 80,000 square miles of the northeastern United States and two large areas of Canada, with an overall population of 30 million people, were suddenly plunged into darkness because of a massive power failure. This led to some bizarre, tragic, and comic results. An estimated 800,000 homebound unfortunates were trapped in subways at rush hour, and thousands of office workers were caught in elevators where they had to remain packed together in intimate contact for hours. Desperate home lovers walked up or down countless flights of stairs and braved other perils to join their families. Several people suffered heart attacks from the unusual exertions. Many slept in offices on makeshift beds. Couples, unable to watch television, read or go out, went to bed early; exactly nine months to the day later a startlingly large number of babies were born—power failure babies, they were called.

The blackout persisted into the following day, and power officials admitted they were mystified by the cause of the breakdown and the extent of its effects. Only after much detective work was the difficulty finally traced to a tiny and apparently insignificant device that had broken down owing to a power overload. Power officials were shaken up. It couldn't happen! Precautions had been taken against that very type of accident. But it did happen. And it had repercussions that were to echo long afterward. For the first time we learned how completely dependent we are on electric power. We had been given a dramatic lesson of how narrow is our margin of safety in the complex world we have created where men and machine coexist in a life-and-death symbiotic weld. We are all at the mercy of an imperfect technology administered by human caprice. The smallest mechanical defect or human failure can produce profound effects. Will the tiny relay function correctly? Will the warning light go on? Will the automatic cutoff valve show its valor in crisis? Will the attendant be drunk or the super show up? Is the repairman competent? Was the night watchman listening when told what to do to avert disaster? Will the key employee be on strike in our moment of peril or need?

A small device costing only pennies can knock out an entire grid across the northeastern United States and part of Canada and cause panic. Virtually the entire system we live by depends on the steady supply of electric power. Since 1930 the demand for electric energy has doubled every decade. In the past twenty years the average American home has quadrupled its use of electric power—from 345 billion kilowatts to 1.5 trillion kilowatts. Our 6 percent of the world's population uses 36 percent of the world's electricity. Energy consumption for each person reportedly rose 25 percent from 1960 to 1968, with a 76 percent increase forecast for the year 2000. A "partly facetious" extrapolation of the present energy-use curve to the year 2100 shows the United States radiating into space as much heat from fossil and nuclear fuels as it now receives from the sun. Nearly half the homes in America today have at least one air conditioner; 98 percent have at least one TV set. We are virtually helpless without the appliances with which we surround ourselves. We are continually urged to buy more of these gadgets that make us progressively helpless and bound to the tyranny of contrivances. At the same time we are warned of massive power famines and the need to conserve electricity.

This is another of our schizophrenic dilemmas. Electricity is essential to our way of life. Directly or indirectly it is behind almost every product and service we have. It also threatens to be our undoing. The production of electricity, its products and by-products are relentlessly bringing about the destruction of the environment. Many feel this is why the pollution problem is virtually insoluble: because our source of wealth and potential destruction are one.

The three primary sources of electricity are the fossil fuels, water power, and nuclear energy. Each produces vast environmental damage. Hydroelectric dams, which account for about 4 percent of the nation's energy, flood surrounding lands, bringing secondary damages to streams, preventing flushing and the carrying away of wastes; as the pollutants build up they kill water life. Nuclear fuels, accounting for less than 1 percent of our energy, emit low level radioactive wastes. The fossil fuels (coal, oil, and natural gas) together supply more than 90 percent of the power and cause the greatest damage. All the sources together release vast amounts of heat that kill off marine life and are causing worldwide atmospheric changes.

The fossil fuels form one of nature's mysteries. They were deposited in the earth hundreds of millions of years ago, but no one really knows why they took their forms. Why did coal remain solid? What chemical or heat action made oil liquid? (Was it, as some believe, composed of vegetation of the sea?) Coal and oil are believed to have produced the natural gas that is trapped in deep beds of porous rock. Gas is considered the cleanest fuel. It is also considered the most scarce. How scarce? No one is sure. Industry has refused to give figures and the government is content to leave it at that.

THE COAL PROBLEM: "LET 'EM EAT COKE"

The origins of coal are more easily accounted for than oil's. It is composed primarily of carbon, plus varying amounts of minerals and moisture. The carbon is believed to have been taken from the air millions of years ago and consolidated into plants and trees. This vegetation sank beneath prehistoric seas, bogs, swamps, and lagoons. Whole forests rose and sank. Some of the matter remained intact and its impression can be seen in coal today—leaves, seeds, stems of plants, whole trees. The rest was broken down by bacteria and microorganisms; the organic debris accumulated, and was compressed by intense pressure from overlying water, rocks, and sediments into the firm mass of carbon we call coal. For man's purposes, the greater the pressure exerted, the higher the grade of coal. Coal can be burned as fuel or distilled into multiple coal-tar products. It provides many of the additives so blithely put into food—dyes, flavoring, and preservatives (some of them are known carcinogens); drugs, perfumes, paints, stains, and synthetic resins. Leftover pitch is used in paving, roofing, waterproofing, and insulation.

As environmental contaminants, coal and oil are together responsible for approximately 74 percent of the sulfur dioxide in the air. In 1967, according to the National Air Pollution Control Administration, factories, power plants, oil and coal heaters discharged into the atmosphere 31.2 million tons of sulfur oxides—the main contaminant identified in every killer air episode. Many attempts have been made to control these emissions. The isolated success is the exception. Indeed emissions are getting progressively worse as more fossil fuels are used. By 1980 emissions are expected to soar 50 percent. By the year 2000, at the present rate of use, they will exceed 100 million tons—three times the present dangerous levels in the air.

There is no real excuse for the fly ash in the air—the soot that spews in black clouds from smokestacks. Technology exists to trap from 95 to 99 percent of the fly ash, but only a comparatively few utilities have installed the necessary devices. No proven method now exists to eliminate sulfur emissions, as noted earlier, but under the Clean Air Act of 1970 the Environmental Protection Agency has set a national limit to be met in four years. Sulfur oxides are a greater health menace than fly ash because they are unseen and can trigger or aggravate respiratory illnesses. A few of the nation's largest power companies have reduced sulfur emissions—usually prodded by local laws—by burning low-sulfur coal and oil.

Sulfur is not an easy contaminant to deal with. It is found extensively in nature. It is part of protoplasm. It is used in a variety of manufacturing processes. It has been with us a long time, and was probably the "brimstone" of the Bible. It appears in coal and oil, ranging from negligible amounts to as high as 6 to 8 percent; it averages around 3 percent. In oil it can be largely removed by refining techniques. There is no known technique to reduce it in coal, although it is believed to be technologically feasible. Once sulfur escapes into the atmosphere in gases there is no way to control it. The alternative is to try to reduce the amount of sulfur burned. A few states have done this by setting legal limits on the amount of sulfur permissible in fuel. Inevitably this has wide economic and political repercussions.

Restrictions on the amount of sulfur in fuel virtually eliminate most coal as fuel. It practically mandates the use of oil. But low-sulfur oil is in short supply, which means that supplies with a higher sulfur content must be refined. That, in turn, means higher prices. Competition for limited supplies tends to further drive up the cost. Coal interests have bitterly fought all restrictions on sulfur content in fuel. They have been supported by industries that are directly affected by coal's fortunes, and others not directly affected but opposed to any controls that could interfere with corporate profits.

New Jersey was one of the first states to pass a law restricting the amount of sulfur in coal. The action was brought on by air so foul that it was becoming unbreathable. The state in 1968 limited sulfur content in fuels to 1 percent by weight; subsequently it was reduced to 0.3 percent in oil and 0.2 in coal. This, in effect, practically eliminated coal. The coal industry recognized it as a death blow if other states followed suit. Other industries saw it as a threat to profits as the costs of other fuels rose.

The new law was immediately challenged in New Jersey courts by a powerful coalition: several major coal producers, five of the nation's biggest railroads, the United Mine Workers of America, which has been identified by a Nader publication as "the nation's foremost 'sweetheart' union," and the New Jersey Chamber of Commerce. A similar suit was started by E. I. DuPont de Nemours & Company. Anxious spectators were the utilities, petroleum companies, and other industrial groups.

There was never any real doubt about the legality of the statute. Even the combine bringing the action allegedly did not take it very seriously. Its only purpose was to delay enforcement for almost a year, according to John C. Esposito in *The Vanishing Air*, a Nader report.* He tells about a perplexed observer who asked a coal trade association executive why his group bothered with the suit when they were bound to lose in the end. "Lose?" the official said. "We didn't lose. Look at how much coal we sold in the last year." In that period the industry sold an

---

* From *The Vanishing Air* by John Esposito, Copyright © 1970 by The Center for Study of Responsive Law. Reprinted by permission of Grossman Publishers.

estimated 8 million tons to New Jersey customers—most of it of the high-sulfur variety.

The biggest single user of fossil fuels is the energy establishment. Coal and oil are the major polluters. Their consumption in utility boilers accounts for half the nation's total sulfur oxides pollution. At least 25 percent of all particulate emissions and nitrogen oxides are traced to fossil fuel power plants. About 5.6 million tons of fly ash and over 20 million tons of sulfur oxides are emitted into the nation's air each year from electric power plants. In New York City, Con Ed alone is held responsible for 42 percent of all the sulfur dioxide emitted. By far the largest amount of the pollutants comes from coal-fired steam boilers. The threat promises to become greater as more fuel is consumed to produce more electricity.

There is a mistaken notion that coal has ceased to be an important fuel. Despite the increased use of oil and natural gas, enormous quantities of coal are still used. Annual coal production now amounts to some 575 million tons. In the next 30 years it is expected to soar to 1.1 billion tons a year. More than half the present production is used by utilities, which have a voracious appetite. Unlike oil, coal reserves are almost unlimited. But four fifths of the coal burned in utility boilers is of high or medium sulfur variety—and this situation is expected to get worse rather than better. The average sulfur content of utility coal will rise from 2.7 percent in 1969 to 3.5 percent at the end of the century, according to studies.

The reasons are complex. As usual they focus on economics. The heart of the matter is that the related and powerful coal interests are deeply committed to the status quo. For coal companies that means not opening low-sulfur mines before the present high-sulfur-content mines are exhausted. Utilities resist change because converting to new fuels or switching boilers to handle new combustion characteristics of low-sulfur fuels would cost money. Railroads do not want any change that would upset profitable hauling contracts. The coal companies have tended to merge or sell out to oil interests, eliminating even a semblance of real competition and thus consolidating power within the energy complex. All the related interests stand shoulder to shoulder to fight off any environmental legislation that would interfere with the present profitable situation of dirty fuels, dirty skies, and dirty politics.

All the various coal interests have worked together under the

banner of the National Coal Policy Conference*—an odd assort-
ment brought together in the early 1950s by John L. Lewis, who
headed the United Mine Workers. The conference fought all progressive legislation that would have in any way interfered with
what *The Vanishing Air* calls "the cozy relationship of its members." NCP consistently blocked or perverted corrective pollution
legislation. First it denied that any health hazard existed from
coal emissions. When action was no longer avoidable "the cheap,
quick and dirty expedient" favored by the energy establishment
was tall smokestacks that disperse the pollutants higher and farther. The third tactic has been to call for control devices in
stacks—but since the perfect device is not yet available, nothing
can be done until it is invented.

The great breakthrough does not seem to be just around the
corner. The coal industry has contributed only token amounts
for research—a total of $107,000—one penny for every 96,000
pounds of coal mined per year. In 1969 the entire electric power
industry spent $90 million for advertising and only $40 million
for research and development, according to S. David Freeman,
director of the Energy Policy Staff of the Office of Science and
Technology. The Federal Power Commission says that overall
the investor-owned industry has spent only about 0.25 percent of
revenues on research and development to meet its various problems.

In 1968 the coal industry asked the government to ante up
$100 million for research, but the government would put out
only $15 million. This, however, represented no small victory for
the coal interests. It subtly put the burden of responsibility on
government rather than industry to find a solution. At the same
time, Esposito points out in *The Vanishing Air*, the level of funding was so low that the electric companies had little reason to
fear new breakthroughs that would eventually put them to added
expense.

Nader's investigators spent months trying to dig out the hidden truths underlying air pollution. They found the going tough,
especially when it came to coal. Government and industry did
everything possible to preserve the status quo. There was a lack
of information. Often there was refusal to provide such information as existed; in one case an industry official refused to tell
facts already on the public record. Coal companies were very

* The National Coal Policy Conference was formally dissolved in 1971.

secretive about their operations; they would not even supply their own trade association with data it requested. There was a lack of research. There was a lack of planning to meet predicted coal shortages. Much of the research done was meaningless and directed more toward public relations and ceremonial use than finding solutions. Some bureaucrats were more involved in trying to control funds and enlarge their empires than in dealing with the problem. Industry had just one objective: to sell more coal.

Government did nothing; coal and its allies were too powerful to tamper with. It could have fixed tough emissions standards to protect the people. It could better have used its meager resources on tough enforcement. (Instead, in 1970 enforcement funds were cut by the Nixon Administration.) Such enforcement would have stopped polluters from ducking out of tough jurisdictions into more permissive areas. It would have forced industry to mine some 95 billion tons of "clean coal" said to be within reach of eastern markets. It would have compelled industry to spend the necessary funds to develop the technology to control emissions. Instead government played industry's game. Here's how Esposito put it:

> The entire edifice of government subsidies, research in place of enforcement, tall smokestacks, and blanket denials of health hazards was erected by the coal-utility coalition to dodge the one course of action that could lead to limited control of sulfur oxides emissions: the prohibition of fuels containing high percentages of sulfur.

In other words, coal used the cheapest control device known: burn and breathe.

OIL: THE UNIVERSAL POLLUTANT

Oil is the lubricant of the modern, industrialized state. It is probably the single most influential commodity in our economy. It is a keystone in domestic and foreign policy. It regulates our standard of living and dying. Directly or with subtle obliqueness it supports or is locked into most of our wealthiest and powerful industries; it is the primary driving force of giants such as the automobile industry, tires, petrochemicals, highway and road-building, and the vast military complex with its voracious appetite for oil. During World War II some 60 percent of all military supplies shipped to American forces overseas were petroleum derivatives.

From oil come diverse items such as plastics, drugs, pesticides, cosmetics, dyes, nylons, food additives, paints, synthetic rubber, fertilizers, explosives, detergents, and napalm that fries human beings and consumes native villages. Oil makes the wheels of American industry turn. It also lubricates our political machinery. It elects federal, state, and local legislators. It dictates Cabinet appointments and diplomatic posts. It supports and frustrates revolutions and wars. It buys whole countries, forms and topples governments. Often oil is the real prize behind the ideological goals we profess in our foreign adventures. It has been called the "invisible world government."

Of all the pollutants unleashed by human activity, none makes so wide and devastating an impact as oil and its byproducts. It gives us the automobile that causes 60 percent of all air pollution, as much as 90 percent in some places. The automobile is a creature of the oil well, and their destinies are interwoven as inextricably as petroleum and pollution. Oil provides the fuel, lubrication, and even the road the automobile runs on. It makes possible the plastic, certain fittings and, in a pinch, even the tires. Cars consume two thirds of all the crude oil produced. This combustion spills into the air the untold millions of tons of carbon monoxide, sulfur oxides, nitrogen oxides, hydrocarbons, lead, unidentified gases, metals, as well as asbestos dust, rubber dust, asphalt and cement dust, and all the other particulates and vapors, known and unknown, with the multiple and unpredictable mixtures they produce.

This might mean pollution to you and me. To the producers and handmaidens of cars it means hard wealth. Seven of the top ten moneymaking corporations in America are involved in the production of automobiles or their fuels, according to *Fortune*. Their profits in 1966 were $6 billion. They represent assets of $70 billion. These giants are among the aristocracy of the modern corporate state: General Motors, Standard Oil of New Jersey, Ford Motors, Chrysler, Mobil Oil, Texaco, and Gulf.

From deep in the earth we bleed the crude petroleum banked there millions of years ago by great chemical and thermal fermentations and titanic geological convulsions. Its wealth is unlocked through a refining process known as "fractional distillation." This is based on the fact that different components of petroleum have different boiling points; they can be separated from one another within certain limits by applying heat and collecting the different products that boil off within varying temperature ranges. Many of the 80 billion pounds of synthetic organic chemicals produced annually by some 12,000 U.S. companies have their origins in these fractions. The most valuable component, economically, is gasoline, in its various grades. The great dream of refiners has always been to convert the crude entirely to gas. But this has eluded the processor. As gas comes off the "top of the barrel," the heavy liquid part that remains in the bottom is fuel oil. This is the "resid" (residue). When the sulfur is largely removed it is known as "sweet resid."

Many of the fractions of crude oil are toxic. Some cause cancer. Some become carcinogenic under the influence of heat. Many of these are emitted into the air following combustion in oil burners and cars to interact with other air pollutants. Benzopyrene, a powerful carcinogen, for example, is released in gasoline emission as a hydrocarbon. Other carcinogens may remain in the asphalt that is used for roads. Ultimately the asphalt and its components will be worn loose by car tires; they will wash down sewers to find their way to the embracing sea, to be absorbed by plants and fish, possibly to poison or cause unknown effects. Or perhaps this toxic dust is blown by the winds to contaminate food crops and fields where animals graze that produce our milk and meat. Through the various avenues—land, water, air—these components of oil enter the food chain.

To keep the industrial machine going requires oceans of oil. The faster the machine drives forward the more petroleum is required. In 1965 the world burned 11 billion barrels of oil. The rate of increase is 6.9 percent a year. This means a doubling in 10 years—22 billion barrels by 1975, 44 billion barrels by 1984, 123 billion by the year 2000. That is based on present consumption and does not allow for increasing use.

Half of the oil produced in all of history was taken in the past 13 years, according to the National Science Foundation. Most of it has been consumed in the United States. The United States demands now exceed 588 million gallons a day, much of it shipped over water. After 40 years of industrial development Russia still consumes barely one-fifth as much. A Japanese uses one-tenth as much oil as his American counterpart.

Oil has the distinction of being the universal contaminant. It degrades and pollutes at every stage of development: extraction, transportation, refining, and use. Some of the worst damage occurs in transportation. Usually we hear only about the massive spills of oil at sea: the notorious *Torrey Canyon* that in 1967 smashed into a reef in broad daylight, because the skipper was reportedly trying to save time and went into shallow waters, spilling 35 million gallons of crude oil into the sea. The oil marred the beautiful beaches of Brittany and France, killed marine life and wildlife and outraged the entire world. There have been many other accidents and incidents. The *Arrow,* carrying 3.8 million gallons, broke up off the coast of Nova Scotia. There was the infamous Santa Barbara oil leak and the disastrous Gulf of Mexico spill and fire in March, 1970, followed by a second in December.

These are the headlines. The more insidious damage takes place daily beyond the glare of public scrutiny. It occurs in thousands of mishaps, spills and normal operations. Up to 10 million tons of oil are spilled into the sea each year. Usually it goes unnoted and unrecorded, but bears terrible testimony in the continuing degradation of the sea.

Some 48,000 ships of more than 100 gross tons now ply the seas. They pollute the ocean with up to 5 million metric tons of oil a year. The 44,000 passenger, cargo, military, and pleasure ships account for about half of the total by flushing oily wastes from their bilges. The world's tanker fleet of 4,000 ships adds the other half of petroleum pollutants from bilges, cargo tanks, and accidental spills.

A prime cause of ocean pollution is the traditional maritime practice of flushing oily wastes into the sea—"blowing the ballast." When a tanker discharges her cargo at port and returns to sea, she must add weight or ballast to restore her seagoing stability. Tankers ballast by refilling their oil tanks with seawater. Eventually this must be pumped out so the tank can again be filled with oil. Until 1964 about 80 percent of the world's tankers flushed their oily ballast directly into the sea. An average 40,000-ton tanker discharged about 83 tons of oil with each flushing. But since 1964 about 80 percent of the world's tanker fleets have adopted a procedure ("load-on-top" cleaning) that reduced the figure to 3 tons. Most of the oil collects on top of the ballast. Under the load-on-top procedure this oily mixture is pumped into a slop tank, but the rest (containing the approximate 3 tons of oil) still ends up on the sea. The other 20 percent of the world tanker fleets do not use this method; it is time-consuming and costly.

Under present international agreements tankers are not supposed to discharge oily mixtures within 50 miles of shore. Ships other than tankers are required to discharge their bilges (the bottoms of ships) as far from land as practicable. This protects the shore somewhat but not the sea. The United States seeks to eliminate both practices by 1975. If enacted, the measure could compel changes in ship design and shore facilities that would cost, on a worldwide basis, an estimated $1 billion to $5 billion. Every day the reform is delayed the sea gets dirtier.

In early December, 1970, two Navy barges dumped their sludge off Florida—a naval equivalent of a tanker blowing its ballast. The result: an oil slick covering 760 square miles only 23 miles off north Florida. The slick was said to consist of more than half a million gallons of waste oil. The Navy said the procedure was done "about twice a quarter over 50 miles from land."

The menace is increased by multiple accidents. In the past 10 years tankers have been involved in 553 collisions, some vessels colliding three or four times. More than three quarters of the accidents occurred close to shore or in ports—the vulnerable coastlands where most of our fish come from. The threat increases as the number of ships and the amount of oil soar. In the past 30 years seaborne oil commerce has risen from 84 million tons to 893 million tons. Most of the oil is shipped as crude, the most toxic form. The new supertankers, some longer than a football field, will carry 300,000 tons of oil. These giant vessels are structurally weak for their size and difficult to maneuver, according to a Presidential report. This means greater risk. Already two of the new supertankers have exploded and sunk. By good luck they were empty.

The experts say there is no such thing as an absolutely clean oil operation. In the first nine months of 1969 some 544 spills were recorded. By 1970 the figure rose to 950. In the first nine months of 1971 it had climbed to 1,240. These figures are believed to represent only a fraction of the actual number of spills. One seriously increasing source of oil spillage is offshore drilling. It now accounts for some 15 percent of the total U.S. production. In recent years drilling rigs have sprung up in American coastal waters like steel mushrooms. The Gulf of Mexico has had 6,000 wells sunk in it since 1960, becoming such an impenetrable forest of rigs that shipping lanes had to be cleared.

Offshore oil operations are rooted in economics and politics, as is all oil production. The continental shelf, beyond the first three miles claimed by the coastal states, is administered by the Department of the Interior. The department permits oil companies to bid for lands they want to lease for exploration and drilling, then gives them very favorable terms. There are further tax writeoffs through which government underwrites much of the cost of exploring and drilling for oil. Industry makes its own surveys that are accepted by the government without challenge. The findings are secret. Often they are not even shared with government. For industry the arrangement is enormously profitable and government profits in a lesser way. In 1969 offshore leases brought the treasury $1 billion; only 0.06 percent of the return was spent for management, hardly enough to do an effective job, as later events dramatically demonstrated.

Federal control of offshore leases puts the Department of Interior in an awkward position. The department is responsible for conserving public lands. At the same time it acts as broker for the exploitation by private industry. The Santa Barbara blowout revealed how poorly this conflict of interest works for the people. Long before drilling began in the channel the risks were well known: geological formations created a serious danger of blowouts. Secretary Stewart Udall had misgivings about granting the leases. He was supported in this by his departmental advisers. But the Johnson Administration needed money for the Vietnam War. Udall's reservations were overridden by Johnson, the Bureau of the Budget, and oil industry pressures. A half-dozen oil companies paid $603 million for leases covering 363,181 acres of offshore lands. Here we have the ugliness of a government exploiting the environment and endangering the sea to support a foreign war unwanted by the people. There were public protests against the drilling. Government and industry assured the public that it was perfectly safe. In return for rents that generally totaled 12 percent of industry's profits off of public lands, the government "remained silent about what it did know and complacent about what it didn't," according to *Earth Times*.

The predicted disaster occurred. On January 28, 1969, a well was brought in at 3,479 feet "crossing two big and innumerable small fault lines." Gas and then oil spurted from the hole. By mid-May the ocean floor was still bleeding. Some 3.5 million gallons of oil had surfaced. False figures given of the continued leakage. Reports of damage were suppressed or minimized. The government said the oil would affect at most 1,000 birds and survival should be in the neighborhood of 25 percent. Actually more than 1,500 oil-soaked birds were rescued; nearly 90 percent died. The state of California said it was "virtually impossible" that the oil could reach San Miguel Island with its vast marine animal population. Three months later *Life* featured a picture of the oil-drenched shoreline of San Miguel.

What was the government's response to this? To keep drilling. A federal panel headed by President Nixon's science adviser, Dr. Lee DuBridge, recommended more drilling in the area of the leak to release the pressure. Another company was also permitted to resume its disrupted drilling. The people of Santa Barbara were outraged. Public fury mounted. What was industry's response? To call in its public relations battalions. A series of ads was paid for by a $50,000 grant to the local Chamber of Commerce. This further infuriated the people. They charged that the ads contained distortions and made the "city's plight sound far less serious than it is," reported the *Wall Street Journal*.

The following December the same well that had caused all the previous trouble sprung a new leak. Once more beaches were fouled. More wildlife died. Tempers flared. Fritz Springman, a spokesman for Union Oil Company, owner of the offending well, was asked by *The New York Times* how the new leak could occur in light of assurance from the company that it would not happen again:

> "We don't have an answer to that," said Mr. Springman in a tired voice. "There's not a great deal we can say because it did happen again. The drilling regulations are much more stringent than they used to be, but this new leak—it's a most unfortunate occurrence."

In February a new disaster erupted. Twelve of 292 wells owned by Chevron Oil Company in the Gulf of Mexico exploded into raging fire. The blaze continued for more than a month, the damaged wells gushed 25 to 40 barrels of oil an hour into the Gulf. Investigation revealed that the company was guilty of 347 violations of regulations. Interior Secretary Walter Hickel, who replaced Udall in 1969, temporarily halted drilling in the Gulf, but soon the companies were back at it. The following December an even more disastrous fire broke out on a Shell platform off Louisiana—"the worst fire in the business." Oil that gushed from the well without burning left a slick three miles long and a mile and a half wide.

Louisiana, which does a nice business from offshore oil, was outraged. Not about the fires and spillage but at Hickel's suspension of drilling. The action was called "disastrous" by Lieutenant Governor C. C. Aycock, in a *New York Times* report: "We have had pollution from oil in South Louisiana for 20 years. But now we are suddenly beating our breasts about it. We are, in effect, killing the goose that lays the golden egg."

The people of Louisiana seem to share Aycock's view. They too are more interested in the golden goose than the egg that hatched it. For years Gulf beaches from Texas to Florida had been fouled with "blobs of oil-like substances that have baffled chemists and other investigators." There was an occasional newspaper headline or TV report. But the most notable reaction to the pollution outcry following the Gulf fire was a rash of bumper stickers that proclaimed: "Oil Feeds My Family."

The upshot of the whole thing was that President Nixon asked for legislation to cancel twenty oil-drilling leases off Santa Barbara. The three companies already drilling were permitted to keep taking the pressures off those shaky wells that they had already gone to so much expense to tap and tame. No steps were taken to cancel leases or prevent future drilling outside the proposed sanctuary where similar dangers await. But it all came to nothing. A congressional committee killed the proposed legislation to cancel the leases. While the latest Gulf of Mexico fire was still raging and oil was still leaking from the wells at Santa Barbara, the Department of the Interior, in December, 1970, auctioned off drilling rights to another 500 square miles off the coast of southwestern Louisiana for bids totaling $850 million— the largest offshore oil lease sale yet. All opposition was ignored or steamrollered by the cry that 100,000 jobs were at stake. Studies are now being made with the probability that offshore drilling will be permitted along the Atlantic coast.

Oil exploration is also being stepped up in freshwaters without real precautions. Drilling is going on in the Florida Everglades and more extensively in the Great Lakes. From 1913 through 1955, just 38 wells were sunk in Lake Erie. Since then the total has soared to 600 and is increasing rapidly. There is danger not only of oil contamination but of releasing brine that could destroy the freshwater above which is used for drinking.

From all over the nation and the world, fantastic amounts of oil gravitate toward the sea. Voluminous amounts of crude are left in the wake of tankers. Huge amounts of oil come from a multitude of other sources: used motor oil (350 million gallons annually), oil sold to refiners for reuse but now often flushed into sewers; from industrial discharges; from accidents to 200,000 miles of high-pressure pipelines lacing the nation; from mishaps to railroads, tank trucks and vessels plying inland waterways; it is spilled during transfers through coastal and riverside terminals. Shipping and port activities alone release an estimated million tons of crude annually. The grand total is likely to be 10 to 100 times higher, according to Dr. Max Blumer, a senior scientist at Woods Hole Oceanographic Institute. Vast quantities come from sewage treatment plants that discharge into streams and rivers. In June, 1966, an estimated 20,000 gallons of used oil was discharged from a Colorado sewage treatment plant into the Colorado River, reports the Presidential Oil Pollution Report. Each spring extensive duck kills are caused by waste oils in the Detroit River.

Thirty percent of all "influx of oil into the world's waters" comes from automobile crankcase oil, dumped by local service stations into sewers, according to the Society of Naval Architects and Engineers. The oil then flows into smaller estuaries and eventually winds up in bays and the ocean. "We are creating a sort of marine wasteland," says Dr. M. Grant Gross, associate director of the Marine Sciences Research Center at the State University of New York at Stony Brook. Dr. Gross spent the summer of 1971 cruising the waters of Long Island Sound and New York Harbor studying these petroleum sediments. He said:

"In those basins with repeated oil spills, you don't even find hydrogen sulphides, that rotten egg smell, building up. Not only are there no large organisms, no fish, but no bacteria—nothing to break down even the petroleum accumulating there. The bacteria can't keep up with the influx of the petroleum. The result is a continuing buildup of this black, gelatinous mass."

Then eventually, the channels have to be dredged, Dr. Gross said. "The dredged material is dumped offshore in areas such as the New York bight (a deep-ocean area at the base of the Hudson River) and becomes as damaging and probably even more damaging than raw sewage sludges. So what starts out as a small local problem ends up as a regional problem and who knows where it will go from there?"

Oil pollution is now universal, says Dr. Blumer: "It is not only on the shores, where it is very evident, but is spreading worldwide so the entire world ocean is affected." Thor Heyerdahl found "floating lumps of solidified asphalt-like oil" in mid-Atlantic. Similar lumps were found by investigators from Woods Hole. The lumps were graphically described by *The New Yorker* as "gooey golf-ball-size black globs . . . trailing sticky-looking streamers." Lumps the size of a pencil eraser were found in heavy concentrations on every square yard of sea surface in some places, said Dr. Richard Bachus. Even the Sargasso Sea, long famous for drifting plants, is now afflicted. Oil residue on the sea surface already equals the amount of its surface plant life, explains Dr. Blumer. This is the more remarkable because the Sargasso lies in mid-Atlantic, far from heavily traveled shipping lanes, fed by the great swirling circular currents of the sea; some of these currents originate in the Gulf of Mexico, which is abused so ruthlessly.

What is the effect of such universal contamination? How is it to be measured? So far damage is considered only in terms of massive spills: the cost of a *Torrey Canyon* breaking up, the value of its cargo, the cost of cleaning up fouled beaches, lost tourist dollars, damage to resort hotel property, windrows of dead fish washing up on beaches and stinking for weeks afterward, fish and shellfish that must be thrown away because they taste of oil. There is no measure for dead wildlife and birds that founder helplessly in oil scum. Most of those that are rescued die. When the *Torrey Canyon* went down, 10,000 oil-soaked birds were treated in England and France. Of 7,849 treated in England, only 450 were alive two months later. What dollar figure do we put on those creatures and others yet to die?

Only now are scientists beginning to understand some of the more subtle long-term effects of oil spills. These effects are never isolated. Nature is not so simple as we would like her to be. Effects are invariably linked in series and chains; they echo mysteriously throughout whole communities in small biological explosions that we are not keen enough to catch even with the most sophisticated instruments. One delayed effect is reduced bird populations, the virtual disappearance of whole species. This is believed to be caused by damage to marine life. Oil in water cuts down the penetration of sunlight, reducing the photosynthesis on which the ocean depends for its critical supply of oxygen. It also eliminates or reduces the minute plants that are the base of the food chain for all marine life. A single cubic foot of seawater contains 20,000 plants, 120 animals or eggs, and plankton. Most of these organisms are not visible to the naked eye. Many of the tiny creatures are extremely sensitive to pollution, even to slight changes in temperature and salinity. They can be easily damaged, undermining the foundation of life in the sea, without the damage being apparent until there is some ultimate and obvious effect, such as birds disappearing. Then who is to relate brutal effect to oblique cause? At a 1971 conference in Washington on oil spills oceanographer Jacques Piccard warned that if oil spillage becomes heavy enough to kill marine plants, the animal life of the world could suffocate. "What happens when oil penetrates all horizons of the sea is beyond measure," says one researcher.

In midocean it is very hard to assess the effects of anything, states Blumer. "Currents move the stuff around. The animals move. However, the effects cannot be doubted." Until recently it was believed that the toxic volatile fractions of oil evaporated quickly and that the rest, the tar, was inert, though messy. This is not so, he says. Tar samples that had been in the ocean from two to six months were found to still contain the toxic fractions —paraffins, benzenes, toluenes (hydrocarbons of the aromatic series).

He too emphasizes the recurrent theme that most damage takes place, near the crucial shore. Oil thus joins the whole concerted chemical-waste onslaught pouring into the coastal waters and estuaries to damage and degrade that environment and its life. Unlike most other contaminants, however, oil invades from the sea as well as from the shore.

A spill can cause biological damage to marine life to an extent never realized until recently, Dr. Blumer says. He and other Woods Hole scientists made that observation after intensively studying the effects of a small spill (650 tons) on a marsh area near West Falmouth, Massachusetts, in 1969. Their evidence indicates that a single wrecked supertanker could probably destroy the entire coast of Maine or a fishing ground such as the famous George Banks.

The most toxic parts of oil, the most soluble, cause enormous biological damage "at the very moment of the spill," says Blumer. This affects all organisms touched, from phytoplankton to basking sharks. Oil can kill marine organisms directly on contact or it can coat and asphyxiate them. The initial kill at Falmouth was enormous, Blumer told *The New Yorker*. "The population went from two hundred thousand animals per square meter of marsh area to maybe two animals. And they say oil is not toxic."

The oil continued to kill long after the spill, contrary to expectations. Windrows of dead fish kept washing up on the shore. Dozens of species were counted. The victims also included shellfish, marsh grasses, and bottom-living eels and worms. This too was a surprise. It had been believed that oil did not affect the bottom but just floated on the surface. "But those bottom dwellers were apparently coming out of their burrows in hordes and being washed ashore or dead."

Nor did the oil break down as it was supposed to; it left a tarry residue that animals could not eat. It saturated the bottom sediment to a depth of at least a foot. Five months later virtually the same conditions existed as at the time of the spill. Every time there was a storm the oil appeared to get stirred back into the water to produce a new kill of fish. "The river, the beaches, the bottom out to a depth of about forty feet are completely dead," observed Dr. Howard Sanders, a marine ecologist. Dr. Sanders observed that strange things happened in the marsh. After five months one species of animal life was returning—a worm that thrives only in polluted waters and had never been found here before. But perhaps most disturbing of all were the animals that were affected but not killed.

> Their behavior became cockeyed. We found bottom fish on the surface—you could put your hand down and pull up cod and eels. It was grotesque. Their escape reactions weren't working at all. Fiddler crabs normally put up a claw and scoot away if you approach. They got the claw up, but they just stayed there. And they were in mating coloration at a time of year when they shouldn't have been. It's questionable whether they could reproduce, or whether offspring could survive at that time of year. The contaminated shellfish have not—as they're said to be able to do—purified themselves, and probably never will.

Blumer has pointed out other hazards as reported by *Environmental Action*. Young organisms are more susceptible than older ones. Higher species are affected when oil destroys their sources of food. Some organisms may absorb enough toxic materials to affect their resistance to infection or their fertility. Some parts of oil are carcinogenic and may cause fatal cancers. There is a danger that carcinogenic hydrocarbons in oil may concentrate in dangerous amounts, much as DDT does. In this way animals not exposed to oil may be harmed by eating contaminated animals—a hazard that also affects humans.

Blumer does not believe that the vastness of the sea is a protection against lethal contamination. Indeed, he told *The New Yorker* that the ocean is so big, and its circulation so broad, "that it's very likely that any pollution, of whatever degree, is irreversible. Four or five hundred years would be required to balance its effects."

Oil pollution becomes very real when it reaches beaches and touches lives. It is no longer the meaningless abstraction that is the salvation of polluters. Pictures of dead and dying oil-drenched birds and windrows of dead fish are devastating. So are expensive cleanup, lawsuits, bad publicity, aroused people, and possible restrictions and penalties. Each thousand tons of crude oil washed onto a beach 30 feet wide can form a layer half an inch thick for 20 miles, according to Dr. Julian McCaull, a biologist and associate editor of *Environment*.

The preferred treatment now is to burn off the oil. This causes excessive air pollution and leaves a hard asphaltlike residue full of the toxic and carcinogenic fractions already mentioned. Other techniques have been tried with limited or no success: trapping "islands" of oil with floating barriers; soaking it up with "vacuums"; mopping it up with absorbent materials such as chalk; pumping it out of the stricken tanker's hold; injecting a chemical that is supposed to turn the remaining oil into an inert jellylike mass, and burying that which reaches the shore.

Chemical dispersants are used extensively in an effort to sink the oil—a form of sweeping the problem under the rug. Frequently these chemicals are more toxic than the oil itself. During the massive *Torrey Canyon* spill, such chemicals were used along 150 miles of Britain's shore. The results were catastrophic, according to Dr. McCaull in *Environment*. He quotes Dr. J. Eric Smith, director of England's Plymouth Laboratory:

> At and near the surface of the sea, vast numbers of the minute flagellates which form the bulk of the tinier elements of the phytoplankton were dead or dying, and pilchard eggs were disintegrating or developing abnormally. . . .
> In the exposed middle reaches of the shore, there developed within hours a scene of progressive devastation, and within a few days virtually nothing remained save for tufts

of dead and dying algae. The rock surfaces were utterly
bare of animals and littered at their bases with cemeteries
of shells.

The effects of detergents that are not toxic in themselves are
considered equally injurious. They break up the oil into fine
droplets that will sink. In dispersion the oil is more harmful to
marine life than on the surface, according to some scientists.
Some remains on or near the surface, reducing sunlight penetrating the water and hampering photosynthesis, and some sinks
to the bottom to affect bottom feeders. The breakdown also apparently accelerated the exposure of marine life to the toxic hydrocarbons in oil, including the carcinogens, reports Dr. Ira N.
Gabrielson, a biologist who heads the Wildlife Management Institute. This can start the potentially dangerous compounds
moving up the food chain toward man.

All of the threats outlined here will be aggravated when oil
starts moving out of Alaska. Many experts have warned against
this new frozen Pandora's Box. This time the target is the delicate tundra, frozen to the strength of iron but easily damaged
and incapable of healing its wounds in the polar chill. Alaska's
permafrost covers 85 percent of the state and reaches as deep as
1,600 feet.

Many of Alaska's creatures live off the sea that will be increasingly and inevitably polluted with oil. The plant and wildlife depend on the thin carpet of vegetation and soil that covers the
frozen ground. In the short polar summers this thin living layer
thaws, insulates the ground below, and prevents it from thawing. Traffic over it could cause terrible wounds, explains the
Christian Science Monitor:

> Rip away that layer, as vehicle tracks do in summer, and
> the sun begins to melt the exposed ground. Alternate freez-
> ing and thawing, plus enhanced erosion, can widen a pair
> of car tracks into a chasm. So one of the main considera-
> tions in Alaska conservation is to keep the permafrost fro-
> zen. And that is just what a buried pipeline carrying hot
> oil wouldn't do.

Ecologists warn of other disruptions and catastrophes if the
proposed pipeline is built. It would transport the hot oil 800
miles across the state from the oil strike on the North Slope to
the shipping terminal at Valdex (the oil must be heated so it will
flow easily). Most of the pipe would be underground. This would

melt the permafrost. The earth would be unstable, it could shift and break the pipe. There is the further danger of earthquakes. Any breakage could pour millions of gallons of hot oil over the vulnerable frozen earth, setting in motion unpredictable ecological upsets.

In some places the line would be elevated a few feet. This would block migrating herds of caribou which the Eskimos depend on for food. Ecologists note that there would be a further threat to wildlife from the effects of air pollution due to burning off gas wastes at the oil wells. This, in turn, could damage the lichens that are the food source of the caribou. Lichens are extremely sensitive to air pollution. The 500,000 remaining reindeer of the Lapland herds are said to be finding lichen forage more and more scarce because of worldwide pollution from industrial centers.

Still another risk would be from the new supertankers that must fight their way through a thousand miles of Arctic waters. Often they have to crunch through ice floes 40 feet thick. When the *Manhattan*—more than 1,000 feet long—"opened the Northwest Passage" in 1968 it stove in its hull bulling through the ice wastes. Ecologists point out that a wrecked tanker could be horrendous for marine life in polar waters. What, they ask, will be the effects of normal oil spillage from bilges and exhausts of these vessels? What will the overall effect be on the whole ecology of Alaska? The Department of the Interior's draft statement on the proposed oil pipeline conceded that there would be damage to fish, wildlife, wilderness, vegetation, and the entire cultural pattern of the region; but the government insists it would be less than that which many conservation experts predict. Despite this admitted damage, the department says the project should be carried out for reasons of national defense. The same reason for going ahead has been advanced by Secretary of Commerce Stans. We often seem so intent on protecting the nation from our enemies that we are blind to the harm we are causing ourselves in the name of national defense.

The big Alaska venture was supposed to be under way long ago. For more than two years millions of dollars' worth of road-building equipment and other supplies have been awaiting the signal to go. Supertankers are being built for the Alaska run. The show is momentarily stalled by a court order imposed on the Secretary of the Interior following legal challenge by conservationists. There's no doubt, however, that the permit will be granted as soon as the legal formality is dispensed with. Hickel was eager to get that oil rolling, as a *Christian Science Monitor* dispatch indicated: "He seized the ironic occasion of Earth Day at the University of Alaska to announce flatly, 'I will issue the permit for the pipeline right of way.' "

This bureaucratic arrogance is frightening when weighed against its possible consequences. The secretary made his pronouncement before studies were completed on the ecological effects of the project. He as much as said he is going through a ceremonial ritual. "I will issue the permit . . ." We are given an insight by David Hickock, associate director of the Federal Field Commission for Development Planning in Alaska. Barry Weisberg, in *Ramparts,* says that with the massive supertankers already in the works, "Hickock notes there still exists an almost complete 'lack of research and investigation in arctic waters on oil pollution, coastal processes, phytoplankton, marine fisheries and mammal populations, and on programs for the development of new technologies for port facilities in the arctic.' "

The long-overdue environmental impact statement recognized some of the more serious problems. It took a more measured view of the pipeline's biological and social impact, considering ways of minimizing some of the damage and admitting that other damage was inevitable. The authors recognized that "the original character of this corridor would be lost forever." Conservationists opposed the project so bitterly and vocally that the government took a more cautious stand. William Ruckelshaus asked the Interior Department to hold up the right-of-way permit for the oil pipeline pending a study of environmental advantages of an alternate route through Canada. Before the request the new Secretary of the Interior, former Representative Rogers C. B. Morton, during congressional hearings on his confirmation as secretary in the spring of 1971, announced that no permit would be forthcoming during the year.

For the men who have advanced the pipeline only one fact exists. Oil. Fortunes to be made, power to be exerted: 100 billion barrels of oil waiting in the arctic icebox to be tapped—almost as much as has been drawn from the whole of North America during the past century. Weisberg tells us a bit about the officials responsible for what he calls tampering with the "great harmonies" of nature: Thomas Kelly, "the state official who presided over the big Alaska lease, was surely moved by hubris when he proclaimed, 'to say that it is tundra today and should be tundra forever when tundra has no economic value doesn't make sense.' " Or listen to a speech in Fairbanks by Ted Stevens, who was appointed to the U.S. Senate in 1968 by then Governor Hickel of Alaska. Stevens is said to have attacked ecologists who came to Alaska to discuss the oil development as "carpetbag conservationists." He referred to the dictionary definition of ecology as dealing with the relationships among living organisms, and then gave ecology a brave new dimension with this triumphant declaration: "But there are no living organisms on the North Slope." His audience was the American Association for the Advancement of Science.

Delegations of businessmen, unintimidated by the prospect of social dislocation or environmental havoc, visited Hickel in 1970 to urge him to get the show rolling. It wasn't expected to take much urging. Weisberg says it wasn't altogether luck that brought Hickel to his crucial Cabinet post where key decisions are made on oil; indeed, Weisberg holds, the oil interests handpicked Hickel for his sympathies and familiarity with their Alaskan problems.

> It is generally accepted in Washington that Atlantic Richfield's chairman, Robert Anderson (who . . . encouraged the opening of Alaskan lands to private development) was most responsible for President Nixon's appointment of Hickel to Interior. Certainly Hickel, with his celebrated oil connections, and his financial interests in the copper of the Brooks Mountain Range and the Yukon River Delta—another potential oil reserve—was not appointed for objectivity, nor for public relations finesse on conservation, given his plans to "build a Fifth Avenue on the Tundra."

Hickel hardly seemed the perfect choice as Secretary of the Interior—a department that regulates much of the nation's natural resources: public lands, oil and mineral leasing, wildlife, and the national parks. It also had jurisdiction over the pollution in the nation's waters. Hickel's philosophy was to subordinate conservation to development. The apprehension of conservationists and environmentalists seemed well founded, *The New York Times* stated. In the weeks before his confirmation hearing, Hickel produced a number of what became known as "Hickelisms." He said he was opposed to "conservation for conservation's sake." He said he feared that "if you set water [quality] standards too high, you might hinder industrial development."

And he observed that "a tree looking at a tree really doesn't do anything, but a person looking at a tree means something."

This was the man President Nixon picked to guard the environment. It appeared that Hickel's only qualifications for the office were political services rendered to Nixon, said the *Times*. But Hickel turned out to be a surprise. He had many tough decisions to make and often he ruled on the side of conservation, even cracking down on the oil industry after the spillages and blowouts in Santa Barbara and the Gulf of Mexico. The latter involved such outrageous violations that Hickel brought charges that led to $500,000 in fines against three major producers—a blow that must have shocked and rankled.

It was not long after Hickel's crackdown on oil and mercury polluters, and conservation orders that struck at various industries, that he was fired. Certainly he had angered Nixon by criticizing him for alienating American youth with his policies. The President summoned him to the White House and, without warning, told him he was through. Nixon said it was over the issue of "confidence." Others offered a different explanation. "Some observers believed he may have turned out to be a more vigorous conservationist than the White House had bargained for," commented the *Times*.

Following Hickel's ouster, in November, 1970, a "White House aide in his twenties" was sent to Interior to fire a half-dozen Hickel aides. Half were professional career conservationists. They were ordered "out of the building by five o'clock." Among those fired was Dr. Leslie L. Glasgow, Assistant Secretary for Fish, Wildlife and Parks, a highly respected biologist on leave of absence from Louisiana State University. Dr. Glasgow said he had been pushed out of the Department of the Interior by political and business interests in a shakeup that represented "a definite step backwards for environment." He said that under the Nixon Administration the environment had deteriorated and that decisions were based "first on politics and second on the environment." Before the sun set on the purge, some of Hickel's recent conservation rulings were being reversed by the Nixon Administration—a ban on the importation of whale products to protect the endangered whales and a regulation barring commercial billboards on federal public lands. Both cancellations met with a public outcry and were later rescinded.

Hickel's firing was seen by the *Times* as "essentially a blow at conservation and only incidentally the settling of a personal ac-

count." Dr. Glasgow was a "mild-mannered scientist whose prime
offense, it seems, was to speak out against the oil companies that
were despoiling the coastal waters of his native Louisiana." A
Times letter writer offered an interesting view of Hickel's departure: "Now the one-eyed man has been banished from the kingdom of the blind."

Nixon replaced Hickel with Representative Rogers Morton,
chairman of the Republican National Committee. During Morton's early months in office he gave little indication of what his
policies would be toward conservation, but he did lift the suspension on exploratory drilling for 14 of the 72 oil and gas leases
in the Santa Barbara channel, to what the *Times* called "the intense nervousness of the region's inhabitants." (More new drilling is now recommended in the channel by the Department of
the Interior, although it concedes the possibility of "serious impacts on the environment" as a result of accidents.)

Was Hickel's ouster dictated—all or in part—by oil? Certainly
he had deeply offended the industry by cracking down on it following the offshore blowoffs and fires, and that $500,000 in fines
rankled. He had also dragged his feet in granting permission to
get out the North Slope oil. It would not be the first time oil
imposed its will on government. A half-century ago the muck-
rakers exposed Standard Oil's machinations in corrupting the
Harding Administration and shaping official policies. Today the
industry is not so crude. But it is no less efficient in getting what
it wants. The techniques of persuasion are more refined. No
longer must accommodation be sinister or surreptitious. Public
relations has made acceptable that which was once unacceptable. Private interests have been repackaged and sold as public
interests. Lawmakers no longer have to be bought off. Now they
accommodate voluntarily. The public, through stockholdings,
has a share in the profits and is no longer so queasy about where
they come from.

Oil's might is so vast, its profits so huge, that even its critics
speak of it with awe. The $20 billion a year oil industry "is fa-
mous for its unsurpassed political and economic power in
America," observes Weisberg. "Its lobbying muscle in Congress
is legendary." Oil enjoys the lowest effective tax rate of any U.S.
industry (7 percent for the 23 largest companies). It has liberal
depletion allowances, "a prime symbol of corporate privilege." It
has a handsome subsidy in the form of a government-granted
quota system; this so-called rigging-mechanism permits the sale

of foreign oil at the same price as domestic which costs about
twice as much to produce. The quota system was a gift to the
industry from another champion of corporate interests. President Dwight Eisenhower granted it in the sacred name of "National Security." The quota system bilks American consumers
out of $5 to $8 billion a year in excess prices, according to a
Nixon Presidential commission; the commission recommended
that the system be replaced by a less restrictive and more equitable tariff program. It wasn't.

There is no doubt that the oil industry will get its way in
Alaska. As Weisberg observes, "An industry that has been able
(notably through the Rockefeller-Standard Oil complex) to treat
the U.S. State Department as a subsidiary headquarters, and at
whose bidding America brings down sovereign governments (as
in Iran), should not expect to have much trouble making its way
in Alaska."

The price of "progress," as they call it, will be high. Already
the Eskimos have been cheated out of land promised them, just
as the Indians were cheated out of their land by our pioneers.
Many Alaskans fear for the future of their land. The frontier will
expand and swallow the earth's last remaining large wilderness.
They see greedy foreigners invading the land and leaving behind
undreamed-of ravages. Bleak landscapes dotted by oil rigs and
barren earth. The night made grotesque by flaming gases being
burned off in flaring orange torchlike flames. Wildlife reduced or
destroyed. The Eskimo further corrupted and estranged from his
culture by the comforts and artifices of the invaders. The crisp
arctic air will become dirtier, and streams will turn into cesspools like ours. No one can say what will happen to the great ice
packs that are keystones in the "great harmonies"—whether they
will be melted further by this new activity, whether they will
make the world ocean rise disastrously, what changes will come
in weather patterns in this frigid land where so much of the
world's weather is born. Everything will happen by chance and
not by design. There is no real policy, no real code for uniform
oil or mineral development. The people will have little say about
their fate; they will accept an illusion of control for real control.
There will be no real brake over what happens to the earth. Nor
is this accidental, according to Hickock: "Both industry and government are deliberately preventing the operation of a public
forum until after the important decisions are made." The problem, then, is not that our current situation results from no plan-

ning, says Weisberg, for clearly the oil companies have a very keen sense of plan and purpose. "It is rather that the plans which do exist are created and executed without public scrutiny or control."

THE UNFRIENDLY ATOM

In this skittish new atomic world we have created, all of us live on borrowed time. We have pointed at the Communist stronghold enough nuclear missiles to wipe them and ourselves out several times over. Their nuclear arsenal is similarly aimed at our heads. But we have the satisfaction of knowing that we can kill each Russian three times for each time they kill us. For the patriot this is a source of pride. For the people of the world it is a nuclear tinderbox. Each year, as the experts see it, as nuclear stockpiles go up, the odds against their not being used go down. Sanity is rarely an effective deterrent against self-destruction.

The damage from these warheads goes far beyond their threat of annihilation and monstrous cost. (The U.S. has spent in excess of $3.5 trillion on nuclear development—two thirds on weapons—between 1940 and 1969.) We have almost desensitized ourselves to the horror of mass death. We are no longer revolted by our own systematic brutalization by a defense establishment that is fed on paranoiac fear maintained by propaganda, not unlike a controlled nuclear reaction fueled by public monies. This insensitivity is part of our nuclear heritage, a progressive inability or unwillingness to feel, walling ourselves off psychologically from our deteriorating world.

The engineers, good technicians that they are, have not let the technology that produced the deadly bombs remain unused. They have evolved the "peaceful atom." Over two decades $15 billion has been spent on its development. But the real price has been a gradual, progressive poisoning of the environment. By the time the extent of the damage was even partly realized, the commitment was great. It was too late to back off. Instead an effort was made to silence critics with massive public relations activity that emphasized a nonexistent safety, as critics of the peaceful atom pointed out.

The primary use of the peaceful atom has been to produce nuclear power generators. These were supposed to help resolve the "energy crisis" that was created as new uses for electricity were found and prompted. Atomic energy was heralded as the miracle fuel. It would revolutionize the production of electricity. Atomic energy, unlike the fossil fuels, is unlimited, we were told. It came along just in time to save us as the fossil fuels were becoming "scarce" or prohibitively expensive. Electricity would be so abundant and cheap that it would not be worth the bother of metering. Such optimism was understandable. A single ounce of radium can provide as much heat as 10 tons of coal.

Several plants were rushed into operation. Of some 3,000 power stations in the nation today, 21 are nuclear-powered; 54 more are under construction, with orders for an additional 42. Those nuclear plants now operating have a combined generating capacity of around 63 million kilowatts—as much as the entire nation was using in 1955. Yet this represents only about 1 percent of the total output we are using today. By 1980 some 200 nuclear plants are expected to supply one third of the nation's electricity.

The AEC projects the construction of 600 large nuclear power plants by the year 2000. No one knows with certainty how much radioactivity they will release or with what results, according to Senator Mike Gravel of Alaska. He points out that two AEC commissioners testified "that they cannot reduce the AEC's permissible effluent levels because they have no idea what the big plants will actually put out," according to the *Times:*

> If this country actually does build 600 nuclear power plants, those plants will produce each and every day about as much long-lived radioactivity as 500,000 Hiroshima bombs. If just one-tenth of 1 percent of it escaped into the environment annually, that would equal the contamination from 500 such bombs every year.

The public has never been adequately informed about the tremendously complex problems and risks of nuclear energy. Even experts disagree. The AEC has added to the confusion by presenting one side. The commissioners "invariably gloss over the hard questions," says Gravel. "We simply should not expect to hear more than the rosy side from the commissioners of an agency charged with a promotional mission."

Much of the optimism that underwrote nuclear power has proved unfounded. Nuclear power is in trouble today on many fronts. Several of the plants are out of operation because of technical difficulties. Others that were ordered have been canceled or postponed. Nuclear plants have proved more costly to build and operate than was anticipated. Operating costs have risen beyond those of conventional plants burning fossil fuels.

But there is a more compelling problem: the lack of safety and widespread damage to the environment. The peaceful atom is now recognized as a greater hazard to world health than bomb tests, according to a group of scientists, representing 15 countries, that reported to the United Nations in 1970. The primary health threats are said to be genetic damage and cancer due to radiation.

For several years there was a running debate over the safety of radiation standards. The AEC insisted the levels set were adequate to protect public health. Many experts challenged this. Among them were two maverick physicists who worked for the AEC itself, Dr. Arthur R. Tamplin and Dr. John W. Gofman. They said the permitted limit was ten times what it should be to protect people and could lead to 32,000 more cancer and leukemia deaths a year if everyone received the AEC approved "safe dose."

The Gofman-Tamplin findings were supported by many independent scientists, as well as others inside the AEC. For a long time the agency refused to revise levels of exposure. It said such revision would interfere with the development of the peaceful use of nuclear energy and raise costs substantially. Gofman and Tamplin charged that the AEC was more concerned with the economics of nuclear development than its health aspects. They charged the AEC with attempts to dissuade them from speaking out, of trying to discredit their work, and of being less than forthright. They said the AEC "persistently refused to make an honest estimate of cancer risk, simply because they have been assuring the electric power industry and other potential nuclear energy users that there is no significant hazard" from then-existing radiation guidelines. They said the government had access to the information on which they based their conclusions. If present standards were allowed to stand, they said, it "must be understood to mean that the government is willing to trade off thousands of cases of cancer and leukemia in return for peaceful atomic energy activities."

Others were even more outspoken in criticizing the AEC. Leo Goodman, an expert on atomic energy for the United Automobile Workers, said the atomic interests had a nervous, defensive attitude. "They feel their whole professional careers are at stake for the lies they have been telling over the years. That's the heart of the matter." Lieutenant Governor Mark Hogan of Colorado charged the AEC with "talking in ways meant to deceive." The more temperate said the clearest fact that emerged was the uncertainty, the experimental nature, of the nuclear program.

Historically, the AEC has not had good marks in anticipating and recognizing radiation dangers. Repeatedly it has had to lower its guidelines for permitted doses as new hazards were revealed. Usually these were uncovered not by government scientists but by independent researchers using government studies that had been the basis for safety claims. In practically every case the government resisted taking a more cautious stance to protect the public. This began with the postwar testing of nuclear bombs—a due-bill that still hangs over our heads. The government insisted that testing was safe. Then it got two nasty shocks: not only was the unanticipated fallout highly dangerous, it was contaminating the entire earth. This miscalculation is reflected in the radiostrontium in the bones forming in children during the massive bomb-testing of the 1950s. Those reckless episodes and their aftermath are recounted by U.N. science adviser Ritchie-Calder. "Every young person everywhere was affected, and why?" Because those responsible for H-bomb testing miscalculated:

> They assumed that the upthrust of the H-Bomb would punch a hole in the stratosphere and that the gaseous radioactivity would dissipate itself. One of those gases was radioactive krypton, which quickly decays into radiostrontium, which is a particulate. The technicians had been wrongly briefed about the nature of the troposphere, the climatic ceiling which would, they maintained, prevent the fallback. But between the equatorial troposphere and the polar troposphere there is a gap, and the radiostrontium came back through this fanlight into the climatic jet streams. It was swept all round the world to come to earth as radioactive rain, to be deposited on food crops and pastures, to be ingested by animals and to get into milk and into babies and children and adolescents whose growing bones were hungry for calcium or its equivalent, strontium, in this case radioactive. Incidentally, radiostrontium was known to the biologists before it "hit the headlines." They had found it in the skin burns of animals exposed on the Nevada testing ranges and they knew its sinister nature as a "bone-seeker." But the authorities clapped security on

their work, classified it as "Operation Sunshine" and cynically called the units of radiostrontium "Sunshine Units" —an instance not of ignorance but of deliberate noncommunication.

The problem of radiation was resolved in a familiar way: by setting standards of exposure—another expression of the "philosophy of poison." This was defended on the same grounds— that radiation is safe in small amounts. It is justified because radiation is found in nature. But critics of the program point out that because small amounts are natural, producing more artificially does not make it safe. It merely increases the hazard. It means that greater efforts should be made to keep public exposure to an absolute minimum to avoid unnecessary ill effects.

The so-called safe dose levels are further undermined by the treacherous nature of radiation. Like DDT, it is cumulative. The buildup in specific organs increases the likelihood of cancer or mutations. Seemingly insignificant environmental concentrations can be transferred quickly through the food chain to deadly levels.

The cell has a "memory" for each insult. Each exposure takes us a little farther along the road to catastrophe. No one now knows where the threshold is; indeed it differs with individuals and all the variables in their life and state of being: age, physical condition, sex, health history, etc. Radiation damage is considered irreversible. Biologists almost unanimously agree that there is no threshold below which damage does not take place. The greater the exposure the more damage. In laboratory experiments each added exposure increases the mutations. Some of the most extensive studies on low doses were done on fruit flies. They can withstand large jolts of radiation (in excess of 25,000 rads) and survive; mutation frequency increased after a radiation dose of only 5 rads (radiation absorbed dose). Plant studies show observable cell damage after as little as 0.3 rads, *Environment* reports. Human cells were observably damaged after an exposure of 5 rads, the lowest dose attempted. In studies of large populations of mice in Las Vegas it was found that those subjected to low-level radiation over a long period had half the lifespan of the controls; fertility was reduced by one third.

These findings have been confirmed in a larger laboratory where humans are the experimental animals. Women receiving low-level radiation in diagnostic studies during pregnancy gave birth to children with more mental and physical abnormalities than did women who had not been subjected to radiation. These abnormalities included mongolianism, leukemia, and disease in general. Curiously, there was a greater than normal variance in the male-female ratio in their grandchildren.

Scientists agree that at very low levels of exposure the risk of damage is small. But when large numbers of people are exposed the chances of one of them suffering injury grows. The risk becomes proportionately greater as the population increases and exposure from nuclear power plants and other sources increases. Each new stress, regardless of origin, adds to the submicroscopic, sublethal poisoning of the earth and its creatures. Ultimately it is reflected in bad health or premature death.

Little is really known about the atom. Its physical properties are better understood than its biological effects. While remaining invisible, it retains a terrifying reality. There is nothing abstract about a cancer or a deformed child. It asks too much of imagination to visualize the cataclysmic fury of unseen molecules under bombardment. They become deranged and smash into one another to create havoc in the orderly world of matter.

Different substances have different properties and powers, explains Sheldon Novick in *The Careless Atom*. Some can be stopped by a sheet of paper. Others can charge through several feet of wood, concrete, or even metals. Many easily invade flesh to lodge deep in the cell with profound adverse effects. The form the disruption takes is said to depend largely on which bodily cells are affected. Lodged in the thyroid gland the radioactive invader may produce cancer. In the gonads it can cause mutation and malformation in later generations. Atmospheric testing through 1962 brought about an estimated 4,800 defective births in the United States. The fetus is most vulnerable to the effects of radiation damage. The fallout from hydrogen bombs that got into milk was particularly hazardous for children.

Biologists repeatedly emphasize the fragility and incredible complexity of the cell, how easily it is disrupted in its normal functions. A radioactive particle piercing the cell is compared to a heavy bowling ball crashing into delicately balanced pins. But the atom's targets are the intricate life processes that we do not even begin to understand. The cell is a complex of interdependent processes, says Novick: "Disturbances of one process may resonate through the whole system, causing a chain of tiny disturbances in the cell which our knowledge is simply not yet adequate to trace."

A nuclear particle, upon smashing into the cell structure, not only can cause cancer and mutations but also has other profound effects. Among them is believed to be premature aging, which is to say, early death. This process is thought to be related to a series of strange events in which the cells attack themselves for unknown reasons. A variety of poisons, perhaps most, almost certainly all of them loose in the environment and acting in concert, can produce this same premature aging process that Dr. Harvey Wiley warned against more than a half century ago. Since radiation is known to produce mutations in all cells, Novick observes, it is reasonable that constant low levels cause repeated damage to the cells, and this hastens the complex process of change that leads to aging.

The process of splitting the atom is enormously complex. Briefly, the chain reaction is initiated in the reactor. There radioactive substances, under bombardment, are rendered unstable. In the process of changing or decaying from one form to another, they emit radiation, explains Novick. Radiation can be in the form of tiny particles, actually fragments of the atom, or as penetrating rays similar to x-rays. This radiation may pass through seemingly solid substances. Some particles, such as the heavy and slow-moving alpha rays, can penetrate only a thin sheet of paper; lighter beta rays may pass through several inches of wood or flesh; some penetrating gamma rays may pass through several feet of concrete.

Novick explains that when one of these particles or rays crashes through some material, it collides violently with atoms or molecules along the way:

> Such collisions are violent enough to tear away some of the electrons which form a cloud around every atom. Stripped of some of its electrons, the atom becomes a positively charged particle, or ion. Such a particle quickly undergoes violent chemical reactions with surrounding molecules as it returns to a neutral state. . . .
>
> Radioactive substances in food are incorporated into tissue just as their stable twins would be; imbedded deep within a cell, the radioactive atom may emit a particle which then streaks away, leaving a track of ions in its path. These atoms then undergo sudden and violent chemical combinations with surrounding molecules. In the delicately balanced economy of the cell, this sudden disruption can be disastrous. The individual cell may die; it may

recover. But if it does recover, its self-regulatory powers may be affected in some way we do not yet understand. After the passage of weeks, months or years, it may begin to proliferate wildly in the uncontrolled growth we call cancer.

Ultimately the radioactive elements (isotopes) produced break down. Some vanish in split seconds. Others are stable, lasting millions and even billions of years. Each has its own rate of decline, which cannot be altered or shortened. The danger of exposure lasts as long as the decaying process continues—that is, until a stable element is produced and the excess energy has been released. The rate of decay is measured by the substance's "half-life."

An element's half-life is the time required for its radioactivity to drop to one half the original value—that is, for half of the atoms of a radioactive substance to disintegrate. Cesium-137, for example, has a half-life of about 30 years, concentrating in the soft tissues. Strontium-90, considered one of the most dangerous radionuclides because it lodges in the skeleton and may cause bone cancer, has a half-life of almost 30 years. Man-made plutonium's half-life is 24,000 years. Iodine-131, which collects in the thyroid gland, has a half-life of about 8 days; but iodine-129, produced in reactors and also absorbed by the thyroid gland, has a half-life of 17,250,000 years. Others are even more persistent.

The term "half-life" can easily be misunderstood. Because a substance such as cesium-137 has a half-life of 30 years, this does not mean that the other half will obligingly disappear in another 30 years. Far from it. After a period corresponding to approximately ten times the half-life, the radioactivity of a substance drops to about 0.1 percent of its original value. Thus, a fraction of radioactive cesium-137 would linger even after 300 years. Much of the radiation produced by the bomb tests during the 1950s is still with us and will remain for centuries. Equally persistent and potentially as dangerous are the peaceful atoms released in "harmless" small amounts. Day by day, year by year, they contaminate the environment.

The hazard of the small daily dose is often overlooked because of the fear of a disastrous accident. This is true of most pollution. The insidious small amount of a contaminant is more dangerous than the rare spectacular incident that makes headlines. When nuclear power plants began to appear, many people had visions of Hiroshima-type explosions. The AEC could easily and correctly reassure them that this was virtually impossible. The greater danger is of an accident that releases enormous concentrations of radioactive materials that build up as a by-product of the fission process. Billions of curies of mixed isotopes are trapped in the thin stainless steel tubes that contain the uranium pellets—the fuel, explains David Pesonen in *Medical Opinion and Review*. "Should there be a loss of coolant and a 'meltdown,' even a modest leak in the surrounding structure would release this stew into the atmosphere."

Several mishaps have occurred in reactors. One was during a fire in the AEC weapons plant near Denver. The AEC at first denied that any plutonium had escaped the plant's boundaries, but investigators from the Colorado Committee for Environmental Information found evidence that soil seven miles away was contaminated with plutonium. The AEC then insisted that this posed no hazard. Independent scientists, however, pointed out that the dangerous plutonium could easily be stirred up by a strong wind and carried off to be inhaled by people in the area. Particles of the right size could remain active in the lung for "up to two years," emitting radiation that could damage tissue, according to the Colorado Council.

In a moment of reckless candor, in 1957, the AEC estimated the possible effects of an accident in a typical nuclear power plant. The study projected 3,400 killed, 43,000 injured, property damage and costs at astronomical levels. Agricultural restrictions alone on a 150,000-square-mile area might have to be placed at a cost of $4 billion, according to the report. That study was based on an accident in a plant about one tenth the size of the reactors now being built. A new study of potential damage due to an accident in a nuclear plant was made in 1964. It was never published. AEC critics charged that the undisclosed findings would have been even more damaging to the cause of nuclear energy, as ever larger plants were being built in more densely populated areas and the threat was becoming greater.

The possibility of accidents increases as our industrial society becomes progressively dependent on complex technology and beset by social unrest. There is the threat of mechanical breakdown and human failure, of sabotage, riots, failing airplanes and even earthquakes. Several of these potential dangers are listed in the *Disaster Handbook*, a disquieting little handy reference guide for the medical profession, which observes: "Contributory causes may be the over-optimistic statements and press releases of AEC officials on the safety of their operations."

The AEC has not had good marks on promoting safety. It is often accused of failure to enforce its own standards. Water radiation levels in Cattaraugus County, New York, are almost ten times above AEC limits, according to a New York State Public Health Department publication. Many scientists charge the agency with ignoring property safety precautions in developing a new "fast breeder" reactor designed to produce cheaper fuels which has a potential for fantastic damage. Some of them say these things do not belong on the face of the earth. In a highly publicized incident the AEC gave the Pacific Gas and Electric Company approval to construct a nuclear power plant at Bodega Head, California—a lonely promontory described as "the world's most treacherous earthquake fault." Only after a public outcry did the AEC begin to doubt the wisdom of the site, and the company reluctantly withdrew.

Insurance companies—the most sensitive barometer of safety —have not looked upon nuclear plants as a good risk. Originally they refused to take them as clients under any conditions. This forced the government, after putting out millions to develop the peaceful atom, to underwrite insurance coverage. But even government has limits. It would insure only to a limit of $500 million per accident—a pittance in view of the possible damage. This was an odd sort of free enterprise, says Novick, "the taxpayer assuming the risk and private industry accepting the profits." The government was confident it would have to remain the insurer for only ten years and in 1957 Congress passed legislation. In 1965 an effort was made to renew the insurance act for another 10 years, to 1977. Meanwhile private companies had been prodded by the Congressional Joint Committee on Atomic Energy, AEC's big brother in Congress, to form pools to provide coverage. Even then the firms took only token amounts spread paper-thin among many companies. Their timidity did more to underscore the risk of nuclear energy than to reassure its safety.

Neither doubt nor criticism has ever deterred the AEC. It has boldly continued to envision and promote magnificent triumphs over nature and its works. It fearlessly conducted underground explosions, loosening vast amounts of scarce natural gas at bargain rates. But there was a slight backfire: the explosions cost more than regular drilling procedures and the gas was tainted with radioactivity, which the AEC insists is not dangerous. There is, in addition, the possibility of artificially induced earthquakes.*

Atomic visionaries dream of ever more daring projects. AEC seriously examined the feasibility of exploding a string of 250 nuclear bombs to dig a sea-level Panama Canal. The blasts would have exceeded 250 megatons or more in energy, a power greater than all the nuclear bombs ever tested, a force equal to two or three San Francisco earthquakes, a cataclysmic thrust that would have sent clouds of radioactive dirt spewing 40,000 feet into the atmosphere. This was justified on grounds that it was cheaper than conventional digging methods. The scheme was abandoned only after opponents pointed out that radioactive debris and worldwide fallout would violate international agreements, that radiation remaining in the great cavities would be washed out by waters flowing into the sea, and there would be unknown biological and mechanical effects of joining the Atlantic and Pacific in an unbroken sea-level canal. And who was to shed tears for the tens of thousands of Central American Indians who would have to be evacuated from jungle villages and coastal fishing hamlets for two years, or possibly forever?

Even this is not the outer limit of nuclear vision. One of the AEC's "wilder schemes" is recalled by Dr. Peter Metzger, a bio-

---

* The arrogance of the AEC and public officials was displayed in the November, 1971, nuclear blast with the odd code name of Cannikin (meaning "little drinking vessel"). Although the atmospheric testing of nuclear weapons was largely banned by the Nuclear Test Ban Treaty of 1963, underground testing has gone on unabated. The latest and largest blast man ever directed against the earth was unleashed with the direct approval of President Nixon, a mile below the surface of Amchitka Island in the Aleutians, despite protests from all over the world. The five-megaton hydrogen bomb had an explosive power equivalent to the force of five million tons of TNT, 250 times greater than the bomb dropped on Hiroshima; it cost $200 million for a project that many scientists insisted had no military or defense value and was potentially disastrous for the environment. After the last rippling shudder, with no visible catastrophe, the AEC exulted that the blast was perfectly safe. The death of large numbers of animals was barely noted. The AEC's long-delayed environmental impact statement estimated that the explosion might kill 20 to 100 otters, a figure boosted to 20 to 240 at a later press briefing. The actual count was between 900 and 1,100, plus some birds and seals. The AEC suggested that the missing otters might have died in the Aleutian storm at the time of the blast and been swept to sea. Scientists hired by the AEC said there was no evidence to support that theory. The AEC, undeterred by previous experience, is now studying the use of nuclear shocks to loosen (for commercial use) natural gas deposits some 5,000 to 7,000 feet under Colorado and Wyoming.

chemist and president of the Colorado Committee for Environmental Information, in *The New York Times Magazine*. The plan, put forth in 1960, proposed utilizing nuclear explosions to generate electricity: *explosions*—not a controlled nuclear reaction. "It began with the explosion of a one-megaton nuclear bomb underground," says Metzger:

> Water would be pumped into the cavity formed by the blast and converted to steam by the intense heat held there. The steam would be pumped back up to the surface and into a conventional power plant for the generation of electricity. Multiple firings would follow, at the rate of a one-megaton blast every 10 days. . . .
>
> How a power plant, built directly above the point of a continuing series of massive earthquaking blasts, could survive was not explained. The steam, highly radioactive, would be recycled and thereby presumably contained. A steam leak in this power plant would be a catastrophe. No mention was made of how an intricate network of high-pressure steam lines could remain intact so close to a series of megaton-size nuclear explosions. Furthermore, the assumption was made that a blast every 10 days can take place in the same hole.

Fortunately, nothing came of the scheme, says Metzger: "The depressing thing is that it could have been proposed before the Joint Committee on Atomic Energy—and listened to without challenge."

Another bold scientist stumped for exploding a peaceful nuclear bomb on the moon to study the reaction. He was restrained by more reasonable men who flinched from such irresponsible experimentation, and others who objected that such a blast would ruin their own lunar studies. Long ago our nuclear pioneers fired a bomb into the Van Allen Belt following its discovery in 1957–58—another piece of stupidity that shows how much we are at the mercy of ignorant men pretending to be knowledgeable, says Ritchie-Calder:

"Every eminent scientist in the field of cosmology, radio astronomy or physics of the atmosphere protested at this irresponsible tampering with a system which we did not understand." They were unheeded. An American statesman glibly dismissed the experiment, designed to see if a brilliant pyrotechnical display, an artificial aurora, could be produced. He proclaimed that Van Allen had a right to set off a bomb in the belt. "It's his." The

British Prime Minister of the day, called upon in Commons to intervene with the reckless Americans, casually asked what all the fuss was about. "After all they haven't known the Van Allen Belt even existed a year before," observes Ritchie-Calder. The Americans joyfully gave the experiment a name, "Rainbow Bomb," and went ahead: "They exploded the bomb. They got their pyrotechnics and we still do not know the cost we may have to pay for this artificial magnetic disturbance."

A warning against such mindless experimentation comes from Metzger, who specializes in atomic radiation. He recently gave up teaching to campaign for sanity in the use of the atom. A risk accompanies every release of radioactive material, no matter how small or how peaceful, he cautions. Scientists are not agreed on the degree of risk; therefore it is necessary to evaluate whether the potential benefits compensate for the risk. "Such an evaluation is not a scientific but a moral question," says Metzger, who puts the issue squarely where it belongs. It should be decided by those who must take the risk. But there is the rub: the public is denied its right to complete information about the risks it is being asked, or more often compelled, to take. Decisions are made in secret, observes Metzger. "The moral and political questions of whether the benefits provided by a new technology are worth the risks to be suffered by all can no longer be decided in secret by a technological elite," says Metzger. "They must be decided by all." The ability to ascertain a few scientific facts does not guarantee the wisdom to determine public policy in using those facts.

The new morality does not yet exist. We are still not permitted to decide our own fate. We must accept official assurances that the risk is worth the benefits. Meanwhile we continue to take our daily dose of radioactivity. One California nuclear plant alone released 900,000 curies of radioactive gases into the air in 1967 and virtually the same amount the following year, according to the Public Health Service. The amounts accounted for about two thirds of the AEC's then permissible limit.

But what happens to the radioactive gases so freely discharged by the California plant and all its nuclear companions? We now know that they take to the winds, much in the manner of DDT, fossil fuel wastes, and other pollutants. They all merge and interact. But the radioactive isotopes lend their unique characteristics to the poisons and substances with which they collide, spreading their infection. Soon they circumnavigate the world, disintegrating, ionizing, and lodging in our tissues to detonate in silent biological explosions, writing a new chapter in the great human experiment that we are conducting on ourselves.

Radionuclides are also discharged in water used to cool nuclear plants. As usual the water is counted on to dilute the poison to harmless amounts. But dilution is no cure for pollution, notes Metzger. "It merely spreads a lower dose to more people." The cumulative nature of radiation makes sport of theories of dilution. Animals can accumulate these compounds in amounts thousands of times greater than the equivalent measurable level in the surrounding water or air. Such organisms become biological time bombs, in the same way as those that eat and magnify DDT in their cells. Or they may contain both radionuclides and DDT, plus some mercury, arsenic, and other dangerous compounds.

A study of the Columbia River in Washington illustrates how low concentrations of radionuclides can build up. The radioactive wastes came from a nearby reactor. Microscopic plants and animals (plankton) had 2,000 times as much radiation as the surrounding water. Insects concentrated 350,000 times the amount. Young birds fed by their parents had concentrations of 500,000. The egg yolks of water birds soared to nearly a million.

At the end of the food chain stands man—a repository of pesticides in his fat, heavy metals in various organs, asbestos in his lungs, a host of transient and permanent pollutants in his blood, respiratory tract, liver, and other organs of detoxification. In addition, Novick observes, all of us carry varying amounts of strontium in our bones, cesium in our muscles, carbon throughout our body tissues. No one escapes. Even the Eskimos in remote Antarctica often have greater concentrations of radioactive materials in their flesh than people in more temperate regions. Each of us is a walking chemical factory. We are embarked on a path from which there is no turning back. There is no expelling these things. We have cast our lot and can only await the results.

From the beginning of the peaceful nuclear program, radioactive wastes have been dumped into bodies of water with no knowledge of biological effects. The few studies made were not undertaken until years later. One radioactive sink eventually studied was White Oak Lake at Oak Ridge, Tennessee. For many years it provided the diet for many creatures, inhabitants of the lake as well as insects and migratory birds that carried its radioactive contents elsewhere. "The highly diluted wastes dumped into the water were being reconcentrated and then neatly packaged and dispatched all over the continent," says Novick. Eventually the offending lake was drained. But what happened to the radioactive chemicals before and after the water was emptied? No one knows. Many are on an eternal journey, as far as we are concerned. An alarming travelogue of a freed radioactive fraction is given by Novick:

> A given atom of radioactive cesium-137 may have passed from water to plankton to fish to bird, been transported thousands of miles; the bird may have been shot down and eaten. This would not be the end of the cesium-137's career. After being retained in the muscle tissue of the man who had eaten the duck, perhaps for days or weeks, the cesium-137 atom would be excreted, picked up once again by plankton in a river, and perhaps pass again through fish, bird and men. Endless other paths might be followed. From White Oak Lake it might have been carried by an insect for a mile and then with the death of the insect become part of the soil organic material which would cycle through plants and back to the soil for years or decades, until perhaps being ingested by another insect as part of a plant leaf and carried elsewhere, picked up by a bird and finally appearing in the human food supply a generation after its creation. Much of the lake water may have passed to the sea before entering the biological community.

The ubiquitous atoms must ultimately reach the suffering sea. "Every living thing on and under the sea is being poisoned with radioactive wastes," Dr. Jerold M. Lowenstein, a specialist in nuclear medicine at the University of California Medical Center, told a 1970 Malta Conference of atomic experts studying the problems of the sea. A Soviet scientist reported that amounts of radioactivity in the Black Sea are high enough to cause embryo fishes to develop deformed backbones, and radiation effects have been found in many other marine organisms, according to *The New York Times*. The experts called "fervently for international controls on radioactive wastes."

They pointed to the increasing number of nuclear plants being built throughout the world and the accompanying hazard for marine life. They noted that many of the new installations lack waste-discharge controls. They recalled the 1963 sinking of the United States nuclear submarine *Thresher* in the North Sea and the large amount of radiation it released. They noted the armada of nuclear-powered ships and submarines now being built and the probability of many accidents and ever greater contamination from them. They warned against the practice of dumping solid radioactive wastes into international waters in sealed containers, because of the accompanying "perils for deep-sea trawlers." Finally, the experts said, there are no international controls to stop the present continuing poisoning of the sea. Unhappily, no agreement was reached on how such controls could be effected. Too often experts are better at conducting postmortems on disasters than preventing them.

In one respect the radiation hazard is not unlike other massive assaults against the sea. It too is directed primarily toward the coastal waters and estuaries that contain most of the sea life. Consolidated Edison in New York and the New Jersey Public Service Electric and Gas Corporation have both proposed building vast nuclear power stations in coastal waters. Charles F. Luce, board chairman of Con Ed, said power demands were expected to grow so much by the 1980s that the utility would have to use the Atlantic Ocean as a site for new generating stations because Con Ed would have used up all the cooling water available on inland sites by the end of the decade. Con Ed consultants proposed that the plants be built on giant concrete barges that would be towed 10 miles offshore and either anchored there or sunk in place, with the generating plants linked by cable to the shore. Earlier Con Ed proposed building the plants on man-made islands. The New Jersey Public Service Electric and Gas Corporation announced that it is considering construction of at least two nuclear-powered generating stations off the Jersey coast within the three-mile limit. The plants would be built on massive concrete and steel barges, rising 175 feet above sea level and anchored to the sea floor.

Builders of nuclear plants almost invariably seek out some body of water that can be used as a cheap coolant. Water from the outside is drawn into the plant to cool the reactor, which produces fantastic heat. When the water is returned to the lake, stream or estuary, it carries the plant's heat with it, along with particles of low-level radioactivity. Water from a proposed reactor on Lake Michigan would be 28 degrees warmer than when it was taken in.

Thermal pollution is not limited to nuclear power plants. Enormous amounts of heat are released by conventional fossil fuel plants as well—all of it vented destructively into water or air. But a typical nuclear plant uses half again more water than a comparable coal-burning operation. Such heavy use of water becomes a threat in itself. The power-generating industry is the nation's heaviest user of water. Almost half the total amount used in the United States is for industrial cooling. By the year 2000 the need will equal two thirds of the nation's natural daily runoff of 1,200 billion gallons. That means the present pollution of half our water will soon rise to two thirds. In densely populated areas 100 percent of the water may be used for cooling power plants.

Thermal discharges are usually dramatized in the mass media by pictures of fish kills. But a dead fish tells only its own story of ultimate catastrophe. It does not reveal the slow, steady degradation that leads to death, what scientist Peter Koringa calls "the general trends, the stealthy deterioration of environmental conditions in sections of the sea of vital importance for its living resources."

The effects of thermal discharges are dependent on many variables: the location and size of the receiving water, the flushing action, the season. Heat affects the physical properties of water —density, viscosity, vapor pressure, the solubility of dissolved gases; it consequently influences the settling of particulate matter, stratification, circulation, and evaporation (which affects weather).

Raising water temperatures strains an entire ecosystem. An increase of 10 to 30 degrees can be calamitous. Only 3 or 4 degrees can have serious effects on fish and other aquatic life. Higher temperature decreases the ability of water to dissolve oxygen, on which most aquatic organisms depend for life. Temperature regulates metabolism, growth, reproduction, and other physiological processes. It can cause fish eggs to hatch prematurely in the spring before natural food organisms are available. It can stop certain fish eggs from incubating and slow down fish in their pursuit of food. It can kill normal vegetation and stimulate the wild production of algae to bring on oxygen depletion and eutrophication. It can disrupt signals for fish to migrate and spawn. It can interfere with the plant-fish-insect-bird relationship, between all organisms and their environment.

Heat affects the rate at which chemical reactions in water progress, and thus it can accelerate and heighten the effects of other pollutants. It can speed the formation of undesirable compounds or change dynamic chemical equilibriums. It can bring about synergistic actions—increasing the power of separate agents so the actions are greater together than the sum of their individual effects. It can increase the toxicity of some compounds and heighten the virulence of fish pathogens. Heat can affect other aquatic organisms, plants and animals, and the microorganisms that together form the pyramid of marine life. It can reduce some species and stimulate the growth of populations of others to nuisance conditions. The intense heat of the reactor can cook and kill microorganisms passing through the facility. In periods of low flow, in small bodies of water, or in heavily populated areas, the entire water supply may pass through a generating plant, soon rendering it almost sterile. Death alone becomes the stabilizing force. Heat can decrease water's recreational value for fishing, boating, picnicking, and scenic beauty, impinging on physical properties and aesthetic appreciation alike. Almost nothing is known about the effects of heat or radioactive substances in water. One of the greatest areas of present ignorance, according to *Environment,* is the nature of the threat of radiation to units of the food chain. The threat is not so simple as death. A mutation that would change the character of plankton in such a way as to alter its position in the food chain might be harmful to its role in the environment and to man, even though it increased the organism's ability to survive. Studies also indicate that higher temperatures may induce aquatic plants and animals to absorb radionuclides more rapidly from the water. Nor are biological effects of heat limited to water. The survival rate of hamsters exposed to radiation at higher temperatures was reduced by 50 percent.

Prudence has been almost nonexistent when it clashed with economic interests. Virtually every river in the nation now shows the effects of thermal pollution. The lakes are expended. Coastal waters are being affected. The culprit is not only nuclear plants but also those powered by conventional means. Two oil-fueled generating plants of the Florida Power and Light Company near Miami discharge about 10,000 gallons of water a second at 95 to 105 degrees temperature. Investigators say this can raise the temperature of adjoining Biscayne Bay to 100 degrees or more. A peak of 103 degrees in June, 1969, caused a major fish kill. The company's discharge exceeds both federal and state limits. Limits for coastal waters are a 4-degree increase in winter, 1½ degrees in summer. The company has been operating under a series of variances granted by Florida authorities. The utility was taken into court by the government in its first action to block thermal pollution.

In 1968, a year after the two plants went into full operation, the complaint stated, the thermal pollution had created a "barren area" of 300 acres. Within a year this had more than doubled. The situation is expected to become much worse when two nuclear reactors now being built are finished. To cool them will require a water withdrawal from the bay of 4.5 million gallons a minute. The water equivalent to all of Biscayne Bay "would pass through the plant and its works in less than a month," the Justice Department stated.

A federal judge subsequently denied the government request for an injunction against the utility, stating that the present warm-water discharge "causes some damage but it is minimal and retrievable." The suit was finally settled in September, 1971, with a compromise agreement under which the utility agreed, as part of the settlement, to construct a $30 million closed system of canals to handle cooling water from two 760-kilowatt nuclear generators then nearing completion at Turkey Point, about 35 miles south of Miami.

Several methods have been tried to resolve the problem of thermal pollution. All have disadvantages. Cooling towers discharge the heat into the atmosphere; in the process it can cause local weather changes—condensation, fogging, and icing of highways during winter months in northern latitudes. Closed-circuit systems reuse the same water, cooling it artificially; like cooling towers, they are very expensive and also ultimately release minute amounts of radiation. Artificial lakes and cooling ponds require huge tracts of land and leave exposed concentrations of radioactive wastes. The Florida Power and Light Company originally proposed to build a 6-mile canal 20 feet deep and 227 feet wide to let the water cool off before entering Biscayne Bay. But ecologists said the trench would cause severe ecological damage, and it would not relieve the basic problem of cooking to death plankton and other small sea life that passed through the plant. Nor would it remove the low-level radiation from the water. Several states have become concerned about these radiation discharges. Minnesota has attempted to impose stricter legal limits on them than the optimistic AEC standards. The state was promptly hauled into court by the Northern States Power Company. The utility is backed by the AEC. Several states have supported Minnesota in its determination to protect its people. "We have gone beyond the point where a deal can be struck with the forces that destroy our planet," says Governor William G. Milliken of Michigan: "We have gone beyond the point where we can justify further pollution for the sake of the economy. What's bad for the air, the water, the land, is bad for people—whether it's good for business or not. We cannot have prosperity at the expense of posterity."

Ultimately a federal judge ruled that Minnesota had no power to require stricter standards for radioactive pollution than those set by the AEC. He said Congress had given the agency this authority and only Congress could change it. Minnesota said it would appeal the decision. But this became unnecessary. In June, 1971, the AEC drastically cut the amount of radiation any nuclear plant could emit beyond its boundaries. The new limit was described as at least 50 times as stringent as that which it replaced, and seemed to justify the complaints of critics.

The following month the AEC's imperious posture was somewhat deflated in a landmark decision by a Federal Court of Appeals judge in Washington. He held, in effect, that the agency had failed to meet the intent of the National Environmental Policy Act. In the past AEC had insisted that its only obligation was the safety of nuclear plants and their radioactive emissions. The court ruling, involving the construction of a nuclear plant near Chesapeake Bay in Maryland, held that issues posed by non-radiological hazards were also the agency's concern and had to be settled before construction and operating permits could be granted. The AEC, to its credit, decided not to appeal the decision. This means it will now look into the question of thermal pollution as well as radiation damage, and review permits and licenses it granted earlier to 96 nuclear power plants. The court ruling plus the new safety regulations caused consternation within the nuclear industry.

During this period of upheaval Dr. Glenn Seaborg resigned as chairman of the AEC to be replaced by Dr. James R. Schlesinger. Two months after taking office, Dr. Schlesinger, in October, 1971, announced a new role for the AEC. No longer would it devote itself solely to promoting atomic energy but in the future would protect public interest in nuclear affairs—an abrupt shift that further upset the industry and left conservationists waiting, somewhat skeptically, to see what the effects would be.

Despite the long-needed reform promised by the AEC, the atom remains unconquered. It has been split and domesticated but still refuses to be tamed. Its by-products and wastes remain behind to bedevil us. Where uranium ore was mined in Colorado, giant piles of leftover tailings from uranium mills mock those who dug them. The miners have been cut down by high death rates from cancer. The tailings now bake in the sun. They wash away in storms to contaminate nearby streams. They leach into groundwaters. They blow away on the wind. The piles refuse to stabilize and are not likely to do so soon. Uranium has a half-life of 5.5 billion years. Radium incorporated into the mass of tailings has a half-life of 1,620 years.

Years ago the fine gray sand was given away free in Grand Junction, near Denver. Builders happily used it to level ground for pouring concrete slabs and backfilling around basements for homes, public buildings, offices, and stores. It was used in about 3,000 sites, but only 10 percent have been accounted for. The tailings, 8 to 12 inches thick beneath concrete slabs, give off a radioactive gas called radon, a product of radium decay. The radon seeps through the concrete. In some homes radon levels may be above those permitted in uranium mines, according to public health officials. Radon in unexpectedly small concentrations can cause lung cancer. It has cut down many men of Grand Junction in the past. Now the people there are not sure if they are next. They can only wait and see.

The tailings are just one hazard. There are also those "hot" radioactive liquid wastes from reactors to be disposed of. They are part of the servicing of nuclear energy plants. Fuel must be delivered to the reactors and the wastes removed, transferred, processed, reclaimed, and the leftovers put somewhere. Some of these fermenting nuclear wastes will remain active for hundreds of thousands of years. They are so deadly that a single gallon loosed into the environment could kill an estimated 2 to 3 million people. About 90 gallons of these murderous juices are now stored in the United States, in burial grounds euphemistically called "farms." One is above the Snake Plain Aquifer in Idado, and beneath it flows the vast underground reservoir that feeds rivers and streams through southern Idaho and the Pacific Northwest. The major one is at Hanford, Washington, in a tract covering 575 square miles.

Here, during World War II, enough plutonium was produced, Novick says, to "end the world in one incandescent flash." Now all but two of the nine reactors are silent. No one is allowed to live nearby. No one can trespass. It costs fortunes to maintain this nightmare with its ghastly possibilities. Listen to Ritchie-Calder:

> There in the 20th-century Giza, it has cost more, much more, to bury live atoms than it cost to entomb the sun-god kings of Egypt. The capital outlay runs into hundreds of millions of dollars and the maintenance of the U.S. sepulchers is more than $6 million a year (add to that the buried waste of the USSR, Britain, Canada, France and China, and one can see what it costs to bury live atoms). And they are very much alive. At Hanford they are kept in million-gallon carbon-steel tanks. Their radioactive vitality keeps the accompanying acids boiling like a witch's cauldron. A cooling system has to be maintained continuously. The vapors from the self-boiling tanks have to be condensed and "scrubbed" (radioactive atoms removed); otherwise a radioactive miasma would escape from the vents. The tanks will not endure as long as the pyramids and certainly not for the hundreds of thousands of years of the long-lived atoms. The acids and the atomic ferments erode the toughest metals, so the tanks have to be periodically decanted. Another method is to entomb them in disused salt mines. Another is to embed them in ceramics, lock them up in glass beads. Another is what is known as "hydraulic fraction": a hole is drilled into a shale formation (below the subsoil water); liquid is piped down under pressure and causes the shale to split laterally. Hence the atoms in liquid cement can be injected under enormous pressure and spread into the fissures to set like a radioactive sandwich.

There are said to be 140 enormous buried vaults, each capable of holding possibly 50 million gallons of the boiling wastes. In a single large tank there is as much radioactivity as in all the fallout produced by all the weapons tests since 1945, according to Novick. There is the nightmarish possibility of accidents: of leakage into water supplies, of gases escaping into the air. Some of the Hanford tanks have sprung leaks "of different degrees of severity." Only recently, long after the deadly casks were buried, the area was surveyed by geologists. Why at this late date? God knows. Geologists found the site subject to earthquakes, heavily laced with faults. They said an earthquake could occur directly beneath the reactors or under the storage vaults. There may have been one nearby as recently as 1918. "The waste storage tanks and plutonium-production reactors are not designed to withstand an earthquake in the immediate vicinity," states Novick.

take a man with a mule 160 hours to plant, cultivate, and harvest an acre of cotton. The same work is now done with as little as 13½ hours of human labor on the more efficient farms, according to a *New York Times* dispatch. Farmers who plowed one row at a time at the end of World War II "now ride mammoth tractors and cultivators and sweep across the fields at five miles an hour, turning the earth six and even eight rows at a time—and shooting weed-killing poisons into the soil as they go."

The new chemical-mechanical technology brought a new kind of revolution to the land and its people. The effects are even more profound than those following the Industrial Revolution, and the price is even higher. Today we pay for our prosperity with universal biological impoverishment, a brutalization of all nature. A poignant picture of the effects of growing cotton in Mississippi today was portrayed in the *Times* in December, 1969:

> This is the season in the cotton country when the air turns clear and sweet and when the fish sicken and die. The chemical pesticides that hung heavy in the air and mixed with the rich black soil during the growing season are now being washed into the bayous, lakes and drainage ditches by the fall rains. People can breathe deeply again. But here and there in the sluggish water lie the bodies of fish, their white bellies turned up as they float among the cypress trees, victims of the chemicals that make the cotton prosper.

Mississippi closed two major lakes to commercial fishing in 1971 following an FDA report that fish from the lakes contained pesticides as much as three times above the danger limit. Fish tested from Lake Mossy and Lake Wolfe ranged from 10 to 79 parts per million of pesticide. The Game and Fish Commission attributed the pesticide pollution of the two lakes to misuse of farm chemicals and to drainage ditches the United States Soil Conservation Service dug for nearby farmlands. The commission said four other state lakes were near the danger level due to pollution by DDT and other pesticides and might soon have to be closed down.

This new technology, often underwritten with poisons and almost inevitably by brutalization of the land, has brought broad social and cultural changes.

Now it is cheaper to use machines and pesticides than pay labor. In 1960 Mississippi cotton farmers spent $46.80 an acre for labor and $13.50 an acre for pesticides and chemicals, the *Times* reports. In 1967, when the minimum wage law took effect, the figures were reversed: farmers cut their labor cost to $13.50 an acre and raised the expenditure for chemicals and pesticides to $24.00 an acre. Thousands of workers were no longer needed on the farms where they and their parents had spent their lives. Many migrated to the cities, particularly in the North. Many had no urban occupation and had to go on welfare to exist. Those who remained behind on the farms complain that the chemicals, spread ten to twenty times a season, make them sick. The land has been so damaged that it is no longer as productive as it once was. Many bugs have developed resistance. New pests have emerged. A few years ago the boll weevil was a problem for cotton and the bollworm was not, said an agricultural official. But the chemicals that killed the weevil and brought it under control also killed the insects that preyed on the bollworm, permitting the worm to flourish. "So you create a whole new can of worms, so to speak," he said.

Biologists fear that the same chemicals that are causing resistance and death may also be bringing about cellular changes that could result in cancer and mutations that could become apparent too late. Scientists have warned that pollution is not only shortening man's life but also leading to genetic defects. Cases are being documented in which a variety of fetal defects are believed to be produced by environmental problems, Dr. Cecil B. Jacobson, chief of the reproductive genetic unit at the George Washington University School of Medicine, recently told a symposium of physicians. The *Times* reported:

> He listed ionizing radiation, viral infections, chemicals and drugs, environmental contaminants and aging as mutagenic factors. He believes that older parents produce more defective offspring because a person is aging in a mutagenic society . . . capable of producing mutations.

The problem is potentially disastrous in view of 80 billion pounds of synthetic chemicals now produced annually. One way or another all of it enters our living environment.

For a long time scientists have been trying to link cause and effect in assessing pollution. It was long assumed that pollution was directly related to population growth and consumption. But three scientists recently pinpointed technological processes as the primary culprit. The study was made by Dr. Barry Commoner, Washington University ecologist; Michael Corr, executive secretary of the American Association for the Advancement of Science's Committee on Environmental Alterations; and Paul Stamler, a researcher for the committee.

Their study, reported in *Environment*, noted that from 1946 to 1968 the nation's population increased 48 percent. But various measurable pollutions, such as algae in Lake Erie, bacteria in New York Harbor, and smog in numerous cities, shot up from 200 percent to 1,000 percent. In the same period the gross national product increased only 59 percent; it and other barometers of general consumption were not enough to correlate with massively increased pollution.

The major factor, according to the researchers, appeared to be certain types of production: Consumption of plastics, which have various adverse environmental effects, rose in the 1946–68 period by 1,024 percent. Mercury use in some major industrial processes increased 2,150 percent. Consumption of synthetic organic chemicals went up 495 percent, detergents 300 percent, electric power (a primary source of air pollution) 276 percent. The article stated:

> Synthetic organic chemicals . . . wood pulp and paper products, total production of energy, total horsepower of prime movers . . . cement, aluminum, mercury used for chlorine production, petroleum products—this particular group of production activities may well be responsible for the observed major changes in pollution levels.

Even if populations stabilize, with stable consumption patterns, the nation would still face increasing environmental problems if the environmental impact of production continues to increase, the scientists concluded.

This, then, is where technology has brought us. After more than a century of "progress," we find ourselves threatened with extinction from the same triumphs that once held such promise. Our entire life-support system is under enormous stress and threatens to break down altogether. The machine that brought us this far threatens to become our hearse. We find ourselves in an unanticipated dilemma. The technology we prize so highly is essentially in conflict with nature. Instead of enriching us it is destroying us. "Since even the smallest mechanical process consumes more energy than it produces, how could the sum of all these processes create abundances?" asks Friederich Georg Juenger.

The machine must, in the end, make man poorer rather than richer in real wealth. That is its nature.

In the process of promoting business, human welfare was almost completely overlooked. What was good for people was less important than what was good for business. The criterion of good business was reflected in "growth." The Gross National Product became not just a measure of commerce but the new American religion. Until this is understood American society makes little sense. An SST is irrational if measured against human values. It makes sense, of a kind, if evaluated in terms of the economy. So do most other environmental degradations. This dedication to business and growth demands a terrible price.*

---

* Growth has been used to resolve all social ills, especially unemployment. It has been particularly destructive in small densely populated developing countries. Puerto Rico, once an island paradise, is cited as a case in point. Two professors at the University of Puerto Rico, in a letter to *The New York Times* (February, 1972) note that since the late fifties industrialization has been encouraged there by offering the lowest production costs in the United States, plus tax incentives. They say the hope of curing unemployment with more growth "has thrown our strategists into a literally suicidal race. . . . We are destroying our small tropical environment in this race." The professors state:

"Already there are three large and continuously expanding petrochemical plants on the southern coast and a proposal for a fourth is under study. Mining companies are only waiting for the availability of electrical power in order to establish a copper mining complex. Mining operations would then be extended to include other mineral deposits. . . . Strip mining and large mechanized pits will be used in the mountainous center of the island.

"The copper refining plant will be located on the southern coast next to an aluminum complex. Electrical power is to be provided by several fossil fuel plants and at least two nuclear reactors dispersed along the coast. During the decade these sources will . . . add about 300,000 tons a year of sulfur dioxide to the atmosphere.

"A group of industries, for the most part pharmaceutical houses, have been recently established on the north center coast upon an uncontained limestone aquifer, one of the two major aquifers on the island. These companies, which are producing mainly antibiotics, are using wholly inadequate disposal methods such as injection wells, irrigation and sink holes.

"Heavy pollution of air, land and waters is therefore steadily increasing and grave environmental and public health problems are already present. An extremely small and densely populated tropical island has very limited ecological compensatory mechanisms to cope with these problems. . . . It is clear to us that a different alternative to exponential economic growth must be sought as a means of dealing effectively with our social, economic and environmental problems. . . ."

Ironically, despite all this growth and environmental destruction, the increasing danger of the toxic wastes contaminating the ground-water supplies, and of creeping into a creek that feeds the Delaware River. This meant, in effect, that the contamination that once flowed into the Hudson was simply being transferred to the Delaware River. The problem had not been resolved but merely moved from one place to another.

The ecological problems generated by our dedication to growth, with the technology it entails, are monumental. In simplest terms, ecologists see the world as energy. Energy powers the earth's great forces, and is made available to man, primarily, through green growth; this growth, in turn, supplies the food we eat and the oxygen we breathe. All matter represents a concentration of energy. Our food, clothes, houses are forms of energy; bodily movements are an expenditure of energy. As matter is broken down into other forms we are dissipating stored energy. This energy has been built up over millions of years and is the world's real capital; it is coal, oil, soil, trees, vegetation, and other resources. All of these are believed to have been formed as a result of the sun's rays beating on the earth. The more harmful cosmic rays are filtered out through the atmosphere. Any change in the atmosphere affects the amount of energy reaching the earth, and thus all of earth's life is affected. Some ecologists figure that approximately 10 percent of the energy reaching the earth can be consumed without harming the earth's capacity to sustain life and the life-support systems. But far more than this 10 percent is now being used. We are "overgrazing" our water and land pastures and poisoning and destroying the whole system of delicate balances that regulate the "great harmonies" we live by. This is the heart of the problem—the deficit financing and destruction of our life capital. This is what ecology is about.

Most technology strikes at the roots of ecology by burning up energy without regard to consequences. Ecology comes from the Greek word for "house" or "home." It is the study of the interrelationship between man and his environment—his world home. "When you pull on something you find everything else attached to it," says the ecologist. All over the world these strings are being pulled with no thought of the consequence. The complex, invisible web of life is being gradually unraveled as man's interference breaks nature's delicate threads.

The complexity of the web was demonstrated in Borneo. There the World Health Organization used DDT to kill malaria-carrying mosquitoes, recalls Dr. LaMont Cole. The DDT killed the mosquitoes but did not kill roaches which accumulated DDT in their bodies. The roaches were eaten by long-tailed lizards, called gechoes, which roam the walls and floors of tropical houses. DDT from the roaches reached the nervous systems of the lizards. The lizards slowed down and became less agile. Cats caught the lagging lizards and ate them. The cats died from the DDT in the lizards. Rats then started moving in from the forests, carrying with them the threat of an epidemic of plague. So cats were parachuted into the village to drive away the rats. The cats got rid of the rats. But then the roofs of the houses started to cave in. The lizards, now gone, it turned out had been eating caterpillars that made their meals from the roof thatching.

Everywhere it is the same. The stream is dammed for power production. This prevents pollutants from flushing away, the water warms, stagnation follows, matter decays, aquatic life dies; from the septic water rise foul odors and acid vapors. In the plains of India and the arid flatlands of the southwestern United States, irrigated land is often so saturated with salt deposits that it is worthless—the fate that overtook the great early civilizations in the Near East; the collapse of agriculture led to the ruin of empire.

The massive use of water today lowers water tables, depriving plant roots of water, killing vegetation, reducing transpiration, affecting local climate, further drying out soil, causing more dust to enter the atmosphere and affecting worldwide climate. Loosened soil washes into streams to suffocate aquatic life and change the quality of water. Bare soil speeds up water runoff from land, causing more erosion, further reducing groundwater levels, killing off more vegetation, and accelerating the cycle of destruction.

Egypt's huge Aswan Dam is a classic example of ecological backlash. It was to be the salvation of Egypt. It would provide hydroelectric power, water for irrigation, more acreage to feed the burgeoning population. Instead it is spreading disease with its irrigation waters and has damaged the once-fertile Nile delta by interfering with the ancient silting process as the outflowing water dumped its load of silt to continually enrich the land. Now it is necessary to use artificial fertilizer from factories powered by the dam so crops can be grown. Algae grow wildly in the newly formed Lake Nasser behind the dam, causing eutrophication and excessive evaporation. The once-rich delta farmlands have been damaged by the encroaching saltwater of the sea. A once-thriving fishing industry downstream is now ruined. The displaced villagers have migrated to Cairo and Alexandria, already hopelessly overcrowded. Any benefits from the dam in providing food were devoured long before it was finished by the exploding population. Recent studies, according to the *Jerusalem Post,* have revealed even more serious problems. Lake Nasser is not filling up as rapidly as expected owing to loss of water in the porous limestone soil; 25 percent of the power is being lost because of dust and soil on overhead wires. Silt in the turbines is a problem. Islands that fishermen were to moor their boats to are "overrun by scorpions and snakes" that were flooded out by the lake. There has been erosion of the river bed because of the lost silt, and now the water runs faster and clearer and is eroding the rock bed. To "correct" the situation, engineers are now proposing to build some ten new dams behind the Aswan.

In North America, Russia and Western Europe entire landscapes are being altered by mammoth engineering projects. Rivers are being rechanneled. Ecosystems as old as the earth are being destroyed. The effects are unknown. It is possible that the proposed diversion of great rivers in North America and Siberia could tilt the earth's balance and slow its rotation, says Dr. Raymond L. Nace of the U.S. Geological Survey. This is not likely to happen, he says, but it dramatizes how little thought is given to the consequences of great water projects; it also reflects his own uneasiness over what is being done.

Bold and mindless engineering projects go on with contemptuous disregard of nature. A recent lesson comes from Canada in the delta of the Peace and Athabasca Rivers in northern Alberta. In 1968 the huge Bennett Dam was completed on the Peace River in British Columbia. Scientists say it has brought a "catastrophic ecological change" in the 1,000-square-mile delta 600 miles below the dam. The people of the delta have sued British Columbia for damages that are expected to run into millions of dollars. Biologists had warned against the project. They were ignored by both officials and the public. The dam was built to supply water power and ensure an even year-round flow of water through the turbines. In a 1970 report entitled "Death of a Delta" a panel of scientists and engineers states: "What was once a fascinating and varied natural complex of lake and marsh is fast becoming a succession of isolated mudflats whose communicating streams are drying up."

Only now is there growing acceptance of the idea that a natu-

ral heavy spring runoff is necessary to replenish waterborne nutrients in the delta's lakes and streams to sustain the wildlife
that flourishes there. The delta was described as "an important
moulting area and staging area for the flights south" of a million
waterfowl. "Some 100,000 birds use the delta as a breeding
ground." Few areas of such importance to migrating waterfowl
exist anywhere in the northern parts of Canada, says a scientist.
Other damage is also reported by *The New York Times:*

> The incomes of some 1,300 Cree Indians and other delta
> people who depend on fishing and trapping have been re-
> duced, and the welfare rolls are high. Scientists have ex-
> pressed concern about the effects of declining water levels
> on food for muskrats, wood bison, ducks and geese.

What is the official reaction to these "catastrophic changes"?
Premier W. A. C. Bennett of British Columbia, who named the
dam for himself, was quoted in the *Edmonton Journal* of September 8, 1970, as having said of the delta (600 miles distant in
Alberta), "That is not my problem." Politics have left the people
who suffer the damage without support: "The Alberta Government, which, like British Columbia's, is organized by the Social
Credit party, has refused to support the suit against British Columbia, taking the position that what is happening to the delta is
a federal problem."

In the United States most of the vast and often-disastrous engineering projects are part of the pork-barrel operation of the
Army Corps of Engineers. The engineers build dams, dikes, canals, locks, lakes, ponds; they divert streams and rivers; they
level mountains, disrupt ecosystems, shatter the earth's crust,
and derange its harmonies for political patronage. During the
Great Depression of the 1930s, the Santee River was rechanneled into Charleston Bay to give employment to idle men. Engineers said the project would help the bay by improving flow and
navigation, and thus prevent silting. Instead it made matters
worse. Dredging now must go on full time and there is talk of
reverting the river to its original channel. But it is too late to
restore natural conditions. There is no recovering the millions of
dollars wasted. There is no return to ecological innocence.

The lessons of past mistakes are lost on the valiant engineers.
They continue to plot more campaigns in their unending war
against nature. The entire world becomes a vast challenge, a ball
of clay to reshape and remodel. The waters of the Atlantic and

Pacific will be joined—if not by nuclear bombs then by conventional methods. Long Island Sound, part of the earth's great encircling belt of water, would be enclosed as a vast freshwater lake. The Colorado River would be dammed, ruining much of the Grand Canyon. The mammoth California water plan would rechannel the waters of the state and alter the ecology of the entire Pacific Coast. San Francisco Bay is one-third filled and more is threatened. Jetports would be built in Jamaica Bay, off the Atlantic Coast; they would slash through the unique Pine Barrens of New Jersey and replace the Great Swamp near Morristown, New Jersey. The ocean bottom is about to be ripped open in a marine version of strip mining, an ecological Russian roulette to get at its mineral wealth. Giant icebergs would be floated from Antarctica into foreign temperate zones to provide fresh water and an experimental mixing of foreign organisms. Enormous iceballs would be fired through pneumatic tubes to Australia. The Everglades, that unique and exquisite river of grass, lives on borrowed time, spared from the jet only to be imperiled by stingy quotas of water released through man-made dikes, and the sea of grass awaits the ecological coup de grâce from the oil rig, the fatal spill that will render it just another legend.

One of the most ambitious and reckless plans of all, still in the talking stage, is called the North American Water and Power Alliance. It would take water from the Peace and Fraser Rivers in Canada, move them into the Columbia River System, then to the Central Valley of California, then south to the lower Colorado River basin.*

Grandiose schemes abound. In a materialistic industrialized society where technology is king and "growth" is God, conservation rarely wins the day. It merely stays the inevitable hour of its own execution.

THAT ARROGANT DESTROYER: THE AUTOMOBILE

The automobile has probably done more to distort our lives and degrade the environment than any other single force. It is the perfect example of technology gone amok—a form of cultural insanity.

Above all, the automobile is a creature of the oil well. Their

---

* Sending out for more water rather than wise use of the present supply is what one expert calls "the quick technological fix." It is still in great style in the United States, it is pointed out.

destinies are interwoven. Together they have made the world almost uninhabitable in their symbiotic assault on the environment. The oil well provides the auto with its fuel, lubrication,
and even the asphalt track it runs on. It makes possible plastic
fittings and parts; in a pinch it can even supply it with tires. The
automobile, it was pointed out earlier, produces some 60 percent
of our overall air pollution, up to 80 or 90 percent in some
places, along with multiple other pollutants already mentioned,
which include noise; autos are responsible for 75 percent of the
racket in cities.

Its sins as a polluter are only the beginning. The car has been
the great destroyer of our times—of physical and social environment alike. Nothing can match its effectiveness in disrupting
cultural patterns and social values. In exchange for poisoning
the air, atrophying our bodies and ravaging the earth, it has
given us mobility and speed. We have traveled far and fast, and
yet gone nowhere. We have even traveled backward. We wind up
with useless mobility, useless speed, and a massive sense of dislocation, disunity, and restlessness.

The offending internal combustion engine may be on its way
out. Congress is in the process of outlawing the sale of new cars
by 1975 if they do not eliminate 90 percent of the exhaust contaminants allowed in 1970 cars (which never met the standards
set for them). The deadline can be extended a year if necessary.
The measure was opposed by President Nixon. Detroit also
fought the legislation. It insists the schedule cannot be met. Industry's view is that any demands on it are unreasonable. It
wants to keep the internal combustion engine, tacking on some
antipollution device at an additional profit. And why not? It's
good business. But most experts agree that it won't be possible to
manipulate the old engine to meet the new standards. A new
type of engine must be developed, but little serious research has
been done.

The research that has been done was primarily to prove that
any departure from the trusty old polluter would be unfeasible to
operate or impossible to build. General Motors hastily knocked
out an unwieldy Rube Goldberg steam engine to demonstrate
steam's disadvantages. GM cannot altogether be faulted for its
dedication to the status quo. It is the nation's richest corporation.
Its gross annual sales are $24 billion (more than the gross na-
tional product of most member nations of the United Nations).

The automobile industry is never so ingenious as when it is

finding excuses to keep from spending money not directly related to profits. It has objected to every proposed alternative to the present engine: steam would be too heavy, too cumbersome, lacking power. Electric cars would be slow and require outsize batteries that would have to be recharged frequently (an argument reinforced by piously pointing out that air pollution would be increased by additional energy necessary to recharge the batteries). There have been other proposals. Each is met by vigorous objections. All boil down to one simple fact: the good old gas buggy is too profitable to get rid of.

Motorized transportation involves one out of every six businesses and one out of every five jobs in the United States. The steel and oil industries depend on it. So do construction and related industries such as building roads, freeways, parking lots and structures, service stations, repair shops, and homes for people in those occupations. The auto industry devours one fifth of the U.S. steel output, 60 percent of the country's rubber, one third of its glass, and large quantities of other raw materials. Repair and maintenance bills for passenger cars exceed $20 billion a year.

Car expenses rank third below housing and food in the typical family budget. The auto accounts for 20 percent of income. It costs the American about $88 a month simply to own a car. Approximately $36 billion was owed by Americans for their autos at the start of 1970. In addition there are operating costs. The average driver is putting out from $4.80 to $12 per hour as he tools down a highway. The typical car owner, buying a standard-sized automobile, signs a contract that will cost him $11,000 every 10 years. In brief, we are indentured to our autos, more slave than master.

The case against the automobile as an instrument of destruction is impressive. It is remarkable that it was ignored so long, a tribute to Detroit's ability to enchant. Only as recently as 1950 was the automobile finally pinpointed as the primary source of air pollution by Professor Arlie Haagen-Smit of the California Institute of Technology. His documentation that Los Angeles smog was indisputably caused by cars was bitterly contested by the automobile manufacturers. The manufacturers vigorously denied their culpability before being forced to admit guilt—and then they minimized it.

There have been proposals to ban the automobile altogether. Most are frivolous. We are too deeply committed to the auto in every aspect of our lives to abandon it. Some cities, however, are trying to limit cars. Tokyo, after being almost poisoned into extinction by monoxide fumes, banished all vehicles from its 122 busiest streets on Sundays, the major shopping day. Pollution immediately dropped 80 percent. Blue skies miraculously reappeared. New York City has experimented timidly with banning cars on a couple of Manhattan thoroughfares and prohibited vehicles from Central Park on weekends.

Simple arithmetic defines our problem. The automobile dilemma will get worse rather than better. Cars are proliferating at an alarming rate. We not only have a population explosion but also an automobile explosion. The number of cars, trucks and buses registered in this country is increasing twice as fast as the human population, according to John A. Volpe, Secretary of Transportation: "Every time the Census Bureau's clock ticks off a net gain of one in the population of the United States, there are two motor vehicles added to the nation's roads."

In the same time we added 2 million people during 1969, we added 4 million motor vehicles. Each car is a swallower of resources, a major poisoner of the environment. There are now 110 million. In 1950 there were only 49.1 million for a population of 152 million people. This means, *The New York Times* explains, that in 1950 there was a car, bus or truck for every three Americans. "Now that ratio is about one vehicle for every two Americans." California has the most with 12 million. Texas has 6.8 million. New York has 6.7 million, almost one third of them concentrated in New York City. New cars are now ground out at the rate of 22,000 a day, Detroit is shooting for 41,000 a day by the end of the decade. Since they are progressively more powerful, each one does more polluting. The highway slaughter takes almost 60,000 lives a year. More people have been killed in automobile accidents than in all the wars the United States ever participated in.

Almost 14 million U.S. families own two cars. Another 2 million own three or more. New roads to accommodate these new cars advance over meadow, field, marsh landfill, and through distressed neighborhoods (rarely do the planners displace or inconvenience the affluent). We lose a million acres of green every year to cars and their services. The creeping carpet of asphalt and pavement smothers the earth so it can no longer breathe. All these new roads mean freshwater lost in runoff—an estimated 335 billion gallons a year; by the year 2000 the loss is expected to be 1.5 trillion gallons—and that's a lot of freshwater lost to a drying earth. We achieve little if we cut automobile pollution, only to keep multiplying the number of cars and keep paving over more of the earth.

With suicidal ingenuity we have locked ourselves into this insanity with a self-perpetuating system. On every gallon of gas sold taxes are set aside for building roads. This amounts to about $5 billion a year. It is expected to reach $6.5 billion within a few years. The chain reaction from the fund is inevitable: the more cars sold, the more gas used, the more revenue raised, the more roads built, the more automobiles produced, the more gas consumed, the more pollution. Cars beget roads. Roads beget cars. The tax money goes into a sacred trust known as the Highway Trust Fund. Behind the fund is the powerful highway lobby—including the American Road Builders Association, Highway Users Federation for Safety and Mobility, American Association of State Highway Officials, and a strong faction within the Federal Highway Administration, bolstered by their grateful friends in Congress. Congress is so deeply concerned with highway construction that any restructuring of the Highway Trust Fund must clear at least eight committees. The fund is almost inviolate.

But at last the government has taken what it calls "steps to rescue mass transit." The Nixon Administration announced late in 1970 the formation of a new quasi-public corporation, National Rail Passenger Corporation, to operate a national rail network. At first nicknamed Railpax, it subsequently became known as Amtrak. From the beginning almost everything possible has been done to assure Amtrak's failure. The funding is niggardly—only $40 million plus government-backed loans up to $300 million, not enough to touch the problem. At the same time $17.3 billion was allocated to highway building and Nixon was fighting for $290 million to keep the SST alive. The new rail agency offers only limited service; it is less a network than a series of trunk lines—many major cities are not even included in the schedule and whole areas of the country are not provided with service. If Amtrak does not show a profit by 1973—hardly time enough to get new equipment into service—it can be scrapped.

Many critics argue that Amtrak is little more than a public relations gimmick by the Administration; its true purpose was to accommodate the railroads by killing their unprofitable passenger service altogether. *The New York Times* reported that when the bill to establish Amtrak was being drafted, "a high official of the Transportation Department said candidly in private: 'The purpose of this bill is to get passenger service off the backs of the railroads, run the wheels off the existing equipment, and then put an end to passenger trains in this country.'" The evidence supports him. Critics called the Nixon Administration Rail Passenger Service Act of 1970 the "railroad euthanasia bill."

The original plan was so bad that practically every railroad organization protested it, including the ICC which has long ceased to champion public interests over corporate interests. The proposal that called for a national rail network did not even provide for service in the Southwest and Far West. ". . . Yet the tentative plan disclosed late last year [1970] had no North-South service on the West Coast, and the Sunset Limited from New Orleans through the Southwest to Los Angeles was to be discontinued," the *Times* stated:

> Both these major routes are now operated by the Southern Pacific Railroad. It is an open secret in Washington that both were included in Mr. Volpe's original version but were knocked out at the White House level. It is also well-known that B. F. Biaggini, head of the Southern Pacific, is a big Republican campaign contributor. He does not want these major passenger routes included in the new network because he wants to clear these lines for more profitable freight operations.

By the time the plan went into effect in early 1971, it had been improved somewhat but still had severe shortcomings. Service was bad and equipment was often obsolete and uncomfortable, tending to drive passengers away as if fulfilling the original death wish. Most railroads are delighted to get the expensive passenger service off their back and keep only the profitable freight service. They have long forgotten that their right of ways are on lands given them by the people in exchange for performing a public service. For years official policy has promoted the automobile over the railroad. State and federal administrative bodies have consistently approved cuts in service. In 1929 there were 20,000 trains in the United States. By 1959 there were 1,500. By the end of 1969 the number had dwindled to 500. By 1970 it was 366. Under Amtrak's rescue of mass transit, only 150 trains are being operated. The roads have driven away passengers with wretched service, often a calculated tactic. The lack of passengers is then used as proof that people do not want trains and an excuse to further reduce service. Each reduction aggravates the highway problem. Official policy, in effect, mandates automobile pollution and the insanity of the automobile.

Every effort to divert any of the automobile fund's money to mass transportation is defeated.* For many years mass transportation has been starved almost to death by Congress. In fiscal 1970 the government spilled from its cornucopia Highway Fund a munificent $2.2 billion for urban highways. From the federal piggy bank reserved for desperately needed mass urban travel it shook loose for all programs—bus lines, subways, commuter lines (almost all losers)—a miserly $214 million, less than one tenth the highway expenditure, less than the $290 million sought for the SST that, at best, would serve only a handful of people.

The present allocation, to be spread over several years, could be spent in one year without making a dent in the problem. This imbalance makes no sense except to the automobile interests. A subway or train is a far more efficient way to move large numbers of people. One highway lane can handle 1,200 cars, or perhaps 2,000 passengers an hour, compared with 40,000 passengers an hour on a single railroad track. Trains are also less expensive to run, more efficient, and less damaging to the environment. It takes only a fourth as much thrust to move a railway car on steel rails as it does to move a rubber-tired vehicle on concrete; a modern train requires only about 15 relatively pollution-free horsepower per passenger to perhaps 10 times that for a pollution-belching auto, according to the *Times*.

Even if a pollution-free car is developed it will not solve the dilemma of the automobile. We will still have traffic jams. We will still lose greenery to creeping pavement. There will be fewer parks, trees and places for people to sit and rest. There will be more tension and less tranquility. There will be snarling cars and snarling people with nerves stretched beyond human endurance. More people means more cars. More cars means more fuel sold, more taxes for new roads, more highways to be filled with more cars, more pollution.

* One small victory for sanity came in 1971 when the Oregon Legislature approved a bill directing that at least 1 percent of all state highway funds be used for the construction of bicycle trails and footpaths. The builder will be the Oregon State Highway Division.

JETS AND THE SST—BREAKING THE
INSANITY BARRIER

Even the automobile is not the ultimate folly. That dark distinction is left to the SST—the Supersonic Transport. The SST is now dead in the United States—one of the real victories for sanity—but it is alive and flying in Europe and Russia, and there is no guarantee that it will not be revived in the United States, in the name of economics and patriotism. By now we know that pollution has no national loyalty. As long as the planes are flying anywhere in the world and the concepts behind them remain viable, their threat is with us.

The Supersonic Transport follows the First Law of Population Expansion: the more people assembled, the more services required, the less attention available per individual, and consequently the greater mass decline in comfort. To offset this disadvantage massive advertising and public relations are required. People must be taught to deny the reality of their senses. This begins by proclaiming the loss in human values and comfort as "progress." Poor service and discomfort are discounted by testimonials claiming that the bad is good and the mediocre is excellent. Petty services—movies, music, and gourmet plastic food—distract from the central insult imposed upon the passenger. This is ridiculously easy in a society where individuals have been increasingly conditioned by the mass media to distrust their own judgment and deny their own feelings. The seduction is reinforced by appeals to sex, romance, and other fantasies that beset a restless, discontented, overindustrialized, overstimulated people.

The SST left behind lessons that should not be ignored or forgotten. Rarely has a public relations campaign matched the intensity of that designed to sell the SST. Black was made to appear white. Doubt masqueraded as certainty. Vision was clouded by frantic appeals to mechanistic and monetary values and false patriotism. The chief salesman was the United States government. Usually we have the excuse of walking into our environmental disasters backward with our eyes closed. We realize the damage only after it has been done and we are deeply committed to error; then we must find a justification to continue. This time we did not have that excuse. Our leaders marched boldly forward with eyes open, bidding the people to follow.

The SST was sold as a symbol of our dedication to the idea
that what is allegedly good for the economy comes before what is
good for people. Even its zealous promoters never went quite so
far as to claim that the plane would be good for people in terms
of human values. They kept preaching the plane's economic role
and nationalistic implications—and even those allegations were
proved false.

The facts can be briefly stated. The SST would have carried
some 300 people in a tube almost as long as a football field (300
feet) at a cruising speed of 1,800 miles per hour—three times
that of the present jumbo jets and more than double the speed of
sound. In exchange for this the public was asked to spend huge
sums of money and accept environmental damage, health risks,
survival risks, and thunderous blasts of noise that could have a
shattering impact on nerves and cause property damage. The
case was stated succinctly by Representative Henry S. Reuss of
Wisconsin:

> We are being asked to spend $1.3 billion, or $4 billion be-
> fore we are through, on a plane that will serve only a mi-
> nute fraction of American taxpayers, while millions of oth-
> ers pay the penalty for this folly in the form of increased
> airport noise, sonic booms, air pollution, and potentially
> harmful weather changes.

Other warnings abounded. A Presidential SST committee recommended that the plane be dropped because it was economically wasteful and environmentally unsound. A congressional
subcommittee (of the Joint Economic Committee) urged that
the project be shelved because of "hidden cost burdens to both
government and consumers and because of imponderable hazards to the environment," reported *The New York Times*. Fifty
scientists, including some of the world's leading environmental
experts, urged that large-scale operation of supersonic transport
planes be delayed until serious questions about the potential for
environmental contamination could be answered. Russell Train
and Dr. Gordon J. F. Macdonald, of the President's White House
Council on Environmental Quality, warned against it. So did
others. All went unheeded. One of the most dedicated supporters
of the plane was President Nixon. Despite his advisers' adverse
findings and warnings, despite his professed environmental concern, despite his vetoes of public welfare programs in the name
of economy, the President repeatedly called for the expenditure

of millions and even billions to help favored private interests at the expense of the environment.

The plane had a unique history. Its development was paid for with federal funds. The project was initiated by President Kennedy with the pledge that the government investment not be permitted to exceed $750 million. The sum had already reached $737 million when President Nixon asked for another $290 million to continue the work. President Kennedy had promised that government participation would end with the prototype stage. It was believed that private industry would then take over. But even industry did not have enough faith in the project to invest in it. The government then proposed to go far beyond its original pledge; it would spend up to $4 billion more to complete the project. The government claimed it would recover its initial investment from royalties on the first 300 planes sold. But the planes were expected to cost more than $60 million each, with few purchasers in sight. A transportation official said the SST could bankrupt not only the airlines that buy it but also the aircraft companies building it.

The proponents of the plane offered two basic arguments: it was big and fast. They also said it would be necessary to American prestige and that it was essential to maintain a favorable balance of payments and without it America would lose its lead in aviation. These arguments were dismissed by experts as so much nonsense. One enthusiast argued that any aircraft that could reach any place on earth within twelve hours "is good for mankind, truly an instrument for peace." This was typical of the logic behind the SST. How would getting from one place to another a few hours faster lead to peace—unless the mission was to drop a few bombs on some scoundrels? And even then missiles could do the job faster.

Proponents dragged in the almost-inevitable red herring: Russia is developing a supersonic plane and would be ahead of the United States. England and France jointly have the experimental Concorde flying. It was largely overlooked that already the Concorde has run into trouble. The first 10 flights brought 578 complaints and 332 claims for damages. Many protests have been filed against its thunderous noise. People near French airports are filing legal actions. Classrooms reportedly have to stop lessons every few minutes. French cathedrals dating back to the Middle Ages are said to be suffering fissures and cracks from shock waves set off by the plane. Stones in the exquisite Strasbourg Cathedral, one of the heirlooms dating back to the eleventh century, are reported to have "split, flaked or burst" due to damage from air pollution and vibration from supersonic craft.

The most vigorous salesman for the SST was William M. Magruder, a former test pilot who became the program director. He expressed confidence in the giant aircraft and its environmental harmlessness. At the same time he conceded that atmospheric sciences in general are "primitive," and it is "impossible" now to calculate the environmental effect of the plane. This would seem reason enough to stop it until answers are available. Instead, Magruder proposed a $15 million research program over four years to find out the SST's effects on weather and climate; meanwhile development would proceed. Magruder said that if the problems proved insoluble and would cause intolerable environmental effects, he would recommend cancellation of the program. Nothing was said about all of the billions of dollars that would have gone down the drain by then. Nor was it explained why such basic research was not proposed until after four years and almost $700 million dollars had been spent on the program.

The entire SST program was full of mysteries. One was the longtime absence of a statement outlining the plane's environmental impact, as required under federal law. When the long-awaited statement was released, Senators Proxmire and Nelson charged that the Administration suppressed critical comments from the Department of Health, Education and Welfare.

A parade of scientists warned that it would release large amounts of water vapor into the stratosphere, possibly increasing the water content an estimated 50 to 100 percent. In the rarefied air of the stratosphere the contrails would not disperse as they would at lower levels; this could cause clouds that would act as an insulating shield to keep the sun's rays from reaching the earth. A secondary effect, warned the Presidential committee, was that "a significant increase in relative humidity would alter the radiation balance and thereby affect the general circulation of atmospheric components." That means adverse weather changes.

An even greater threat was to ozone. Scientists feared that the water vapor released into the stratosphere would undergo complex chemical changes that might destroy part of the ozone that encases the earth and thereby jeopardize continued life on this planet. Ozone exists ten to thirty miles above the ground; it isolates the thin, dry envelope of air that composes the stratosphere from the lower atmosphere, and it absorbs the sun's lethal ultraviolet rays. If ozone did not exist, the ultraviolet rays would blind humans and animals, burn land plants, and possibly make higher forms of life on the planet Earth impossible.

Dr. Harold Johnston of the University of California, a leading authority in atmospheric chemistry, said the SST was a threat to life because the nitric oxide in its exhaust would convert the ozone into oxygen gas which is not a complete shield against ultraviolet rays. Dr. Johnston estimated that a fleet of 500 SST's, operating an average of seven hours a day (which had been predicted by 1985) could reduce the ozone content of the atmosphere by half within less than a year. As a result, he said, "all animals of the world (except of course those that wore protective goggles) would be blinded if they lived out of doors during the daytime." He said this would occur even for those who never looked up, because atmospheric molecules strongly deflect, or scatter, ultraviolet rays, which during daylight would therefore strike the eyes from all directions.

Some of his associates at Berkeley believed the searing ultraviolet would also kill all plants except those underwater, explained Walter Sullivan in *The New York Times*. "Without land plants it would be difficult for land animals, including man, to survive." Others predicted that with less ozone protection against ultraviolet rays there would be an increase in skin cancer.

Scientists further cautioned that sulfur particles from the fuel could absorb enough heat to warm the stratosphere by as much as seven degrees centigrade. What would such a warming mean? No one has the faintest idea. But one scientist warned that "when you change something on a global basis you had better watch out."

Russell Train, chairman of the White House Council on Environmental Quality, pointed out that even on the ground the SST would add another hazard: subsonic speeds necessary for landing and takeoff would result in insufficient fuel combustion "with a resulting heavy discharge of pollutants into the atmosphere. Both atmospheric pollution and ground contamination seem likely to result." Train concluded that "justification for proceeding with the program is not now apparent."

The most obvious and dramatic hazard was ear-shattering sonic booms. The Presidential committee said the effects of the sonic boom would be "considered intolerable by a very high percentage of the people affected." It would be the ear-splitting equivalent of 50 jumbo subsonic jets taking off simultaneously, according to Dr. Richard L. Garwin, a leading physicist and long-time Presidential science adviser. He urged the government to immediately terminate all direct or indirect support of the SST. "Each flight, in saving a few hours for 200 or 300 people, will jolt every person, animal, building, and fragile archeological relic in a fifty-mile-wide path throughout its length," said Dr. Bernard D. Davis, a professor at Harvard Medical School.

Advocates gave assurances that the SST would never be used over heavily populated areas because of the noise. But Dr. Davis pointed out that they "strenuously resist" having this written into law. This is the inevitable tactic of the polluter: offer assurances that cannot be enforced. Then, when it is convenient or necessary, ignore them. By then it is too late. The public has adjusted to the initial abuse. Each subsequent loss can be inflicted with even less resistance. Step by step there is an erosion of standards. Each new degradation is barely discernible from the last. The important step is the first compromise. That's the one that corrupts; the others that follow are merely an extension. President Nixon, in a television appearance with Art Linkletter, told schoolchildren that "in 1980 you will go from Washington to Los Angeles in an hour and a half—that's how fast we'll be moving with the new planes that will be available then"—in the present technology that means the SST. And it means directly over land, despite the Administration's official position that the SST would never fly over land. Ultimately, the SST proponents, in desperation, promised legal guarantees that the SST would not fly over land. By then it was too late, and there was still no assurance that any law written could not be changed.

There were other disadvantages that helped cool off the airlines as prospective SST investors and buyers. It was found that the plane would require longer runways than were originally believed necessary. This would mean the added cost of new construction or new airports that would be even farther from cities. It would mean even longer delays at each end of the flight. Time saved in the air would be lost on the ground. The noise level would be three times that anticipated. The ban on land flights, however long it lasted, would make the planes of limited use to the airlines. The Environmental Defense Fund was suing for noise standards before the plane left the ground. SST proponents promised noise modifications. This meant more delays, higher costs; and there were questions about how much the noise could be cut.

Every disadvantage posed by the SST would be added to those already inflicted by commercial jets. The average arrival and departure leaves a trail of 141 pounds of particulate matter, oxides of nitrogen (some believe they may leak into the stratosphere and damage the protective envelope of ozone), hydrocarbons, noxious odors, and sulfur oxides—not an inconsiderable amount in view of 500 to 2,000 landings and takeoffs daily at the nation's larger jetports. In Los Angeles, jets produce as much pollution in a day as one million cars, according to estimates. The aviation industry insists its pollution is insignificant, amounting to only about 1 percent of all pollution. The percentage is small, the total is not; it amounts to 3 to 4 billion pounds annually. Further, the particles emitted are small; they would therefore tend to be easily airborne and to lodge deep in the lungs of those unlucky enough to inhale them.

In normal flight a jet emits about two and two-thirds tons of carbon dioxide and one-third ton of water vapor every 10 minutes—a considerable sum in total. Scientists warn that it may have certain effects on the radiation balance of the air and consequently on world climate. Every transatlantic flight also burns up about 30 tons of oxygen, with whatever effect that may have on atmospheric conditions.

Jets also have another adverse effect that only came to public attention recently. When flying they accumulate kerosene in small holding tanks. In landing or takeoff, pilots dump or blow this out of the tanks. The kerosene either lands on runways and makes them hazardously slick, or takes to the wind as another air pollutant. About 6,700 tons (2 million gallons) are dumped into the sky each year by domestic carriers. Washington National Airport alone is doused with an estimated 110 tons of runoff fuel a year, enough to have "significant effects on the health and welfare of people who live and work near the airport," according to Air Pollution Commissioner John T. Middleton. Some 500 gallons of the fuel are dumped needlessly over Miami daily in peak seasons, says former airline Captain William L. Gutherie. Captain Gutherie became concerned about the environmental and safety hazard of dumping kerosene. He celebrated his thirtieth anniversary as a pilot for Eastern Airlines by refusing to dump the fuel in the air over Miami. He was promptly fired. The airline issued a statement with a familiar ring: only a "small amount" of kerosene was involved and it was not a hazard. There is no present authority to make the airlines stop the practice of dumping kerosene.

Airplanes have taken us down a long primrose path. They began as simple conveyances that had a pleasant sound as they droned overhead. But the jet was another technological kettle of fish, with its intolerable whining roar. It was inflicted on us before we knew what was happening. It grew into the huge 707 and the massive 747 that can fly the population of a small town—and wipe it out in a single crash. Flying is now an unpleasant experience, all advertising and public relations to the contrary. A passenger is so much meat to be moved as swiftly and economically as possible. The SST as originally conceived would have taken us another step in that direction. Then, belatedly, drawbacks began to emerge. But the political and economic commitment was too great to back off, even when it became apparent that it was an environmental disaster. It was easier to minimize the damage and rationalize the risk.

Among those who stood to profit from the SST were the Boeing Aircraft Company, who manufactured the plane, and the General Electric Company, who was building the engine. Another was the state of Washington where Boeing is located. Washington is one of the few states with two senators who are both chairmen of influential standing Senate committees. Senator Henry M. Jackson, Jr., and Warren G. Magnuson, both Democrats, are among the most powerful men in the Senate. They battled hard for the SST, even making personal appeals among their colleagues for support. The President did not wish to offend such powerful Senators who can often make or break Administration policies, especially in a Congress controlled by the opposition party. President Nixon, whatever his failings, has never been accused of not being an adroit politician.

"Scoop" Jackson, as he's known to his friends in Washington, has been especially congenial with the Administration. He has supported Nixon's Vietnam policies, and he was permitted to introduce important environmental legislation that was joyfully signed into law by the President. Jackson declared himself against noise pollution while beating the bushes for the SST with its renowned sonic boom and other environmental liabilities. One of the tricks of good politics is to be able to compartmentalize yourself. The first rule of politics is expediency.

Jackson is known among his colleagues, "jocularly," says *The New York Times*, as "The Senator from Boeing." It was accepted that the prosperity of Seattle, which practically means the state of Washington, would suffer serious consequences if the SST program was canceled. This would not be good for Washington or Boeing or Senators Jackson or Magnuson. One solution to their dilemma was put forth by Representative Henry S. Reuss of Wisconsin: "Why in the name of common sense do we not put Boeing to work making a mass-transit vehicle, and GE [the co-contractor] to work producing a pollution-free engine for it?" But that was too sensible to be taken seriously. Other critics, facetiously, said it would be more logical to give Boeing and Washington State a direct subsidy to forget the SST and write off the cost to environmental protection. That made more sense than the SST.

There was an even stranger aspect to the SST episode: the public acceptance of any new abuse inflicted upon it. Occasionally there was a brief expression of outrage in the letters column of the *Times*. But, generally, the public could not have cared less. The SST served one useful function: it epitomized our loss of reality. It spotlighted our misshapen values. The President in his State of the Union message proclaimed the urgency of combatting pollution before we are all wiped out—"it's now or never." A couple of weeks later he asked Congress for $290 million to push ahead with the SST, a known polluter and a possible threat to life. There is not enough money for schools, housing or antipoverty projects, but veto follows veto for social welfare funds. There is not enough money to clean up the nation's air or water or to find means to properly dispose of our wastes without contaminating the earth. But there is money aplenty for special interest groups and political advantage. We are already overwhelmed by pollution that cannot be avoided or controlled. Yet fully warned of the health risks and potential environmental damage, the U.S. government used its vast power to try to force on the people yet another source—one that was wasteful, unnecessary, and potentially disastrous.

The showdown fight over the SST was played out behind the scenes in Congress in early spring of 1971. Leading the forces against it was Senator William Proxmire, with Nixon and Magruder trying to beat down Senate opposition. In the final hours almost every legislative trick known was used, including pairing by opponents who canceled one another's vote by not voting so their constituents and powerful forces would not know how they stood; there were absences from roll-call votes, trading, mysterious changes in loyalties, broken promises, and curious party splits. The House continued to support the SST. The Senate wavered in its opposition but mustered enough votes to win by the narrow margin of 51–46.

It is tempting to describe the SST's defeat as a victory for conservation. In a limited sense it was. Certainly it demonstrated an important lesson: for the first time conservationists had shown that they were developing political muscle and could influence public policy.

Unfortunately, many who voted against the SST did the right thing for the wrong reason. They were more concerned with economics than the environment. And the victory was only a local triumph at best. The world's ozone does not stop with America's borders. SST's are still flying over Europe and the Soviet Union, with the promise that they will soon be in commercial service around the world with whatever damage to the cosmic fabric of life that they bring—and the threat of reviving the project in the United States.

noise . . . noiSE . . . NOISE!

One of the more unpleasant by-products of progress is noise. Only recently has noise been recognized as a pollutant and health hazard, in the same way as noxious gases, chemicals and wastes that befoul air, water, crops and soil. Like other pollutants it interferes with nature's harmony.

Noise is measured in decibels—the term "bel" derived from Alexander Graham Bell's name. A decibel is one tenth of a bel. A decibel represents the lowest sound detectable by the human ear in quiet surroundings. A noisy office, for example, registers about 100 decibels. A noisy sports car or truck, 90. A riveting machine, 130. A jet plane at takeoff, 150. A power mower, 114. Neurological damage—damage to the nervous system—is said to begin at about 85 decibels. The threshold of pain is about 120 decibels.

The decibel is logarithmic in its effect—that is, 60 decibels of sound is not just 10 added to 50; it is actually 10 times more intense than 50 decibels, 100 times more intense than 40 decibels, and 1,000 times more intense than 30 decibels, according to the publication *Plain Truth:* "A jet taking off at close proximity (140 decibels) is *one billion* times more intense than 'normal' background noise in a suburban neighborhood (50 decibels)."

This sounds fantastic. But it is explained that *intensity* is not quite the same as *loudness*. "The best example of this is a 5,000-watt radio station signal versus a 50,000-watt signal. The signal (intensity) is 10 times—10 decibels—as powerful, but the volume (loudness) is only about *twice* as powerful."

What does all this mean to people? To those living in cities it has a very real meaning. Generally they exist perilously close to the threshold of damage and pain. Many work under conditions that are intermittently, or even almost continually, above it. Most urban dwellers cross the threshold of pain or damage for brief periods day in and day out: the shriek of a siren in the middle of the night, a sudden excavation blast, jackhammers that rattle the brain, huge air compressors at construction sites, buses grinding away to disturb sleep, rock groups cracking the sound and sanity barrier as they shriek of tragic events and forlorn young lives. Music being what it is these days, many youngsters of seventeen are said to hear no better than men of seventy-one. This amounts to a speeding up of the natural aging process.

Any sudden loud noise, such as the sonic boom from an SST or a military supersonic jet, can cause violent reactions and possibly serious effects. The sonic boom occurs at the point an aircraft exceeds the speed of sound. The air in front of the plane simply does not have enough time to get out of the way in a normal manner, explains one study: "Every cubic inch of air remains motionless until the very last instant: not until the plane has approached within a half-inch does the air begin to move out of the way. It must then move out of the way in a few *millionths* of a second."

The sonic boom produces in people what is known as the "startle syndrome." Typically, we are told, the syndrome includes hunching the shoulders, pulling the head forward and downward, crouching slightly, the body tense, the heartbeat accelerated. A person subjected to such attack by sound may blink, jump, or cry out. Various stomach symptoms may result, along with feelings of fear, surprise, terror. Sonic booms can produce amazing effects on property—everything from triggering rock slides and avalanches to cracking weak building foundations and shattering windows.

These effects are dramatic and demonstrable. But the greater damage comes from the routine erosive assault on the human nervous system from the noise that is part of daily life. Noise is a stimulant. We are all overstimulated, particularly the young. Possibly by now we are all hooked on noise. Many city dwellers get nervous, even ill, when they go to the country and are forced to endure the silence. Those who object to noise are a persecuted minority. They have almost no rights, certainly no voice.

Gradually this cacophony is approaching what many experts think may reach a lethal level. Noise is a demonstrated killer. It has killed mice at just under 180 decibels, slightly more than a turbo-jet at takeoff and slightly less than the Saturn V rocket at launch pad. Keeping pace with our social and industrial progress, the overall loudness of environmental noise is doubling every ten years, according to the Federal Council for Science and Technology. In some communities it shot up four times between 1956 and 1968—32 times what it was in 1938. Noise is like smog, says Dr. Vern O. Knudsen, chancellor emeritus of the University of California—"a slow agent of death. If it continues for the next thirty years as it has for the past thirty, it could become lethal."

No one really escapes the effect of continued noise at harmful levels. Some are affected directly in physical damage, others indirectly in social and economic loss. More than 7 million persons work at noise levels high enough to damage hearing, reports the U.S. Public Health Service. We are not told how much, or how little, is spent on research to try to control noise levels. Ordinarily technology is not developed to the point where it protects the worker from side effects; it halts when the immediate mechanical function is achieved.

To the exposed individual this has a very real meaning. In one noisy factory workers lost an average of 50 decibels of their hearing range after 15 years exposure to a 100-decibel level. Their normal threshold of hearing (20 to 25 decibels) had been changed to about 70 decibels. Normal conversation (60 decibels) had to be repeated louder (at 70 to 75 decibels) for the workers to hear. Accordingly the government set 90 decibels as a health hazard for an 8-hour-day working environment. Thus, noise becomes part of the "philosophy of poison." We live as close to the danger level as possible without plunging to immediate disaster.

It is literally true that no one escapes. Even unborn babies can be damaged by noise, according to one study. For years it was believed that the human fetus was protected from unpleasant conditions in the outside world "in the comfort of the mother's womb." But recent experiments have shown that the fetus is not so isolated after all. It seems to react to harmful environmental conditions just as the mother does—another example of the invisible and indivisible web of life. The supersonic transport jets may therefore pose a threat to the health of unborn infants in that "the planes will be one more noxious addition to the total assault of an increasingly noisy environment," concludes Dr. Lester W. Sontag, director of Fels Research Institute in Yellow Spings, Ohio.

The emotional effects of noise stress on humans have been repeatedly suggested by animal studies. Rats seem particularly like humans in many respects, psychologically as well as biologically. The effects of cities on people were simulated in rat experiments by Dr. Joseph Buckley of the University of Pittsburgh. He and his associates found a definite correlation between city life and high blood pressure and irritability. They put some 5,000 rats through a stress chamber (a box about the size of an office safe, which is equipped with bright lights that flash alternately from the walls; from loudspeakers come varied sounds of clanging bells, buzzers, and something similar to the roar of a jet taking off). The chamber was buffeted about 140 times a minute to simulate a commuter train or car ride.

Under such pressures the unlucky rats developed high blood pressure. The condition appeared to be permanent even when the animals were removed from the chamber. After a week in the chamber the rats became irritable and dangerous to handle. To test the results, Dr. Buckley and an assistant exposed themselves to the same conditions. They soon noticed a rise in their blood pressure, accompanied by increasing irritability.

Another study found that trees and grass are effective mufflers of noise. Each 100-foot width of trees can absorb noise 6 to 8 decibels of sound intensity. Research reveals that noise can be reduced as much as 65 percent by trees and grass in combination. Used alone they were less effective. As suburbs spread and trees are eliminated and roads cover green vegetation, noise levels inevitably rise. To the individual this means an increase in background levels of noise. But unless he can prove actual demonstrable physical damage he is not likely to be taken seriously. The point was made in a 1963 case in New Jersey. Some Jerseyites complained about industrial and transportation noises. They said these noises were health hazards because they interfered with sleep and eating and "life in general was made miserable." But the State Health Department would not accept this because there was no direct proof of disease. Therefore there was no damage.

The case is rarely so explicitly stated. To prove damage there must be a demonstrable dollar loss: this is calculated in workdays lost per year, estimated economic loss to business or industry, the cost in Workmen's Compensation funds for damaged hearing or other provable afflictions due to excessive noise. Noise is considered a menace only if it interferes with the worker's efficiency and therefore costs the business community or taxpayers money. For those who must have human damage translated into dollar figures, noise pollution costs U.S. employees more than $4 billion annually in accidents, absenteeism, inefficiency, and compensation claims, according to the World Health Organization. Here we again have the risk-benefit formula. So much risk, so much benefit. The workers take the risk, the economy reaps the benefit.

A new noise consciousness, however, appears to be gradually awakening. In a report to Congress in January, 1972, the Environmental Protection Agency asked for noise limits on aircraft and other transportation equipment, construction equipment and internal combustion engines. The agency said that noise was having a harmful effect on the public and would get worse unless action was taken:

"Whereas noise levels sufficient to induce some degree of hearing loss were once confined mainly to factories and occupational situations," the report stated, "noise levels approaching such intensity and duration are today being recorded on city streets and, in some cases, in and around the home."

New Jersey, formerly so tolerant of noise and often called the nation's noisiest state because of its highways and heavy industrialization, finally, apparently, had an earful. At the same time the EPA report was filed, Governor William Cahill signed the nation's first statewide comprehensive noise-control legislation.

The new law empowers the state to establish noise-level standards for automobiles as part of the state-operated inspections; to restrain industries from disturbing surrounding residents; set curfews for specific kinds of noise; prohibit the use of machines that do not have mufflers; and bar the use of ma-

chines and other noisy equipment unless they meet state-estab-
lished noise-level specifications. No longer, points out Ronald
Sullivan in *The New York Times*, will the state allow a sub-
urbanite to be jarred from a Saturday morning's sleep by the
unmuffled racket of a power mower or by the ear-grinding clat-
ter of a garbage truck grinding up refuse. "However, there are
no state provisions for barking dogs."

RESOURCES: DIG, CUT, DESTROY, MOVE

The rape of America began with the Pilgrim Fathers. As John
Stewart Collis tells it in *The Triumph of the Tree*, the Europeans
first fell upon their knees and then fell upon the Indians. The
Indian was easily disposed of. He did not understand the meaning of land ownership as the Europeans conceived it. Land, to
him, was sacred—a gift of the Great Spirit, to be held in trust.
The white man did not understand this reverence. Land was to
be used and exploited as any other commodity. The Indian was
easily bilked out of much of his land, with a few beads and
pieces of wampum, and driven off the rest. Firewater and fire-
arms, in the end, sent the red man on the final thousand-mile
"trail of tears" to Oklahoma. There was no place for the Indian
in the new nation. He was a liability, according to President
James Monroe: "The hunter or savage state requires a greater
extent of territory to sustain it than is compatible with the progress and just claims of civilized life . . . and must yield to it."

And yield he did, what was left of him, huddled into a barren
corner of the land he had loved and cared for. The invaders
found themselves in possession of, as Collis describes it, "a
wildly beautiful land, enormously fertile." After eliminating the
Indians, the white man went to work on the trees with the same
intense ferocity:

> They went out against the forests with the thoroughness
> of an invading army, attacking first one stronghold then
> another. For a hundred years the white pine trees of New
> England held out. Then one day it was found that all had
> fallen on the field. After the white then the yellow. The
> movement of destruction advanced relentlessly onward
> from the forests of Maine to New York. In ten years those
> battalions were defeated and the lumber-troops entered
> Pennsylvania and Ohio and Indiana, from whence they
> moved in turn to Michigan, to Wisconsin, to Minnesota,

and thence again through the Rocky Mountain region on to the Pacific Coast. That was the northern campaign against the trees. There was a similar offensive in the South, from the Carolinas to Texas and on through Arizona and Colorado.

Ultimately practically all of the virgin forests were gone. Not all of the timber was used. Much was burned or left to rot. Huge bonfires burned day and night with such fury that there was said to be no darkness in some towns. The fires raged for three years continuously. Accidental fires were extremely common. Men were hired to fight the fires. This made matters worse. Fires were started by men who needed jobs.

At first, Collis explains, the cutting of trees was a bigger industry than their use. But the steady growth of mill power at length made the lumber merchants very rich and powerful. The first waterpower mill in 1631 could cut only 1,000 board feet a day. By 1767 the new gang-saw cut 5,000 a day. In 1820 the circular saw cut 40,000 a day. In 1830 the steam saw cut 125,000 a day. The figure is now 1 million a day. In Colonial times trees were cut down ruthlessly to feed the devouring lumber mills, to fire iron smelting furnaces (one furnace could consume an acre of forest a day), and to provide wood for many uses in early commerce—minerals and fuel for glass manufacture, to prepare salt brine to cure cod for export (one cord of wood for eleven bushels of cod).

The rape of the forests was no haphazard operation, Collis points out. Several million deforesters took part in the campaign toward the annihilation of the American forests. They worked almost as a coordinated army, not governed by a general or over-all command as much as a philosophy that had the same effect:

> A movement on the grand scale must always be informed with an idea. It must have its philosophy. It must proclaim a doctrine. This was not lacking here. The philosophy of the invaders of this huge and bountiful continent was the philosophy of *inexhaustibility*. The idea was simple: there is no limit to the wealth of this country. The doctrine was pure: we are the masters and lords of this land, and may do as we please. The command was clear: pillage and pass on! There is more beyond.

The use and abuse of technology in America grew out of three primary principles: the philosophy of abundance, the pursuit of comfort, and private ownership of land. These formed the base of the pyramid of growth. There was no restraint, no control. Only one question was asked of any new technology: would it produce quicker and cheaper? The consequences to the land or its people were not considered. The land was there to be exploited—a sanction, a mandate, found in the Bible itself and quickly incorporated into the commercial creed of pioneer America: "And God blessed them, and God said unto them, 'Be fruitful, and multiply, and replenish the earth and subdue it: and have dominion over the fish of the sea, and over the fowl of the air, and over every living thing that moveth upon the earth.'"

Dominance became holy mission. The land was ripe and ready, its wealth waiting to be used. Labor was there to be exploited. Technology, exploitation, and religion became one. Every man was a king. Land was his domain. He could do with it as he liked. No holds barred. Workers, like women, were chattels. The public was a vast consuming maw to make the prudent industrialist and businessman rich. The law of the land was *caveat emptor*—let the buyer beware. At its worst it was a form of commercial anarchy.

Ultimately limited regulations and controls were imposed in some areas. But they were in conflict with the underlying American philosophy of the rights of property—the right to the unrestricted use of men and materials. Not even those who passed the legislation nor those assigned to enforce it wholly believed in what they did. Restrictions were a violation of the American code. The enforcers frequently looked the other way when infractions were committed. Laws were riddled with loopholes to give an illusion of protection when little or none existed. Regulatory agencies, such as the Bureau of Mines and the Department of the Interior, did less to protect the resources entrusted to them than to become partners in their systematic exploitation. The regulatory agencies were generally run and staffed by men imbued with the spirit of frontier America. Their sympathies were basically with the exploiters.

In the great giveaways of the nineteenth century, men with connections bought land from the federal government for as little as 10 and 12 cents per acre for prime forests; railroads and favored industries got it for nothing. First the trees were ripped off and sold; then eager prospectors ripped into the earth's entrails to remove its mineral wealth. The exploitation of resources was underwritten by the Mining Law of 1872, an open invitation to plunder the land. It is still in effect. Under the law anyone may enter federal lands, except national parks and other areas specifically barred by law, to look for minerals, primarily metals: iron, copper, lead, uranium. An individual may stake out a claim of 20 acres or, if he joins with others, 160 acres. The claim is filed with the county, not the federal government. So long as the individual or company searches for minerals the land is his. If he discovers minerals he can file for a patent to the claim: $2.50 per acre for a placer claim (as surface panning for gold), $5.00 for a lode claim (vein or deposit). He then receives from the government full title to the land, including title to the surface resources; often this includes valuable timber.

The same bargain fees established a full century ago when the land was considered inexhaustible are still in effect today. The destructive technology now available makes the land even more vulnerable than it was then. There are no rules governing how a miner goes about exploring his claim, according to James Ridgeway in *Ramparts*. A prospector may cut a road through timber, bulldoze, dredge, or strip the land at will. In some western states "prospectors are encouraged to gouge the ground because state laws require that they dig a pit and engage in the charade of erecting a works of sorts on each claim as proof that they are really working it." Often the effect is devastating. In Wyoming, for example, mining companies bulldoze or plough big strips of land in search of jade. Acres might be destroyed in hopes of turning up a jade boulder. Usually it doesn't. Then the prospectors move on to a fresh piece of land, leaving the mess behind. In Arizona, where half of the United States supply of copper is produced, a huge open pit mine outside of Tucson uses so much water that groundwater supplies have been lowered to the point where crops and vegetation are affected.

Each development in technology has multiplied the damage, always in the sacred name of free enterprise. Greed for gold inspired some of the earliest and most brutalizing devices to ravish the earth. During the California gold strike, gold towns sprung up like mushrooms. A California gold miner named Anthony Chalbot devised an ingenious canvas hose and nozzle that washed gold-bearing gravel into placer pits for processing. This evolved into an even more powerful hydraulic apparatus known as the "Little Giant"—a huge nozzle that could rip away an entire hillside, according to Stewart Udall in *The Quiet Crisis:*

For every ounce of gold collected, tons of topsoil and gravel were washed into the river courses below. With the spring floods, clear streams became a chaos of debris, rocks and silt; communities downstream were inundated with muck, and fertile bottomlands were blanketed with mud and gravel. The town of Marysville, along the Yuba River, was forced to build ever-larger levees that rose higher than the city's rooftops. In 1875 a big storm sent the Yuba over the levees and filled the city with silt. . . . in their reckless effort to extract gold, the hydraulic miners asserted the right to damage other resources irreparably—and the homes of other citizens as well. In the 1870s, the right of each individual to do as he pleased was sacrosanct.

The same feverish drive for quick riches was expressed in ruthless damage from almost every form of mining. The assault on coal was particularly disastrous and permanent. Some of the delayed effects are only now being felt—and paid for. Exploitation of men and material was rampant as deep mines were dug and often collapsed because of the failure to provide safeguards. Hundreds of miners died ghastly deaths. No compensation was given those who became ill. Coal that could be taken cheaply and easily was removed and new cuts were begun elsewhere. Abandoned mines were left unsupported and often uncharted. Today these huge underground caverns and linking tunnels are caving in. Often they cause houses and buildings on the surface to collapse and foundations to crack.

Roads have been known to disappear under motorists. A man in a new convertible, happily eating a banana, suddenly found himself upside down in a 30-foot hole that had abruptly swallowed the highway, car, man, and banana. In some of the forsaken mines great fires have raged for generations. More than 100 coal fires burn in the mountains of Pennsylvania alone and threaten to affect 700,000 people and $500 million worth of property, according to *The New York Times*. The earth suddenly splits open and steam from a hidden cauldron squirts into the air. Nearby residents must walk for blocks to find cool water. Sulfuric smoke from underground infernos blackens the earth, compounded by carbon dioxide emissions, to foul the air. Recently a woman put her dog in the basement of her Wilkes-Barre home; the next day the animal was found dead from the fumes.

Two million acres in twenty-eight states—much of it in Pennsylvania—have collapsed in this form of subsidence, according to the government. A million more acres are expected to sink by the year 2000. But there is no one to sue. The mine operators long ago made their fortunes, and decamped. Now the bill must be paid by the government—a delayed additional bounty to the mine owners from the taxpayers. Millions are being spent to relocate families. Millions more are being spent to fight the fires. The government has already spent over $15 million to control mine fires in the Appalachians. Two years ago Pennsylvania allotted $200 million to fill in abandoned strip mines and fight underground blazes. This is part of the hidden cost of the cheap goods produced from that cheap energy that American industry boasts of.

The crime against nature and the future does not belong only to the unenlightened past. Even now we are making our own contribution to tomorrow. An estimated 92,000 surface acres a year are currently being undermined with little or no planning or control over surface development, according to the Federal Bureau of Mines. Mine production over the next thirty years is expected to advance the spread of potential subsidence to 200,000 more acres a year by the turn of the century. The huge slag heaps that tower over these wastelands, many within the boundaries of communities of up to 100,000 people, some burning for thirty years, are also expected to grow and flame. New mine wastes to be accumulated over the next thirty years—much of it combustible—are expected to add 3.6 million tons to the 18.5 billion tons of tailings already extracted. This adds up to mountains of waste—potential air and water pollution as well as a ghastly scar over the wound in the earth's entrails.

Eventually frugal operators found a cheaper and more efficient way to remove coal—strip mining. This is technology gone rampant. It must be the most destructive force yet devised by man to rip at the land. The earth is gouged out from the top to reach surface deposits of coal. Enormous electric shovels, soaring 100 feet high and burning more electricity than a small town, work 24 hours a day. They creep relentlessly across the earth, obliterating all in their path—woodlands, fields, even towns and highways. Whole mountaintops are torn off and 20,000 "lofty serpentine miles of 'bench' slice across the highlands of nine states," reports the *Times*. When the seams of coal are exhausted the pits are abandoned and the machines move elsewhere to gouge out new cuts in the earth. Virtually no laws exist to control the strippers.

Strip mining has become increasingly popular because it is cheaper than digging deep mines; the latter also involve costly new federal safety regulations. More than 35 percent of the nation's annual coal output of 566 million tons is produced by stripping. By 1980 more than 5 million acres will have been defaced by strip mining, according to a government study. Strip mining is so popular that even the Tennessee Valley Authority—the federal monument to conservation—now plans to strip-mine 50,000 acres it owns in Kentucky. For many years TVA has been buying cheap coal from strip mines. The agency's charter was granted on the premise that it would protect the environment. But TVA has given precedence to another mandate—that it produce cheap electricity. One of its critics, Harry M. Caudill of Whitesburg, Kentucky, in a letter to *The New York Times*, charged that TVA's demand for cheap coal stimulated surface mining because stripping costs less than tunnel and pillar operations:

> Coal suppliers for the agency began taking mountains apart and rolling the rock and dirt into the valleys. TVA got cheap coal and America got bankrupt mining companies, crippled and dead miners, multitudes of widows and orphans on welfare, millions of acres of ruined land and thousands of miles of dead streams. Through it all Mr. Wagner [Aubrey J., TVA board chairman] and his board were able to report profitable operations and the sale of cheap electricity.

Ironically, much of the stripping is on farmlands already ruined by exploitative soil practices. Now, with the discovery of stripping, land prices have soared, from a base $150 an acre to as much as $3,500. Those who hold out are often driven out by block-busting tactics such as surrounding the wanted properties with coal leases to force sales.

Abandoned strip mines are wastelands. The dusty cuts are—as a *New York Times* correspondent described them—"blotched by ugly ponds of orange and rust-colored acid water, standing in chemically 'hot' lagoons of sulfuric mine drainage." These acid waters poison other streams. More than 7,000 miles of Appalachian waterways are already affected and the flow of poisoned streams is growing as the mines proliferate. Another 6,000 waterways are damaged in Missouri, Kentucky, and Illinois. Many streams are so burned from the acids that they cannot support any aquatic life, plant or animal. The Monongahela River, which flows through the coal regions of West Virginia and western Pennsylvania, is so scalded with acids that it cannot freeze over in the coldest weather. Cleaning up the acid water would cost an estimated $10 billion. Officials will not even guess what it would cost to repair the mess of the mines. Many think they cannot be reclaimed at any price, especially those cut out of steep mountain grades. Strip mining desolates tens of thousands of acres a year of some of the nation's most beautiful mountain country, with little likelihood that the damage will ever be repaired. Of 1.8 million strip-mined acres, only 58,000 have so far been restored, according to the Department of the Interior. Reclamation becomes virtually impossible, as erosion by wind and water prevents healing and the damage accelerates year by year.

There is talk of using the scourged land to grow vegetation for wildlife, for cattle, or for Christmas trees. Experts say that generally this is so much pie in the sky. Most stripped areas become disaster areas. An 11-year federal field study was made of a 25-square-mile creek basin in McCreary County, Kentucky, a strip-mined coal region. The project concluded that relatively small-scale strip mining there between 1955 and 1960 killed or reduced fish in streams by filling them with acid and mineral poisons, brought down hundreds of thousands of tons of channel-clogging silt, and killed or stunted trees, and that the devastation was likely to continue unchecked for years.

What befell this region and Appalachia is in store for other states in the West and Middle West if they do not act. The deposits in Appalachia, which have yielded 25 billion tons of coal during the past century, are largely played out and the earth is as impoverished as its people. Now the despoilers turn westward. Some 26 states have been found to have strippable reserves of 128 billion tons of coal, according to the U.S. Geological Survey. The spoil banks would themselves cover 42,000 square miles— an area larger than the state of Ohio, one of the favored targets.

A few states have passed feeble laws to restrict stripping or to minimize its effects. These laws are usually so riddled with loopholes as to be meaningless. An Ohio law was quickly undermined by seventy amendments. State legislatures friendly to coal interests bottle up corrective legislation in committees or pass laws that are a travesty. As coal-producing states fought bitterly over diminished markets, the lawmakers were loath to saddle mine operators with "unnecessary" cleanup costs. Legis-

lators are often controlled by the coal producers. Even indepen-
dent-minded lawmakers can be swayed by threats of mine clos-
ings and lost jobs. State courts are friendly to the coal operators.
Judges are prone to make any regulations attempted unenforceable; they often refuse to uphold such laws as exist or to impose
penalties. Pollution-control boards in the coal states are generally weighted by members of the coal industry, posing a serious
conflict of interest. In Kentucky, for example, acid drainage
from coal mines is a severe water-pollution problem; but the
president of the Kentucky Coal Operators Association is on the
Water Pollution Control Commission, according to *The New
York Times*.

Recently Knott County, Kentucky, balked. It had its fill of the
strippers and passed a law banning the practice. The measure
was promptly challenged in the courts. It is given little chance
of survival in Kentucky's coal-oriented courts.

In West Virginia, Secretary of State John D. Rockefeller, IV,
made a valiant effort in 1971 to ban all stripping "completely
and forever" to save it from "the spreading cancer" of surface
despoliation. Rockefeller said surface mining in the state now
covers more than 65,000 acres and is doubling in area every
year; about 6 to 7 percent of the state's area is already despoiled,
and the devastation could spread to 40 percent. The coal industry reportedly spent $100,000 lobbying and advertising against
the ban and mobilized 2,000 persons at an antiabolition rally. In
the end Rockefeller's conservationists succeeded only in winning
a two-year prohibition on the issuing of state surface mining
permits in 22 of the state's 55 counties where such mining has
not yet begun.

There is little hope of relief from government. The Federal
Bureau of Mines historically "has shown more interest in protecting the coal industry than in regulating it," the *Wall Street Journal*
points out. Following a disastrous mine explosion that cost 78
lives, Congress passed the Federal Mine and Safety Act of 1969.
But the record of enforcement was "dismal," noted *The New
York Times*. During the first year of the new safety act's operation, mine deaths rose 25 percent; 100 miners were killed during
the first nine months of 1970 alone.

The new law was soon undermined by bureaucratic interference. The director of the Bureau of Mines, John O'Leary, a rare
bird in mine bureaucracy, tried to enforce the law. For his pains
he was fired by Nixon, a dismissal reportedly insisted upon by

leading coal industry executives. The Bureau of Mines was then in charge of Under Secretary of the Interior Fred J. Russell, who was Acting Secretary of Interior following the ouster of Secretary Hickel. Russell promptly ordered the Bureau of Mines to require its mine safety inspectors to obtain telephoned authority from Washington before they could shut down unsafe mines. The law stated clearly that violations must be corrected within a reasonable time or the mine closed down, and it made provision for hearings to restrain overzealous inspectors. The law did not provide for some faceless bureaucrat to decide if a dangerous mine should be left open for political or economic reasons.

The law also called for appointment of a thirteen-member technical advisory committee on coal mine safety research, a highly complex engineering discipline. Such members were to be "knowledgeable in the field of coal mine safety research." They were to be paid $100 a day for the part-time posts. Among Russell's appointments were a former airline stewardess (a Republican national committeewoman), two Republican county chairmen, a doctor's widow, and a retired grain executive.

Following demands for Russell's removal, he "resigned." Russell, previously a heavy contributor to the Republican party, returned to the White House staff, so, as President Nixon said, "his talents might be put to use more effectively in some other area of the Administration."

Nixon's replacement nominee as director of the Bureau of Mines had such close industry ties that Congress was embarrassed and protested. That candidate withdrew. Nixon's next choice, Dr. Elburt F. Osborn, an industrial scientist, had better luck. He was confirmed in October, 1970, and must have endeared himself to the industry by promptly coming out against putting strip mining under federal jurisdiction. Control of the mines should remain at the state level, he said. And that's just where it is—where there is little regulation and little likely to come.

Ultimately the victim of the wanton disregard for the land is the little man who is displaced and damaged as the earth is ravaged. It is men like Luther W. Johnson, a 64-year-old Kentucky farmer. He had the bad fortune of getting in the way of the relentless power shovels. Johnson has sued the Bethlehem Steel Company to stop it from stripping out 10 million tons of coal over 3,000 acres near his home. Johnson, born on his farm, has lived there all his life. He shows a visitor the 60-foot poplars that have grown during his lifetime, and points to the land coarsed by water gone bitter from mine operations upstream so his cows cannot drink. He speaks plaintively of his love for this patch of earth. Against its loss "money don't seem of much account," he drawls. Johnson represents a perverse truth that keeps popping up with discouraging frequency in the losing struggle to preserve the earth from the despoilers: since filing the suit he has received threats from the mineworkers, men who are his neighbors and whose lives, like his, are being diminished by the creeping pits that devour anything that gets in their way.

For the uncaring and unfeeling there is no place for sentiment or love of land. It carries no price tag and thus has no value. This is what makes stripping possible. For the despoiler, the conservationist is a despised obstacle to be insulted or dismissed with contempt. The vice president of Consolidated Coal Company has publicly branded conservationists as "idiots, Socialists and Commies," according to a *New York Times* article. The head of the Ohio Coal Association asks this bloodless question: "What are we going to do? Are we going to cut off the electric power because some guy has a sentimental feeling about an acre of coal?"

The damage is not limited to coal mining. Other mining operations also leave dreadful wounds in the earth. In Butte, Montana, the removal of copper ore by the Anaconda Company has had epic effects. Once the mines were underground. In recent years strip mining has taken over. The creeping open pit has literally devoured the town and its life.

Years ago Butte was a prosperous community with stately mansions on "the richest hill on earth." Here the copper barons lived in crude Western elegance. In 1917 the town reached a peak population of 100,000. A fifth of the residents were miners "digging ore in the slopes and drifts that honeycomb the rocks beneath the community." The supply of copper ore appears inexhaustible, and is removed by giant 100-ton trucks filled by 50-ton shovels that have gouged away a substantial part of the town. Over the years the mine has chewed its way relentlessly toward the business district, devouring colorful neighborhoods with names like Dublin Gulch and Sin Town, reports the *Times*. The pit grew steadily bigger and deeper. Now it plunges 600 feet into the earth and is expected to sink to 2,000 feet—more than one third of a mile.

As the hole encroaches toward the business and residential

district, stores have closed and people have fled. Only 23,000 now remain, most of them living directly or indirectly off that hole. Anaconda has kept buying land as the huge trench snakes closer to the city's vitals. Now it is only blocks away. The city fathers are considering leaving Butte to the mine and moving the city eight miles south, the ultimate surrender to progress. Planners expect the federal government to pay most of the estimated moving cost of $150 million. No one uses the dirty word "subsidy."

Few resist when it comes to putting protection of the land ahead of profits from its destruction. One who did was Cecil Garland, owner of a small general store in Lincoln, Montana, a town of 800 population. There the Anaconda Company tried to lease land to dig a $50 million open-pit copper mine even larger than the epic trench in Butte. It would have brought an 800-man weekly payroll up to $100,000, a prospect that excited local businessmen. But the project was opposed by a small band of conservationists who feared the result. Their leader was Garland, a former Las Vegas craps dealer who had come to Lincoln to be close to things he cared for—the land and nature. He went to work for the Forest Service. Soon he found that the woodlands it was supposed to protect were actually being destroyed by the service through the overcutting of timber and building of access roads. Garland quit and opened a store. When Anaconda tried to come in, he formed a protective association that demanded environmental safeguards from the company. Anaconda's reply was to withdraw. Were Garland's townsmen grateful to him for saving the land from despoliation? No indeed. Instead they turned on him and boycotted his store, trying to drive him away so the mine could come in. He too had learned the bitter lesson of becoming "an enemy of the people" when he tried to preserve the land from wanton destruction.

The latest target for mineral mining is the sea. Oceanic mining engineers point out that the sea's vast mineral wealth will become available to new technology just as many land sources are running out. Supplies are almost inexhaustible, they say. This is curiously reminiscent of what was once said of the now-despoiled and exhausted land. The great riches in the deep beckon. Already nations are quarreling over the spoils and maneuvering for advantage. With or without treaties to share they will extract the treasures—nickel, copper, and cobalt, nodules of manganese dioxide described as "big as potatoes and scattered over the ocean floor," just waiting to be gathered. But biologists pose nagging questions: How many of these minerals once taken and used will return to the sea in dissolved form as a source of new marine dislocation? And what will be the effect of the new mining procedures? One company will soon begin experimental mining operations off the southeast Atlantic Coast with a vacuumlike device that will draw materials off the sea bed; halfway up it will screen out fine wastes and spew them into the undersea in a broad fan. An almost certain result will be the smothering of bottom life over a broad area, the biologists warn.

Mining is already under way in the shallower reaches of the sea, on the continental shelf, which supports most marine life. It takes its place with the oil drilling, heat pollution from coastal plants, and the storm of poisons and other pollutants that rain down continually from the atmosphere and pour out of rivers.

One threat in particular bothers oceanographers. Recently the American Museum of Natural History in New York sold for $50,000 the rights it held to mineral deposits of aragonite (a highly pure form of calcium) off the Bahamas. Many conservationists felt the museum—a respected member of the conservation establishment—had shown more interest in turning a profit than protecting the environment. This was probably shortsighted on their part. In a growth culture, even museums must grow. Museums today are big business.

To get at the minerals unlocked by the museum will require tearing up the ocean floor—a form of marine strip mining that could leave lasting scars, according to an oceanographer. Others say the dredging will destroy the fragile balance of one of the few remaining natural areas of the earth—the warm waters at the edge of the continental shelf. They fear that the small particles freed during the digging will cloud the water seriously; it could interfere with photosynthesis, affect marine life, and send the particles drifting on currents to contaminate other areas.

The developer, the Dillingham Corporation of Honolulu, is not intimidated by such fears. It foresees changing the face of the ocean near the Bahamas. Already it has taken such steps. It has built an island with dredgings scooped from the ocean floor; here it expects to stockpile the removed aragonite. By the end of the year the island is expected to grow to 200 acres. Eleven other man-made islands will follow. They will provide a base for new industries to be built. The gorge in the sea floor where the minerals were removed will become a giant deep-water terminal for supertankers. Possibly there will be a complex of refineries on the new man-made islands. The supertankers would use the terminal to transfer oil to smaller vessels that would deliver it to the American mainland. All of this inevitably means more pollution of air and sea.

The prospect of the new marine complex alarms some biologists. One was upset about the threat of disastrous oil spills in this vulnerable area. But the museum apparently does not share this worry. Its director, Dr. Thomas D. Nicholson, told the *Times* that it was "strictly a commercial venture. We thought we had something marketable and we went ahead and sold it. Maybe some fish might be disturbed but they could probably find someplace else."

TIMBER: SLASH, BURN, ABANDON

The destruction of the green canopy of forests that once covered most of America—and most of Europe as well in time long past—led to environmental changes. Such changes are still taking place. Forest, soil, water, and weather are so interwoven, so interdependent, that one cannot be damaged without damaging all. The forest looks to the soil for nourishment. In return its roots hold the soil in place and its leaves shield the protective organic matter from the sun's heat so it does not dry out. The forest's thick leaves deflect driving rains and soften them to gentle drops. The raindrops, bearing the nitrogen and other essential elements collected in their journey through the air, ease slowly into the carpet of living organic matter on the forest floor to provide moisture that enables soil to live. Water helps soil microorganisms break down the "trash" on the forest floor—an accumulation of fallen leaves, dead branches, wastes from living creatures and the corpses of the dead. These various remains are converted into organic nutrients. The nutrients in this form are available as food for the trees and plants of the forest. This growth now provides nourishment and shelter for the multiple organisms, insects, and animals that help protect and sustain the forest in the interlocking web of life. The water, having nourished and refreshed, is purified in its filtration through the soil. It then enters the underground network of streams, finds its way into buried springs and reservoirs, finally joining running surface water to spill into rivers and make its way back to the sea to be evaporated and distributed again as precipitation.

The forest benevolently goes further than merely easing water to the ground. It also helps provide moisture for the community it presides over. Tree roots wander far underground to forage for food supplies of water. This water is tapped and lifted through the trunk, to the limbs and leaves, which release it into the air. A full-grown willow tree can transpire as much as 5,000 gallons of water in a single day. From enough trees a cloud can form. The moisture rises on the warm air, chills and falls back as rain, snow, sleet, or some other form of precipitation. Upon returning to the earth it completes the cycle again and again, helping the forest to maintain its own life. The moisture given to the air by the forest refreshes the atmosphere, cooling it in the summer, warming it in the winter, avoiding damaging extremes at all seasons. The leaves and branches temper the wind. The roots and falling leaves prevent erosion and hold vast quantities of water. The earth itself, with its organic matter, becomes a giant sponge. If all of this were not enough, the leaves, in providing the trees with food from the air, from nitrogen and other elements, give off the essential oxygen that purifies the air and makes life possible. Trees are also "air scrubbers." There is some kind of mysterious antiseptic quality in the volatile substances excreted by a forest; it has been found capable of checking pathogenic fungi and harmful bacteria. The bacterial content of the surrounding air in parks of conifers was 200 times less than that of city air.

The primary wardens of forests provided by nature are birds and other animals and insects. They have handled the job of managing forests effectively without help for millions of years. The forests were in excellent shape when the first settlers arrived. Birds, left alone, are marvelous at controlling pests. A single bird can eat its weight in insects in a day. Various species of birds police their own selected domain. Each is a specialist. Each attacks its own specific insect fare. Different species of birds have different elevations which they police, as if through some cosmic arrangement.

The insects are also necessary to the overall economy of the forest. They help pollinate. They attack and destroy defective trees, thinning out the weakest members so the sturdy that remain have room to flourish. Similarly, these unwitting insect predators help remove dead wood. In this they are aided by lightning and its fires, high winds, ice storms, and snow that breaks off dead branches. These fallen limbs are added to the organic mass on the forest floor to be broken down by the busy micro-organisms.

All the while the animals that dwell in and visit the forest are contributing their part. Squirrels and other creatures bury seeds and aid in starting new growth. The forest is constantly under-going an evolutionary process—destroying, renewing, and re-placing, finding the species best adapted to the environmental conditions that prevail. Once it reaches this stable state it is said to be in climax.

So the forest is not just a group of trees and a few creatures but a living interdependent community, ever changing and yet the same. "Forests are so much more than meets the eye," says John Stewart Collis. "They are fountains. They are oceans. They are pipes. They are dams. Their work ramifies through the whole economy of nature."

The opportunity to visit and experience forests becomes in-creasingly remote as they steadily dwindle. Those that remain are managed like factories (many are called "tree farms"), or they are subjected to slow death or immediate ruination by ex-ploitative practices. Two thirds of the world's forests are gone, according to a U.N. report. The United States now has only about 900 million remaining acres of forest. Of that amount the government owns approximately one fifth (183 million acres). Only about 20 percent of the national forests are in anything like a wild state. The rest have been turned over to logging and graz-ing. Each year the loggers extend their borders, hacking away at the public wilderness that remains and commercial stands alike. A total of 97 million acres of government forest land is now classified as commercial timberland—subject to cutting. Some 90 percent have been opened to loggers; 20 million more acres came under the commercial heading in the past 20 years.

The number of forests promises to decline still more as wider usage is found for wood and stimulated demands become greater. In recent years timber usage has increased steadily as acreage has decreased. By 1985 the use of pulpwood is expected to rise two and a half times over present levels. By the year 2000 the overall use of wood will double, with fewer acres to grow it on. This is accomplished by an industrial trick called "sustained yield" forestry. It means pushing the land to its utmost capacity to produce—and sometimes beyond it.

Sustained yield forestry is the logging industry's rationalization for the overcutting. The term implies putting back what is taken away. This is the exception rather than the rule. It is also the exception on public lands. The Forestry Service is starved for funds, like most government agencies assigned to protect resources. It generally lacks the will to protect and the funds to replace. Some 50 percent more trees now are being cut than are replaced by planting and natural growth combined, according to the service.

Sustained yield forestry means putting the land under heavy stress. Trees no longer grow at nature's leisurely pace. They are mass-produced on a schedule fixed by computers. The land is heavily fertilized, growth cycles are speeded up, superseeds for supergrowth are developed, trees are "harvested" before they reach maturity. Enlightened companies protect their investment by replacing cut trees with seedlings. Others are concerned only with fast profits and do not think in terms of fifty or sixty years from now. The objective is to cut as much timber as fast as possible and acquire new lands to cut more. Such firms often are dedicated to destructive practices. Whole areas may be cut down at once with "slash and burn" methods rather than cutting selectively. That which can be used is hauled away to mills for lumber or pulp. That which cannot be removed economically is burned, adding to the air pollution. This random cutting leaves huge tracts denuded and unprotected. The barren land, especially the hills, is subjected to ruinous erosion. Through erosion the United States is estimated to have lost somewhere between one third and one half of its total topsoil. The Mississippi River alone carries away one cubic mile of rich silt from its mouth every year.

The Forest Service has done less to protect the forests than to aid in their exploitation. Representative Dingell of Michigan has called it "a wholly-owned subsidiary of the timber industry."

The Forest Service engages in many destructive practices. Livestock in excessive number overgraze certain national forest lands. This can be as harmful as any other overuse of land. It wipes out vegetation, leading to erosion and soil loss. Laws have been passed to block overgrazing, but such laws have been undermined by powerful lobbies representing livestock interests, supported by their representatives in Congress. Many of the public grazing lands are now ruined—denuded and desolate as the barren hills of Jordan, as wind and rain wash away the vulnerable soil.

The nation's timberlands are simultaneously being subjected to other abuse. Recreational use increases as populations swell; large numbers of people tramping through forests can be deadly, killing vegetation, causing fires, and chopping down young growth for firewood or in acts of vandalism. Air pollution is taking an increasing toll. Every year some 30,000 acres of forests are cleared to make paths for power lines that carry the electricity produced from the coal taken out of the mines that damage the earth; the power thus produced leads to greater environmental damage in the widening circle of cause and effect. The trees sacrificed no longer can produce the oxygen that formerly helped clean the air of sulfur dioxide from the burned coal. New highways built with the revolving highway fund account for more lost forest acreage, adding their weight to the oxygen deficit. The cars that use the roads replace oxygen with pollution. The circle grows progressively wider and more vicious. It widens as the runover of trees is speeded up and more usage is found for wood products. More pulp is used every year.

The forest is also a victim of politics and economics. Public officials and industry have worked closely to exploit federal timberlands and bring about their liquidation. Raids have been made repeatedly on public forests by modern timber barons who bend the law to their own advantage. These raids are carried out behind a smokescreen of artificial lumber crises, public relations, phony conservation, manipulated timber prices, rigged bidding, kickbacks, and politicians working with industry to sell wood on public lands for the profit of private interests. Since World War II there has been the heaviest wave of clear-cutting (stripping the land bare) in the history of the world, according to Charles H. Stoddard, former director of the Federal Bureau of Land Management. He notes in the *Sierra Club Bulletin* that the scheme was supported by a "slick advertising program (squirrels running over stumps, etc.) designed to lull the concerned public into complacency." Damage was extensive: ". . . roads were gouged out of steep hillsides; skidding with heavy equipment and clear-cutting in overly large patches permanently damaged watershed soils and cover, scraped stream-bottoms and resulted in silted fisheries habitat."

Stoddard documents one of the massive raids in Oregon. The details are complex. Practically every known abuse was found there. The cast included industry, county and federal officials, a Washington lobbyist "who was a professional 'influence peddler' and a close friend of the Assistant Secretary for Public Lands," assorted politicians, and "three former BLM officials" whose objectives were "to give political protective custody over this 'sweetheart' arrangement between the federal government, industry, and local government." The latter kept timber prices low by limiting competitive bidding and the allowable cut high "by a combination of harassment and flattery of federal administrators."

The promoters of the Oregon-California scheme took pains to make sure no harm would come to "this federally sponsored intrigue." They set up the usual "advisory board," composed of "loyal timber company ex-politicians, 'kept' consulting foresters, and county cohorts with a couple of innocent citizen conservationists for window dressing." The advisory board helped pressure politicians and key officials in and out of government.

This is the key point: how regulatory bodies are broken down. Industry exerts such unrelenting pressure that regulators cannot do their work. Leverage comes from many sources: former government employees, lobbyists, politicians, and industry brokers, such as trade associations and others. According to Stoddard: "This constant, almost daily pressure on BLM administrators in Oregon . . . proved to be the determining factor in negating our efforts to broaden the management objective of these lands from timber production alone to include a balanced output of water, wildlife, recreation, and scenic value." In the end the only concern was with lumber production.

Some 2 million acres of public lands were involved. The annual allowable cut in western Oregon was increased from 950 million board feet to 1,190 billion. In the mid-1950s it had been only 600 million. The counties got 75 percent of the revenues from the sales; the remaining 25 percent went to the Federal Treasury. The counties and industry had such "tight control" over BLM that bidding was controlled. This squeezed out smaller competitors and made possible the continuing sequence of evils, including political contributions:

> . . . the artificially low timber prices made possible highly profitable operations for the industry insiders and large payments were made to the counties. Both situations made possible slush funds available for heavy political contributions to the campaigns of the Senators, Congressmen and gubernatorial candidates currently holding office.

"The losers are, of course, the citizens of the U.S. whose lands are stripped, whose Treasury receives only a pittance, and whose political processes are corrupted," observes Stoddard. And who is to change this system? Certainly not the Department of the Interior, "since that unhappy Department is locked in the vise grip of vested interest pressure groups opposed to change."

Everyone is taken care of so no one will want change. For a long time the Bureau of Land Management received no revenue from the timber sale. "To grease the then creaking wheels, the Oregon congressmen and then Senator Guy Cordon [long a faithful servant of the timber interests] arranged for a 25 percent [of the 75 percent timber revenue to the counties] kickback to the BLM for roads, reforestation and administration of timber sales." This arrangement, "which has really accelerated the liquidation of your forests and mine, was actually hailed in Oregon by a former chairman of the O & C [Oregon and California] Board, as a generous and foresighted act of conservation statesmanship." Conservation, like statesmanship, is where you find it.

In 1960 Congress recognized the threat to the public forests. It responded with the multiple-use law, that provided for selective logging in certain forests, but was to be balanced by other demands: water supply, wildlife protection, recreation, and timber production. It didn't take long before new assaults began to undercut the bill and hack down excessive numbers of trees without regard to the other legislated uses of the national forests.

A report by the School of Forestry of the University of Montana strongly criticizes the Forestry Service for paying only lip service to the principle of multiple use, while "getting the logs out" (cutting timber) comes first. The report was based on intensive study of Forest Service practices in the Bitterroot National Forest along the Montana-Idaho border. Multiple-use management "does not exist," it charges. What the Forest Service was engaged in at Bitterroot was not "timber management" but "timber mining."

The forests have been decimated by repeated raids. If you want to rape the earth you begin by declaring a crisis. Whenever we are running out of some precious resource, the raider has license to go where he will, do what he likes, strip ruthlessly, and leave a mess behind, all in the name of public welfare.

One of the most persistent raids was the so-called National Timber Supply Act of 1970. It was written by and for industry. It almost became law with the support of President Nixon. The measure was such an obvious steal of public timber that *The New York Times* editorially observed that "the Administration ought to be fighting this timber grab, not supporting it." The act would have profited only the logging industry and despoiled much of the 19 percent of the country's forest land still owned by the people, the *Times* stated: "In 15 years, old growth would be cut that the Forest Service has been planning to ration out over a century. Logging would be king—and forget about scenery, environment, and everything else."

The raid was prepared in the usual way—by first declaring a timber crisis. This took a bit of doing but industry was up to the challenge, helped by its friends in Congress. It cited the need for more housing provided by the Housing and Urban Development Act passed earlier in the year. Thirty-three members of the Agriculture Committee that sponsored the legislation also used the need for housing to justify it; ironically, only 9 of the 33 had voted in favor of the measure to provide the housing that had suddenly become so vital.

Following the customary public relations bombardment, the bill went before Congress in February, soon after the President's urgent environmental distress cry: "it's now or never . . ." He too cited the need for lumber to meet urgent housing needs. He not only forgot the recently proclaimed environmental crisis, but also ignored the freshly signed Environmental Policy Act that required a statement outlining the environmental impact of the proposed legislation. The statement was never forthcoming.

Several other oddities were curiously overlooked or ignored. At the time virtually no housing was being built—not for want of lumber but, rather, lack of funds due to the Administration's tight-money policies. Even as proponents of the bill pointed to American mills closed for want of timber, huge supplies of prime Northeast logs were being shipped abroad: 4 billion board feet were exported in 1969, more than half of it to Japan. Japan, we are told, was then so glutted with wood that it was exporting plywood to the United States.

Timber famines abound with oddities. The United States, allegedly so hungry for wood, still clung to a 20 percent tariff on plywood imports. And during the crisis the industry's profit ratio doubled, climbing "beyond what should reasonably be expected," according to George Romney, Secretary of Housing and Urban Development. While coveting a 10 percent increase in the allowable cut on public lands, the industry was not touching a backlog of 26.6 billion board feet of national forest timber it had already purchased—equal to twice the annual cut.

One more curio: despite the alleged shortage, industry was making no effort to conserve wood. Instead it was continuing wasteful practices, such as burning small growth remaining from clear-cutting that could have been reclaimed for use in pulp (because burning is cheaper than recovering) and feverishly researching to find new ways to make growth faster and cheaper and develop new uses for it, an effort in which it is supported by government. The forest products industry is the fourth largest in America, points out Malcolm Margolin in *The Nation:*

> It thinks only one brute thought: bigger! The very companies that yell "timber shortage" the loudest spend millions of dollars to develop "disposables"—throwaway evening gowns, tuxedos, tablecloths, and the like. For every scientist seeking to save wood, there are a hundred searching for new ways to waste it:
>
> When the Defense Department stops ordering pinewood ammunition cases, the loggers do not rejoice at the salvation of the Southern pine forests; they demand (and get) a House Small Business Committee conference to find out why. And last year when the price of lumber suddenly jumped 30 percent and plywood nearly 80 percent, Congress held three separate hearings to discuss, not how to decrease the demand but how to squeeze more timber out of the forests.

Conservationists were outraged by the latest grab. They mounted one of the biggest counterattacks in history. Enormous pressure was put on Congress to reject the bill. And reject it did. It refused even to consider the measure—a rare snub. But this did not discourage President Nixon. Only four months later he ordered the Secretary of Agriculture to "permit increased harvest of softwood timber" so the housing goal of 26 million homes by 1978 (provided by the Housing and Urban Development Act of 1968) could be met. He offered the sop of saying the increased cutting should be accompanied by money for manpower to plant new trees. He failed to note that the Administration was already 5 million acres behind in the forestation program and had been unwilling to provide money to catch up. Nor were those exports of logs to Japan mentioned.

To conservationists, the President's action was discouraging. Once more it demonstrated the difference between Presidential prose and Presidential priorities—a dedication to industry that almost invariably outweighs environmental needs. As one conservationist characterized the maneuver: "It's just another administrative end run."

Apparently stung by such criticism, and recognizing the need to preserve the tormented forests, the Nixon Administration in January, 1972, announced that clear cutting would be curbed by Presidential executive order. The step was to be taken on recommendation of the Council on Environmental Quality. It seemed odd, at best; theoretically such restrictions had been in effect right along. As soon as the timber industry learned of the plan it dispatched its top lobbyist to the capital. Three days after the Nixon proposal was announced it was dropped. A *New York Times* editorial lamented the lost reform and observed that the order's swift extinction "suggests that the logging industry was more persuasive with the President than his own environmental advisers."

SICK SOIL, SICK PEOPLE

The loss of the forests has had a disastrous effect on the soil. There is no longer protection against the drying sun. The soil particles crumble and die. There follows water erosion and wind erosion. The same thing happens to prairies when their protective blanket of grass is killed by overplowing, overgrazing, or other abuse, especially when followed by droughts. This is what happened to the Southwest in the 1930s when it came to be known as the "dustbowl." Large areas of Texas and Oklahoma were literally blown all over the world. Hundreds of miles away (I then lived in Missouri) the sky would gradually turn red. You could not see the sun at midday. There was just a red haze with an eerie glow behind it. People held handkerchiefs over their mouths; some wore masks. The dust was everywhere—in your hair, your mouth, your lungs. People tried to shut their houses against the dust. It was in the food, on the furniture, on beds, in clothes, everywhere.

Chastened, the dustbowl had to reform. With federal funds great barriers of trees were formed to block the wind. Grass was replanted to peg down the soil. New agricultural practices, among them contour plowing to trap water and prevent runoff, were introduced. Ultimately the soil was stabilized, the drought ended, and the storms stopped. But there was no recovery from the loss. New soil is not easily come by. One single inch takes 500 to 1,000 years to form. Over the entire surface of the earth there is only a thin skin of topsoil, averaging only six to eight inches deep. This light dressing provides our margin of life. Lose it, damage it beyond its power to function, and we are all done for.

The epic of soil begins two or three billion years ago. It is a vital chapter in the drama we call creation. The molten mass believed to have been hurled off by the sun hardened into solid rock. The first life was probably the lowly microbe, single-celled bacteria or fungi. Using carbon dioxide and the acids excreted, these organisms exerted a solvent action on the rock, breaking off minute fragments. This was the start of soil. The bacteria died to provide the first organic matter. This blended with the rich rock dust. At first, so the theory holds, the process was slow, taking millions of years. Then the elements lent a hand to speed up the process: heat, cold, wind, rain, glaciers, and complex biological actions. Night and day brought temperature variations. The rocks developed fissures, water penetrated the cracks, froze like a wedge and cracked the rock, providing more surfaces for the elements and the bacteria to work on.

Then lichen appeared—part algae, part fungi, host and parasite to one another, with no distinction between plant and stem. In their symbiotic union they covered the exposed surfaces of rocks, excreting acids, digging tiny tentacle roots into the cracks and fissures, aiding the long, slow process of turning rock into soil. The lichen were followed by moss and they perished to add their remains to the accumulating mass. By now there was a thin film of soil. In this base ferns took root, multiplied and died, and other plant forms evolved. Then came the trees, at first small and hesitant, little more than thin reeds; the tree experimented, growing taller, forming a rooted platform, sending up a stalk which thickened into a trunk that exploded into branches and leaves, ultimately initiating the recycling of matter and water to house and feed the community it lorded over. In parts of the world the forest growth did not stop with trees. Dense jungles sprang from the earth, choking it with vegetation, dripping water—great rain forests, vines crawling, clinging, intertwining like snakes, a vast, tangled green stronghold that was impenetrable, probably alive with strange life in a yeast of creation.

Eventually men arrived. In their war on nature the process of creation was reversed. Deserts began to creep across the earth's surface where it was flat, and equatorial rays boiled down. From denuded hills washed billions of tons of precious topsoil that spilled into rivers, blocking navigation, smothering marine life, washing into estuaries and the sea. Eons of evolution were lost in hours or days.

The condition is worldwide. Greece, once 60 percent forests, is now only 5 percent covered by trees and has had serious soil losses. Many of the trees were cut down by Christians to purge the peasant mythology that gods dwelt in these aboreal abodes. In the United States there was a more mercenary reason for cutting trees and ripping up the sod. Soon the unprotected soil was overgrazed, overplanted, and beset by long droughts. Soil began to wash and blow away on a continuing but less dramatic scale than the dustbowl. The nation annually loses about 5.5 billion tons of topsoil, most of it through wind and water erosion. From farmlands alone the loss is about 3 million tons a year, even as the need for greater food production grows.

There are many myths about soil. Often thought of as lifeless and inert, it is, in fact, dynamic and alive. It fairly teems with bacteria, fungi, molds, yeasts, protozoa, algae, insects, and other minute organisms that live in the top few inches of earth. Several billion bacteria can be found in a single gram of rich soil. In addition there are earthworms, those important husbandmen of soil. Good soil supports a worm population of 26,000 per acre. Joining them are shrews, mice, moles, ants, wasps, yellowjackets, and other creatures that burrow, tunnel, and cultivate the earth, loosening it so water and air can flow among the molecules of dirt to perform their vital chores.

Soil is a mixture of oxygen, water, minerals, and its living community. Loam soil is composed of one-quarter water, one-quarter air, and one-tenth organic matter—"it thus swims, breathes, and is alive." It is not solid but a working union of billions and billions of particles, each encased in an invisible film of water, each a tiny world in itself dependent on the other tiny worlds around it. An essential part of soil is the organic matter, a rich compost of leaves, grass clippings, decaying plants, roots, crop residues, manures—anything that lives and dies. All of the soil's elements must be in balance, interacting and supporting one another in a physical, chemical, biological exchange. Too much water releases vital nutrients too slowly. Too much air speeds the release up so the mineral content is depleted too quickly. Overly aerated soil evaporates moisture into the atmosphere so it dries out.

Humus, the decaying organic matter, is necesary to a healthy soil. It keeps the particles from composting into a solid mass that cannot be penetrated by oxygen and water. It turns the earth into a sponge that can absorb enormous quantities of water—an amount equal to its volume—preventing rapid runoff and erosion. It also provides food for the living community in the soil and its dampness stabilizes the temperature.

Soil, air, and water are never separated. They are forever interacting in a great symphony of mutual need and mutual aid. Four chemical elements make up the bulk of living matter, explains Dr. Barry Commoner: carbon, hydrogen, oxygen, and nitrogen:

> . . . they move in great, interwoven cycles in the surface layers of the earth; now a component of the air or water, now a constituent of a living organism, now part of some waste product, after a time perhaps built into mineral deposits or fossil remains.

Nitrogen, he explains, occupies a special place among these four elements of life because it is so sensitive an indicator of the quality of life:

> Nitrogen deprivation is a first sign of human poverty; a certain outcome is poor health, for so much of the body's vital machinery is made of nitrogen-bearing molecules: proteins, nucleic acids, enzymes, vitamins and hormones. Nitrogen is, therefore, closely coupled to human needs. Indeed, as the world population grows, nitrogen will become, increasingly, the crucial element in our effort to avert catastrophic famine.

Nitrogen also is the key to the soil problem. But first the stage must be set:

The early arrivals in this country found superb soil across the heartlands of America. There was no finer soil in all the world. It is lyrically described by Aldo Leopold in *Sand County Almanac:*

The black prairie was built by the prairie plants; a hundred distinctive species of grasses, herbs and shrubs; by the prairie fungi, insects, and bacteria; by the prairie mammals and birds, all interlocked in one humming community of cooperations and competitions, one biota. This biota, through ten thousand years of living and dying, burning and growing, preying and fleeing, freezing and thawing, built that dark and bloody ground we call prairie.

It too was a victim of the false philosophy of abundance. In the early days of the nation this ocean of rich fertile soil seemed endless, just as had the trees and the minerals and the air and water and all the other riches so prodigally spent. The settlers and homesteaders were not good soil managers as the Indians had been; they fenced off units of land and then overworked them; they fragmented the prairie as commercial units—a division not recognized by nature. When one patch of ground was exhausted they simply moved to another. Later, as land became more scarce, prudent farmers began restoring lost fertility to the soil in the form of animal manure and various cover crops—so-called green manures.

The great change came after the mid-1800s, with Justus von Liebig, a German chemist, known as the father of modern chemical agricultural practices. Von Liebig analyzed the ashes of burned plants and found three primary elements: nitrogen, phosphorous, and potash (NPK). He concluded that soil fertility depended primarily on their presence in the soil, and supplying these three elements in soluble form would restore soil fertility. The experiment worked like magic. Crops fairly leaped out of the ground in abundance. Agriculture was revolutionized almost overnight.

The discovery of NPK virtually ended the use of humus. Buying bags of chemical fertilizer was cheaper than transporting and spreading manure on the farm. Eventually manure was considered a liability, a problem waste to dispose of. Once manure enriched the land on the small farm. Now it was concentrated in feedlots. One cow produces as much waste as 16.4 present-day Americans. For each pound it gains, a cow produces 5 to 25 pounds of waste. Twelve chickens produce the organic wastes equal to that of one man. Little of this is recycled into the earth but becomes a contaminant that poisons streams and drinking water supplies. More than two billion tons of cattle manure annually are now taken from feedlots every year, piled up to leach and poison groundwaters. Often it is so full of chemicals that it will not break down. Some of it is dumped into rivers where it uses up oxygen and causes huge fish kills and other pollution problems. The mountains of manure are causing several Midwestern feedlots to go out of business. At least two in Kansas recently shut down.

The shift from humus to chemical fertilizers had far-reaching consequences. The new fertilizers did not replace soil fertility; they only enabled the soil to use up such fertility as remained. Gradually fertility declined. Today it is half what it was 200 years ago. There was no loss in total crop production. Often there were dramatic gains as the hot chemicals jolted soil organisms to feverish activity. As organic matter was used up, however, the structure of the soil changed. No longer was it moist and friable. It dried out to become a claylike hardpan. Continual plowing, disking, and hoeing were necessary to break the crust so water and oxygen could penetrate to crop roots. The barren soil, baked by the sun, was subject to water and wind erosion and to drought. Most of the marvelous virgin prairie, with "its black, spongy soil, the splendor of its many wild flowers and the sweep of its tall grass blowing in the wind," to quote *The New York Times*, is almost gone. Only small plots remain as "museums." Various private interests are trying to buy up tracts to restore them. One small area of near virgin prairie was preserved as a cemetery near Plainfield, Illinois. It is a foot higher than the surrounding land that was farmed almost to extinction.

With the coming of the factory farm, thousands of acres of prairie were planted in wheat. Nature was overwhelmed by this monoculture. It had no experience with large single-crop adventures. Virgin prairie was a varied mixture of dozens of different perennial grasses and native flowering plants; nearly 100 species of wild flowers were not uncommon. But chemical fertilizers and overgrazing have turned the sea of grass into seared hardpan. More than 100 million acres of public rangeland are rented to ranchers by the Department of the Interior for grazing herds of cattle. Cattlemen pay the government only 44 cents a month for each animal, compared to grazing charges of $3.50 to $4.00 per animal per month on privately owned grasslands—a subsidy retained by the vigorous lobbying of big cattle interests. Secretary Udall tried to raise the rent, but Secretary Hickel refused to put the increase into effect.

Despite large government programs, overgrazing has left 30 to

50 percent of the grasslands in bad shape. The return from this overuse is not even enough to pay for the damage by reseeding. Year by year the condition gets worse. "Man has probably misused the earth's grasslands more than he has misused any other plant environment vital to man," says Dr. Robert C. Anderson, director of the University of Wisconsin Arboretum, as quoted in the *Times*.

The loss of soil fertility brought on a train of evils. It changed the character of the land and produced a different kind of plant. Old balances were destroyed. At the same time huge tracts were planted in crops that were foreign to the region. These crops often required irrigation that left destructive salt deposits,* and almost inevitably had to be dosed with chemical fertilizer that further reduced the soil fertility. Most of the nitrogen in the fertilizer washed off into streams or percolated into groundwaters, to degrade the water. Planted fields attracted hordes of pests. Man's answer was not to respect nature's protest but to violate her further. The answer was mass spraying. This killed the natural predators and made the bugs worse. Then resistance developed. Imbalances became greater. Pests previously controlled by predators exploded in number to become a menace themselves. The war between man and nature escalated. Nature sent in armies of pests in greater varieties. Man desperately replied with ever more lethal sprays: enter DDT and the related slow agents of death, enter parathion and the related immediate agents of death, enter 2,4,5-T and 2,4-D that cause mutations and birth defects. Enter all the other correctly designated "economic poisons" designed to short-circuit nature.

Gradually the persistent pesticides build up in the soil. The soil itself becomes poisoned along with the plants. Soil organisms are killed off. The soil is rendered even more unstable and infertile. There is evidence that some pesticides are harmful to the vital nitrogen-fixing bacteria in the soil. "It may well be that the avalanche of approximately seven hundred million pounds of organic pesticides a year will eventually bring disaster to the human race," says Representative Dingell.

Now the poisoned, overworked soil falls under attack from another unexpected foe: air pollution. From the smokestacks of

---

* Some of the most extravagant and wasteful water habits are said to belong to farmers of the Southwest, and they raise a basic national question, according to *The New York Times*: "Should the government pay farmers not to till the soil in states with high rainfall while it subsidizes farm irrigation in states with low rainfall?"

cities pours sulfuric acid that dissolves minerals. Accordingly, the high concentrations raining down from the air into the soil can alter conditions needed for healthy plant life; high acidity can reduce tree growth by as much as 5 percent, along with damaging leaves and other vegetation. Acid in the soil can leach out minerals, such as calcium; they then drain into lakes and rivers, further upsetting the already disrupted ecological balance of plant and animal life.

The attack increases. Now, added to the poisons and stimulants, are biological tricks to deceive nature. The farm tactician develops clever hybrid seeds to increase yields; these high-yield seeds require heavier jolts of fertilizer and pesticides. Soil fertility is reduced even more. It is a process of "mining the soil"— taking from it such fertility as remains. An illusion of progress is created through deceptive bookkeeping—entering the exploitation of soil capital as a profit. The production of huge crops is not proof of soil fertility, but represents exhaustion of the earth's nutrients, warns Dr. William A. Albrecht, former chairman of the Department of Soils at the University of Missouri College of Agriculture, one of the world's foremost soils experts.

For the moment the tricks of technology still work. Mountainous crops are turned out. The more the soil produces, the harder it is pushed. The soil, like the forests, is regulated by the computer and economist and not by the biologist. Government researchers devise new ways to produce crops, and then pay the big factory farms fortunes *not* to produce beyond set limits; some of the biggest growers receive up to a quarter-million dollars annually in federal subsidies—an indirect welfare payment that we hear less about than the pennies so begrudgingly given to the poor and destitute. Despite such restrictions, huge surpluses pile up and are given away to maintain artificial price levels. The smaller grower cannot compete with this big-money game; he is forced out of business and sells his land to developers who put up houses for the expanding population, or he is squeezed out by economic pressures and his land is bought up by the new "corporate farm," also known as conglomerate farming or agribusiness, so enthusiastically endorsed by Nixon's new Secretary of Agriculture, Earl L. Butz.

"This surrender, in one form or another, takes place 2,000 times a week all across America," points out *The New York Times*. "That is the average number of farm sales weekly, according to the Agriculture Department." Many of these family

farms are bought out by huge corporations seeking diversifica-
tion, the article states: "International Telephone and Telegraph
now produces not only transistors but also Smithfield hams.
Greyhound now runs not only buses but also turkey processing
plants. John Hancock now sells not only insurance but also soy
beans."

Farms owned by the corporate giants consist of thousands of
acres. A few corporations will dominate an entire industry.
Twenty corporations are now said to control the broiler industry;
only a couple of decades ago broilers were raised primarily in
barnyards of family farms.

> Today, there is virtually no market for barnyard chickens.
> Instead the family farmer is usually growing broilers under
> contract for one of the big agrigiants. In a shed built with
> a loan from a corporation, he feeds mash produced by the
> corporation to chicks hatched in the corporation's incuba-
> tors. When the birds are mature, the corporation takes
> them away, slaughters them in its own processing plant,
> packages them prettily, then ships them off to a supermar-
> ket—perhaps its own.

The factory farm becomes progressively bigger and more
efficient. The soil is pushed even harder, along with the animals
that live off of it. Chickens are fed antibiotics and other drugs to
make them grow bigger faster; lights are left on all night so they
will be tricked into laying around the clock; they are kept on
wire screens so they can have no contact with dirt or disease; the
sick fowl are quickly "processed" and hustled into cold storage;
only after a long battle were cancerous chickens removed from
the food supply. Cattle are fed the artificial hormone stilbestrol,
a known carcinogen. Residues of stilbestrol remain in some
meat sold on the market. Cattle are also fed a variety of bizarre
substances to cut food costs—plastics, manure, sawdust, and
shredded paper. All meat animals are given drugs—hormones,
antibiotics, tranquilizers, parasite killers, etc. Many bags of feed
contain poisons and carry warnings that the animal should not
be fed the contents for a certain number of hours or days before
slaughter. The plucky consumer is thus at the mercy of the
farmer and slaughterhouse obeying these instructions. He must
hope his safety outweighs their convenience. And he must hope
that even if all instructions are obeyed, the chemicals break

down or are eliminated as they are supposed to be, theoretically.

The measure of success in American agriculture is purely quantity. How much corn per acre? How many gallons of milk per cow? How much butter per gallon of milk? How much beef gain per pound of grain? By those standards it is a booming agriculture. The computer and not human welfare is the referee. But there is another measure: at what cost are the gains produced? Is there a connection between the quality of food and human health?

The basis of life is protein. Protein is made up of the amino acids, those complex and little understood compounds so essential to human health. As soil fertility declines, the protein content of crops shows marked decline. Since the start of the century the protein content of some hybrid corn has been cut in half, according to Dr. Albrecht. Wheat protein fell 5 percent in 15 years, he says. "In its place we have moved more and more to those plants which let them build up with vegetable bulk, carbohydrate, let them fill themselves with starch instead of protein, and the human diet has come down."

Many experiments attest to the direct connection between soil fertility and plant and animal health. Bulk can be grown on almost any soil, explains Albrecht. But merely by changing the composition of the soil the composition of the plant is also changed. This can have profound effects on the animal eating the plants grown in the soil. Merely by subtracting or adding trace elements in the soil, Albrecht grew plants that when fed to rabbits made it possible to alter their reproduction or sexual eagerness. He also demonstrated that soil fertility has a direct bearing on the amount of pest infestation on crops. He and other investigators recognize that pests attack plants and animals that are suffering from deficiencies or disease. Thus, sprays merely treat the symptoms of soil deficiencies without touching the real problem. Often they preserve "sick" plants. In a celebrated experiment, Albrecht demonstrated the resistance of healthy plants by growing spinach in alternate rows of fertile and infertile soil:

> By increasing soil fertility it was possible to give plants 100 percent protection from insects without use of sprays— at the same time adjacent plants suffering from infertile soils had 100 percent attack. Using spinach we had two rows of bugs, then no bugs, two rows of bugs and then no bugs. Not because we sprayed but because we got a different condition under them.

Other investigators have demonstrated the link between soil fertility and animal health. Sir Albert Howard found that cattle fed compost-grown crops had an immunity to the highly infectious hoof-and-mouth disease. His countryman, the eminent Sir Robert McCarrison, proved a direct connection between diet and health. McCarrison fed rats the diets of different classes of Englishmen, duplicating in the rats the same diseases suffered by people. McCarrison also determined that health began in the soil. Man is literally created out of the earth, since it is the earth that supplies, through the agency of plants, materials out of which he is made, he said.

> Impoverishment of the soil goes on apace because we take out of it in the form of crops more than we put into it in the form of animal and other organic manures: This impoverishment leads to infertility of the soil and this, in its turn, to a whole train of evils; pasture of poor quality; poor quality of the stock raised upon it; poor quality of the foods this stock provides for man—meat, eggs, milk; poor quality of the vegetable foods he raises for himself; and faulty nutrition with resultant disease in plants, beasts and men.

Whatever the source of nitrogen, the unsuspecting soil microbes convert it into nitrate. In nature, nitrate is found only in small amounts. It is a "limiting factor" (other necessary elements must be present but cannot be utilized because of the absence of a key nutrient). The sudden release of "instant food" from the chemical nitrogen fertilizer intoxicates crops into wild growth, and we have those miraculous yields. But the land, mined of its organic matter, cannot hold the nitrogen and nitrate. Large amounts are quickly washed into streams and leached into groundwaters, touching off a familiar chain reaction.

In a balanced natural system the amounts of organic nitrogen and nitrate dissolved in the water remain low, explains Dr. Barry Commoner. The population of algae and animals is correspondingly small, and the water is clear and pure. Because the nitrogen cycle in the soil is tightly contained, relatively little nitrogen is added to the water in rainfall or in drainage to the land. But now the nitrogen fertilizer joins with the phosphorus from detergents, sewage, and industrial wastes. They excite normal algae growth into explosive and destructive red tide blooms. The thick growth prevents sunlight from penetrating the water. Photosynthesis is slowed down, possibly hampered further by DDT and other chemicals in the water. The algae bloom, decay, and die, possibly killed by more chemicals applied to them. They sink to the bottom. Eutrophication follows. We are off on another accelerated cycle of environmental destruction.

The huge quantity of nitrogen released into the environment also presents health hazards. It finds its way into human water and food supplies. Nitrogen is relatively harmless itself but can be converted by intestinal bacteria into toxic nitrite. It can also react with certain food or even bodily constituents to form other compounds, some of them powerful carcinogens. Volatilized into the air, nitrates react with waste hydrocarbons under the influence of sunlight to produce noxious smog.

Nitrogen runoff has caused serious contamination of drinking supplies. Many states now consider it a health hazard as well as a cause of eutrophication of streams and lakes. Crops are also picking up excessive levels of nitrates. Baby foods may contain 300 ppm or more. This can lead to methemoglobinemia—a condition in which the blood's oxygen-carrying power is destroyed, and which may be manifested in a condition known as "blue baby," according to medical authorities.

Our question, then, is what does this abuse of soil and its products ultimately mean to human health? "The physical and spiritual health of a single individual and mankind as a whole should be the final yardstick in measuring both progress and problems," says author Isaac Asimov. That seems a fair test. Americans should score high, based on our propaganda. They are repeatedly assured that they eat better than other people, they live longer than their ancestors, they are healthier. But this is not quite the case. Our national health statistics are more dismaying than reassuring.

The nation's health services now cost about $70 billion a year. Of that amount $7 billion goes for treating illnesses directly resulting from the environment; an additional $25 billion a year is lost through missed wages and services attributable to environmentally caused illnesses, according to Paul Kotin, director of the National Institute of Environmental Health Sciences.

The suffering behind the figures is more unnerving. Heart disease is epidemic.[6] Mental health is a disgrace (500,000 mentally ill children alone). Cancer is rampant, even among the young. The diseases of old age strike sooner and linger longer; they make cruel sport of our victory over the infectious diseases that

once cut us down quickly and mercifully. Now we are so concerned with trying to find relief to ease our miseries, having our
organs cut out, taking drugs, seeking palliatives, that we do not
ask the vital questions: What brought about this condition? Was
it avoidable? We accept sickness and premature aging as our
natural lot. We deal with the symptoms rather than the cause.
Forty percent of our young men—the flower of the nation—are
rejected for military service for health or educational deficiencies. The increasingly longer age span we boast of is largely a
statistical fantasy, founded primarily in the dramatic reduction
of birth mortality several decades ago. From 1890 to the present
day only about four years have been added to the life-span of an
American reaching age forty, according to vital statistics for
1967. The figures reveal almost no increase in the life expectancy of Americans since the 1959–61 reporting period. Nor is
our health record impressive when compared with those of other
countries. In thirty foreign nations twenty-year-old men will live
longer on the average than will American men, according to Dr.
Jean Mayer, nutrition adviser to President Nixon. The lifetime of
a twenty-year-old woman in twenty-one nations will surpass that
of American women.

In 1950 the United States had the fifth lowest infant mortality
rate among the nations; eighteen years later that position had
changed to thirteenth. In 1966 Sweden, Japan, the United Kingdom, and East Germany, among others, had infant mortality
rates lower than the United States, according to the U.S. Public
Health Service. Men in each of these countries were also expected to live longer than American men.

Once more the technology that should enrich our lives robs us
of the thing most dear to us, our health. To Albrecht this is almost inevitable. He sees technology as an instrument for utilizing the raw materials of nature. It can move toward its ultimate
perfection only by impoverishing nature. "Life is foreign to the
essence of technology." Like many other soil experts he sees
doom ahead if we do not act quickly to preserve whatever remains of the health in our soils, that irreplaceable heritage. We
are left with this warning:

> We must soon face the dilemma of feeding ourselves on
> paved streets because the rural community is about to be
> the dead victim of a parasitic, technical soil exploitation
> that has failed to appreciate the biological aspects of the
> soils in the creative business of feeding all of us.

THE WEATHER: CHANGING . . .

Profound global changes have inevitably followed the disruption of the earth's great harmonies. The mechanical and chemical onslaught knocked askew the delicate natural balance that regulates the interchange of soil, water, and air. New conditions were created by multiple insults and intrusions on natural life-support systems: heat produced by cities and their power plants, deserts and farms where there had been forests, roads and pavement in place of fields, mountains leveled off, waterways diverted to new man-made courses heavy with silt and contaminants, tidal action and flow impeded and eutrophication, soils damaged or ruined and dust sent flying into the atmosphere, mines scarring the earth and their tailings airborne, radioactive wastes and pesticides pouring forth to be evaporated into the air, newly synthesized chemicals from smokestacks and tailpipes interacting in a bewildering complexity of unknown combinations, the sea's pulse fretted by barriers and change, its coastline altered and fettered by commerce and filled in by developers. The dialogue between man and nature becomes ever more strained and unresponsive.

These new conditions have been reflected by changes in weather. Such changes, in turn, have triggered further changes in the earth's surface and its rhythms. Together they form a progressive round-robin of change begetting change. Weather itself is a result of a series of complex interwoven events. It is made up of such phenomena as the spinning of the planet on its axis as it revolves around the sun, its peculiar tilt and shape, the life breathed into it by the atmosphere, a mighty cosmic bellows regulated, in part, by the slant and intensity of the sun's rays, the amount of water in the atmosphere, which is alternately heated and chilled by day and night, the seasons, the saturation and reflection of the sun's heat, natural electrical charges, chemical reactions prompted by elements in the air, the location and height of mountains, the depth of valleys, deserts, and forests that can speed or retard the winds that are set to churning and racing by enormous updrafts from heated surfaces, and masses of cold air sinking on polar ice caps or stalking invading warm fronts backed by mighty pressures.

The world's winds, set to whirling by the spinning of the globe through space, charge upon one another; from the north comes

icy dry air raging through the vast gorge formed by the Appala-
chians to the east and the Rockies to the west; its counterforce,
the moist air currents, travel north at a more leisurely pace from
the Gulf Coast. They fall upon one another in a wide pressure
front, the warm air heavy with moisture thrust upward by the
heavy cold mass. The chilled moisture condenses and falls as
rain, snow, or ice. Mountain obstructions may also send air soaring into the dry cool regions above, chilling the air, squeezing
out the vapor in precipitation. The spent wind continues now as
a parching dry force that sweeps across the broad flatlands and
prairies, scorching the barren land left unprotected by the plow
and overgrazing. The heated molecules of the air, agitated, push
away from one another, adding to the movement; in Kansas the
sea of wheat turns golden brown and the blistered highways
seem to buckle in the burning rays of sunlight. The moving air
strikes a sunbaked mountain and rises, sinking at night into the
cool valley below, traveling in eternal circles, returning to the
poles and taking off again on new journeys.

That is the scenario for weather. But the expected rarely happens. It is frustrated and diverted in its mission by the small
irregularities of topography, local and worldwide conditions; the
air masses that should clash head on go astray; they may hesi-
tate, retreat, advance, or break off to perform oddities of
weather, capricious designs in fog and mist, storm, freak unsea-
sonable snowfalls or inversions, hurricanes, tornados, sudden
calm, a magnificent cloud formation that breaks off to dissolve
into delicately tinted cirrous clouds or to explode abruptly in a
violent rainstorm, itinerant and unreasonable waves of heat and
piercing bolts of lightning. Weather, unlike climate, is a rogue,
unpredictable and ever changing, capable of outrageous surprises and catastrophic effects. Climate can be a bore; weather,
never.

Around the globe, with its smorgasbord of weather, clings the
protective envelope of atmosphere that breathes life into the
planet. This atmosphere is a shield against the sun's scorching
heat. Ozone in the stratosphere deflects the ultraviolet rays. The
earth's crust of soil and crown of vegetation heat up faster than
the ocean and cool air charges toward the heated surface during
the day; the process is reversed at night, when the earth cools
faster than the water and the air returns to the sea, flowing out
over the warmer ocean surface as if seeking relief.

These are all mysteries of unknown origin. Investigation

slowly unlocks the possible patterns of effect but not the miracle of cause. The riddle is vexed by man's activities, as unpredictable as the weather itself. Human activity has produced the notorious greenhouse effect. This is caused by carbon dioxide emissions that trap the sun's heat and prevent its escape back into space from whence it came; the effect is heightened by the 200 million tons of contaminants released annually into the atmosphere and dust from the soil that blocks the suns rays and chills the earth. Will we freeze to death from the particles that block the sun's warmth or bake in the trapped heat? Scientists are not agreed on which end awaits us. They also pose other tantalizing questions: What unexplained weather effects are produced by the new supervening blanket of ozone generated by fuel emissions? What global changes will follow the sun beating on new deserts where forests once sprayed fountains of water into the air? What meaning is there in altering the physical structure of the world's soil so it holds less water and sends great volumes of dust aloft? What will be the ultimate result of turning the atmosphere into a giant test tube, part of the vast human experiment in which we are all engaged?

Ultimate effects are unknown but many changes in weather are already noted by atmospheric experts; others are suspected. Profound repercussions are feared. In the past six years there has been a 50 percent rise in the humidity of the stratosphere, usually an extremely dry region, raising the water content from 2 to 3 ppm (an addition of some 380 million tons of water since 1964), according to recent Navy tests. Normally, air moves into the rarefied region of the stratosphere to be chilled and have its moisture wrung out, which is then dumped back into the lower atmosphere in the form of falling ice crystals. Any change in the stratosphere would weaken its wringing-out process. Scientists believe the accumulating moisture is due to natural processes. But they may be triggered by some secondary effect from man's activities. No one is sure. Knowledge of the stratosphere is almost nonexistent. Into this almost unexplored and unknown realm nations plan to send the SST and other supersonic craft to discharge their wastes with unknown effects.

Closer to the earth's surface, more research has been done. Scientists know that temperatures over cities rise as much as 25 degrees above those of surrounding areas. This causes an artificial upward rising of heated air currents. The rising heat forces vagrant clouds upward to be chilled in the cooler air above. Rain and snow are squeezed out over the vicinity of these urban centers; snow may fall on cities in such quantities that traffic is paralyzed and fortunes in tax monies are spent removing it. "In addition to producing smoke, noise, and frayed nerves cities also make snow," observed Walter Sullivan, science writer for *The New York Times.* Scientists are now trying to prevent the accumulation of snow in and around cities by seeding clouds with silver iodide crystals from a distance. Theoretically, the water vapor will adhere to the crystals, as it does with a nucleus of dust, to form water droplets; hopefully they will become large and heavy enough to fall as snow or rain before reaching the superheated city. Thus one artificial process is met by another, and normal precipitation patterns are altered and realtered.

Scientists believe that the enormous accumulation of dust, along with DDT and other contaminants, in the air causes far-reaching weather effects. Dust appears in the atmosphere in incredible quantities from loosened and dried-out soil, industrial processes, deserts and volcanoes, mine tailings, coal ashes, and other natural and man-made sources. Effects are believed to be even greater than were previously suspected. One study found that the reddish dust whipped up by storms over the Sahara Desert and borne by trade winds across the Atlantic to the West Indies may cause hurricanes. Dust, it is explained, can cause atmospheric turbulence.

Dust presents an additional hazard because it contains fungal spores, bacteria, and other low forms of life that originate in sub-Sahara Africa. Researchers said dust could also play a "significant role" in the formation of rain, since a major component of dust consists of clay minerals known to be efficient as the nuclei for raindrops. Dust content in the air is so great today that it formed a visible haze over the ocean in recent tests conducted by Dr. Joseph M. Prospero, a University of Miami oceanographer. "It was like flying into New York," he said. The dust was said to travel about twenty miles an hour, taking seven days to reach Barbados, and was detected at altitudes as high as 20,000 feet. Often it reached such concentrations that it outweighed the amounts of sea salt in the atmosphere. It was found to hold "all the expected concentrations of earth minerals—with two exceptions." Dr. Prospero said the flying dust contained unusually large amounts of lead and zinc, "indicating that a small part of the dust was pollution from industrial continents such as Europe and North America"—further proof that pollution recognizes no national boundaries.

There is an interesting history behind the dust study. It was begun by British scientists who went to Barbados in the belief that they could find air so clean that they could measure the concentration of extraterrestrial dust in the atmosphere. Instead, their sample yielded such large quantities of reddish dust that they had to abandon their original project, *The New York Times* reports.

This airborne dust is an ominous threat: it is known to reduce the amount of sunlight reaching the earth by reflecting 15 to 20 percent of it back into space. In such diverse places as northern Arizona, Yellowstone, and the Adirondacks the air is said to be 10 times (1,000 percent) as dusty now as it was 70 years ago. Over the Pacific, dust has increased 30 percent. In central Asia it has increased 19 times in the 40-year span since 1930. How much has this contributed to sudden torrential rains that have inundated "various local areas with disastrous rock slides, changing the course of streams and soil runoff?" asks the investigator. Experts say that dust not only can cause and increase rainfall but can diminish or eliminate it as well, and change its character. During the cane-harvesting season in Queensland, Australia, the leaf is burned off before cutting and harvesting; this casts palls of thick smoke over wide areas. "Downward of these areas, rainfall is reduced up to 25 percent from levels in neighboring regions unaffected by the smoke," reports the *Wall Street Journal*. Changes are also taking place in the United States, reports Dr. Vincent J. Schaefer, director of the Atmospheric Sciences Research Center at the State University of New York: "Instead of the downpours we used to receive, we get fine, misty rains and snows in which the drops are so small that they tend to drift rather than fall. We had more than 20 of these this year [1969]. We used to see only two or three a year."

Such precipitation alters normal patterns of moisture distribution. It brings chemical changes due to the airborne contaminants it attracts and washes down in its path. It reflects weather changes that come from altering the character of the world on a global scale. Gradually concern is giving way to fear. Among the apprehensive is Representative Emilio Q. Dadario of Connecticut, an authority on the effects of technology. He says that recent congressional testimony on science, research, and development points to the need to find out what we are doing "to the natural rhythms of earth and to the environment on it." He is especially concerned about the enormous releases and concentrations of heat which did not exist a few decades ago. "These result almost exclusively from man and from his machines," he says. Virtually all of them gulp oxygen and produce heat at a prodigious rate. "Moreover, much of the earth's natural landscape is being altered in such a way as to intensify unnaturally the thermal mechanics of the atmosphere."

Dadario points out that some scientists are "sufficiently alarmed to call for preventative action to be taken at once." To emphasize the threat he quotes Walter Roberts, director of the National Center for Atmospheric Research at Boulder, Colorado. Roberts told Congress that "international regulation of deliberate and inadvertent actions that change our atmosphere has become a necessity, and major measures should be taken for the welfare of mankind. The problem may soon be even more pressing than that of the A-bomb."

PAVEMENT, POACHERS, GUNS, AND POISON: OUR VANISHING WILDLIFE

As recently as the end of the Civil War there were at least a million buffalo. Even the plainsmen were awed by the size of the herds. One recalled seeing in Arkansas in 1870 a herd that stretched "from six to ten miles in almost every direction . . . looking at a distance like a compact mass."

Then the hunters took over. They too were imbued with the frontier spirit that the supply was inexhaustible. The buffalo, like the forests, was just another commodity to be exploited. The new nation grew and prospered, nourished by the blood of the buffalo. The slaughter is recounted by Stewart Udall in *The Quiet Crisis:*

> . . . the army wanted them killed to supply profitable freight in the form of hides; the market hunters wanted to kill them for their tongues and hides; and sportsmen came to kill them for trophies and for pleasure. For a few years buffalo hunting was the main, grisly business of the plains country. . . . At times the hides awaiting shipment were stacked up in Fort Worth in high rows over a quarter-mile long.

The most celebrated and ferocious of the hunters was William (Buffalo Bill) Cody. He would circle among the panicky beasts shooting them down at close range as fast as he could fire and reload. In a single day he killed 25 to 40 buffalo. In a 12-month period he killed 4,280. Other hunters with special rifles picked them off as they grazed. Passengers on trains shot them for amusement. Within a dozen years the buffalo were almost extinct. Now, like the Indians who had lived off them, a wretched band of survivors were rounded up and driven to protected land.

Then the fury turned against the passenger pigeon. Probably never in the history of the world had one life form been more abundant. In the early nineteenth century there were an estimated five billion passenger pigeons. Mighty flocks would cross a continent in a day, so dense that a shadow followed them in flight. They are recalled by the naturalist John Muir in *The Wildlife of America:*

> I have seen flocks streaming south in the fall so large that they went flowing over from horizon to horizon in an almost continuous stream all day long, at the rate of 40 to 50 miles an hour, like a mighty river in the sky, widening, contracting, descending like falls and cataracts, and rising suddenly here and there in huge ragged masses like high-splashing spray. How wonderful the distances they flew in a day—in a year—in a lifetime.

The accounts of their slaughter are chilling. "Here they come," cried someone, and the pigeons arrived with a noise like "a hard gale at sea . . . a current of air" as the birds passed over. The creatures landed. They were clubbed and shot and their necks were twisted. It must have been bedlam, rendered stark by bonfires that lighted the night. Mighty numbers of pigeons perched in the trees, and branches broke under their weight. The roar of gunfire and shouts that accompanied the bloody business continued throughout the night. At dawn such pigeons as remained flew off. The remainder were cast into huge piles for pickling and eating by people or were eaten by pigs and vultures.

The passenger pigeons depended on huge numbers to excite their mating instinct. With dwindling flocks they ceased to reproduce. The last one died in a Cincinnati zoo in 1916.

For many animals without protection the massacre still goes on. Every spring some 245,000 seals are brutally clubbed to death off the Canadian coast so their pelts can be made into sealskin coats, suede handbags and luggage. The large majority of the victims are baby seals which are beaten and skinned (sometimes alive, it is said) in front of their terrified mothers, who looked on helplessly. "At the end of each day, thousands of mother seals can be seen nuzzling and mourning over the carcasses of their dead pups," according to Lewis Regenstein, coordinator for the Committee for Humane Legislation, in Washington. Until 1970 as many as 500,000 seals were killed annually in the Gulf of St. Lawrence and along the Labrador coast, according to Friends of Animals; in 1970 a quota of 245,-000 was imposed.

A few months after the Canadian slaughter, the U.S. Department of Commerce carries on its annual Alaskan "sealing operation," in which some 60,000 (mostly adult males, but also some females, says FOA) are killed. The Commerce Department sees nothing wrong with this slaughter. Animals are on earth for man's use and profit, the department has stated. Commerce Secretary Maurice H. Stans, after a personal inspection tour of the seal "harvest" in Alaska, denied that the clubbing is inhumane (the animals are first stunned with a baseball-like bat and then bled to death). A different version comes from Tom Bywaters, a film producer:

> Each morning during the killing season, the animals are driven inland for the kill. The men walk them a great distance. Great clouds of steam rise from their bodies. From time to time a seal falls in a quivering mass, exhausted. At the end of the long drive, the animals are broken into pods of five or six. A man rattles a tin can. The seals raise their heads in panic. Crack, crack, crack. Other men smash their skulls. Despite government claims that the animals are quickly rendered unconscious, time after time I saw the animals hit again and again because one blow had not been sufficient to stun them. The seals are smashed in the face, the flippers, the back. Eyes pop out of skulls and the animals shriek in pain . . . flopping and twisting in the tundra for an excruciating amount of time until . . . smashed again into unconsciousness.

Stans claims the number of seals is estimated at 1.3 million, compared with 200,000 sixty years ago, and only the surplus is killed. "There is no present danger whatsoever of extermination of the herd under these policies," he says. His optimism and figures are denied by Friends of Animals. FOA says the present size of the herd is closer to 329,000 than the 1.3 million claimed by Stans. Regenstein insists that the natural size of the herd is over 5 million, according to official U.S. Government figures that one rarely hears cited. Even the 1.3 million claimed, although an inflated figure, represents a depletion in the herd, says Regenstein. Where, then, are the surplus seals the Commerce Department says it is killing? he asks. Further, charges FOA, Secretary Stans' claim that only bachelor seals are killed is untrue. In 1969, according to FOA, hunters invaded the breeding rookeries, "in a desperate attempt to fill the quota, 11,594 female seals were taken. Every one had one pup. Every one was pregnant. Every one of those pups died of starvation."

FOA says that industry's claim that the herd has increased 1,000 percent under the fur seal program is a deception. It cites government data which show that in the last ten years the birth rate has dropped from 643,000 in 1960 to 306,000 pups born in 1970—a 52 percent decrease.

The seal operation, according to FOA, cost American taxpayers $2 million in 1970 "to subsidize . . . the seal kill." As part of the government contract, awarded to a single private firm, "$75,000 is being spent on advertising to obscure the facts in the war of the environment."

A bill introduced by Senator Fred Harris of Oklahoma and Representative David Pryor of Arkansas would protect all ocean mammals—seals, whales, porpoises, dolphins, polar bears, walruses, sea lions, and manatees—from being killed in U.S. waters or by U.S. citizens anywhere. The bill, initiated by Alice Herrington, founder and head of Friends of Animals, would ban the import of the fur or other by-products of the same mammals, removing much of the economic incentive for the killing of the creatures. The United States is also urged to take the lead in negotiating an international treaty for their protection. The Harris-Pryor bill—The Ocean Mammal Protection Act—is, as expected, vigorously opposed by the business-oriented Nixon Administration, the fur industry, and the "sportsmen's" lobby.

The list of threatened species is long. It includes wolves. Although they are now known not to be the vicious predators of legend, they still bring a $200 bounty in the Arctic. The vanishing grizzly bear is still shot for sport. Bears are off limits in Alaska but the oilmen pouring in for the great North Slope strike have found a way to get around regulations and shoot them in "self-defense." In the Arctic, polar bear hunting from planes is a favored sport. The plane pursues the bear until it is exhausted; then the triumphant sportsman shoots the animal down with a high-powered rifle with telescopic sight which permits him to remain at a safe distance. Only 10,000 to 15,000 polar bears are said to remain worldwide, and some 1,500 a year are being killed off, mainly by U.S. hunters in Alaska.

In the sea several species of animals have been "overharvested," as the euphemism has it. The most gravely threatened are the large whales. The blue whale, said to be the largest animal ever to inhabit the earth, is believed to have passed the biological point of no return. The few hundred left are likely to die off from natural causes; the decreasing number makes it difficult for the huge creatures to find one another and mate in the vastness of the oceans. "One of the next to go will be the humpback whale whose intricate seven-octave 'song' has inspired symphonies and has been recorded on a popular album," says Regenstein. The sei, bowhead, gray, sperm, finback, and right whales are also seriously depleted and have been placed on the government's Endangered Species list. "It is estimated that several of these whale species have about one year left before their numbers will be reduced to a point at which their survival will become doubtful."

> At the present time the remaining 70 species of whales can still be hunted, killed, or imported into the U.S. Even porpoises and dolphins—among the friendliest and most intelligent of all the world's creatures—are now being "harvested" in such numbers as to threaten their future existence.

Whaling has long ceased to be the romantic pursuit we once thrilled to in *Moby Dick*. The modern whaling ship is a highly efficient factory. The whale is electronically stalked by sonar and airplane; then into it is fired an electronically aimed harpoon armed with a grenade that explodes inside the creature. The carcass is inflated to float and is located by an attached radio transmitter. Technology enables the hunters to kill three or more whales a day, where once the take was limited to two or three a month. A whale factory ship cuts and "processes" a whale in thirty minutes.

Whale oil and meats are used in a variety of products: lipsticks, shoe polishes, lubricants, and cat food. All could easily be replaced by common substitutes. When the United States put several whale species on the Endangered Species list, two whale-processing companies pressured the government for their removal. The grounds? That whale oil is necessary to "national defense."

Under the International Whaling Commission whales are hunted and killed on a quota system that is a tragic farce. Although the whales are almost exterminated, the commission in 1971 set quotas for all of the world's oceans at the same high level as prevailed in the previous year. Only in the North Pacific did the commission reduce the allowable catch by 20 percent, but even there the new levels—nearly 4,000 sei whales and more than 10,000 sperm whales, for example—are far above the rate at which nature is replenishing them. Judged in terms of the facts about the whaling crisis, *The New York Times* stated that "the commission's actions seem almost grotesque, rather like telling a firefighter to pour on slightly less kerosene."

The Whaling Commission is dominated by the Russians and Japanese, who between them account for about four fifths of all whales caught. In 1971 Secretary of Commerce Stans commendably ordered that the United States no longer would use its kill quota to hunt whales. But under pressure he licensed one company to continue for another year; it ceased operations and ended U.S. whaling in December, 1971. The U.S. action, however praiseworthy, is virtually useless unless the Russians and Japanese can be persuaded to stop their greedy onslaught against the vanishing whales, and as long as the United States permits imports of whale products it indirectly encourages the slaughter. Some of the United States whalers who were forced out of business were bitter because the International Whaling Commission now permits Russia and Japan to hunt whales in the international waters off the state of Washington. They said this makes no sense and largely nullifies the American gesture.

The slaughter and depletion of animals is worldwide. If a creature has economic value it is under attack, directly or indirectly, legally or illegally. Laboratory needs have almost cleaned out many of the big monkey-hunting grounds in India, Kenya, Somalia, Ethiopia and the Colombian part of the Amazonian forests. In Europe birds are caught in fine nets, often illegally, and eaten, even rare songbirds. "The Italians trap and eat everything," complained a Scandinavian. "Nightingales, swallows, even very small birds with only a few grams of flesh." Russian fishing fleets, using devices that suck in fish like a huge vacuum cleaner, leave behind virtually nothing to mate or spawn. Other nationalities use fine nets about as efficiently. Some still illegally explode dynamite underwater to kill fish. Adult fish are being eliminated as their remaining offshore spawning grounds are being poisoned by wastes and chemicals from engineering projects, power plants, and factories. Many species of fish that were recently abundant are now becoming scarce, among them the sea salmon.

The safari to hunt and kill wild animals is still promoted in what the *Times* describes as a "ghoulish species of business literature":

> Some recent examples of the genre illustrate the nature of the appeal: the nearer a creature is to extinction, the more exciting it is to get a specimen for your living room. One pamphlet advertises "the best jaguar hunting possible." Another rhapsodizes about the "search for the noble tiger," the thrill of "the first sight of dark brown stripes on an orange-yellow body . . ." A third boasts of "perfect performance, every one of the company's safaris having "bagged one jaguar." Still others feature the killing of crocodile, blackbuck antelope, and rhinoceros.

A short time later Pan American Airlines announced that it no longer would sell safaris to kill wild animals but would promote photographic safaris, and it was hoped that other airlines would follow their lead in protecting vanishing species of wildlife.

Once there was room for wild creatures in the American West. But now "civilization" is closing in, "even in the high mountain meadows and rushing streams, and the animals are fading away," reports the Associated Press. The number of wildlife species on the official endangered list jumped from 89 to 102 in the last year, according to a 1969 report by the National Wildlife Federation. Throughout the world nearly 1,000 species are considered endangered. Extinction may be direct by outright killing or indirect by destruction of habitat.* Along with the endangered animals, some 2,000 kinds of flora, representing about 10 percent of the 20,000 plant species on earth, are threatened with

---

* Some conservation organizations are now turning their efforts from trying to save individual species to trying to preserve the habitats, or biotopes, in which the animals survive. This, as one conservationist explained, includes their special food and the way they find it, the conditions for breeding and "carrying on their normal social life."

extinction, according to the authoritative International Union for the Conservation of Nature and Resources.

On the list of endangered species are the timber wolf, the whooping crane, and the Montana west slope cutthroat trout. When a species is endangered, it is not enough just to stop killing it; survival is possible only if sufficient numbers are bred in captivity and then set free. Listed as rare—infrequently sighted by hunters or conservationists—are the grizzly bear, arctic grayling, spotted bat, and prairie falcon. The timber wolf and mountain lion have been almost exterminated by guns and poisons of herdsmen and rangers who see them as predators—a label falsely applied to some animals. Along with the poisoned "predators" go many other animals whose only sin is hunger.

In one month of 1971, there were 48 bald and golden eagles (protected species) found dead in Wyoming, according to Lewis Regenstein, coordinator for the Committee for Humane Legislation. About half the eagles had been poisoned by thallium sulfate, a poison spread by the U.S. Department of the Interior throughout the western United States in an effort to exterminate coyotes. Regenstein said there is not a doubt that "the killing of these eagles is part of a deliberate, well-planned campaign, aided and abetted by the U.S. government, to wipe out all the predatory animals which might compete with agriculture interests":

> At the behest of cattle and sheep farmers, the Interior Department has adopted a mass and indiscriminate poisoning campaign designed to wipe out all wild animals which these ranchers consider undesirable. This massive effort involves distributing throughout the western United States tons of grain and meat baited with the deadly poisons, cyanide, strychnine and sodium monofluoroacetate, or 1080. The intent of the program is to "eliminate" such predators as coyotes and mountain lions. . . .

A rancher was charged with various offenses connected with the poisoning of the eagles. On his ranch investigators found seven antelope carcasses baited with enough thallium "to kill every animal in the state." * In Colorado 54 more dead eagles

---

* Public furor erupted over disclosure of the poisoning of the eagles; the irony was not lost that the bird symbolizing the nation is threatened with extinction by poison laid down by the U.S. Government (which also may be an omen for humans). Several conservation organizations sued the government to halt its "predator control" program. Consequently President Nixon, in conjunction with his annual message on the environment, early in 1972 issued an Executive order forbidding the use of poisonous chemicals on public lands, including those leased to sheep ranchers.

were found, apparently victims of high-tension wires, pushing them further toward extinction.

The Department of the Interior has been spending $8 million a year on "predator control," virtually soaking the entire American West with poison. A *New York Times* editorial notes that the department has put out poisoned carcasses on the open range. It has used the "coyote getter"—a baited cartridge on a steel rod stuck into the earth, which shoots cyanide into the animal's mouth (and sometimes these vicious devices get unwary children at play and adults). "From airplanes it has scattered hundreds of thousands of lard-covered strychnine balls like deadly snowflakes. It is a thoroughly shameful, senseless and savage business carried out not by crackbrained individuals but by an arm of the United States Government.

"This hideous practice has resulted in the slaughter not only of coyotes but also thousands of bobcats, mountain lions, badgers, foxes, opossums, raccoons, beavers and porcupines. Every impartial scientist who has seriously studied this predator control program has condemned it as unnecessary, inhumane, and ecologically a disaster."

The Department of the Interior has estimated that 80,000 coyotes are poisoned yearly in Texas, New Mexico, Colorado, Wyoming, Utah and Idaho; the poison is planted in bait at low concentrations and causes a painful death. The chief victims have been coyotes because sheep raisers blame them for the loss of lambs. *The Times* editorial states: "But since lambs are naturally fragile animals and since coyotes primarily feed upon jack rabbits, rodents and carrion, the suspicion has long existed that many sheepmen 'cry coyote' mainly to establish a basis for claiming tax losses."

The government poison program has been administered by the so-called Division of Wildlife Services of the U.S. Fish and Wildlife Service. Despite occasional propaganda to the contrary, the Wildlife Services "exists almost entirely on behalf of the livestock industry," according to Jack Olsen in *Slaughter the Animals, Poison the Earth*. He points out that "something like 50 percent of its budget is paid directly by sheepmen (who pass the cost along to the buyers of wool and lamb) and the rest by taxes."

Aerial gunners also take a toll of wildlife. A helicopter pilot, James O. Vogan of Murray, Utah, testified before the Environmental Subcommittee of the Senate Appropriations Committee that bald and golden eagles were illegally shot from the air in Wyoming, along with antelope, elk, deer and Canadian geese. He estimated that 570 eagles were taken from the air over one Wyoming ranch, and 200 more were shot out of the air over other ranches in Wyoming and Colorado between September, 1970, and March, 1971. Vogan said a Wyoming rancher paid $15,000 for the killing of eagles and coyotes from helicopters.

Van Rennselear Irvine, aged 51, a wealthy and influential stockman, pleaded no contest to 29 counts of violation of state game laws connected with the death of 50 golden and bald eagles. He paid a fine of $679. Casper businessmen told *The New York Times* that the action was too soft and the situation gave the state a bad name. One said, "If I shot just one eagle I'd be in jail in a minute."

Also on the way out is the once-proud mustang. He has been almost dispatched from the prairie he roamed with such grace. Now he graces the inside of a can labeled "dog food." A century ago there were millions. Only 17,000 now remain and they are disappearing fast. Ranchers and herders consider them pests because they too must eat. A journalistic tear is shed for them by the *Wall Street Journal:* "Trapped and sold for six cents a pound on the hoof, shot and left to rot on the range, driven off cliffs, the mustang is a slowly vanishing symbol of the American West."

The carnage of wild creatures is increased by cars. They kill wildlife and domestic animals alike, along with the 60,000 humans they rack up annually. In the southwest, where highways are wide and cars go fast, it is a local joke that a cow has to graze at 20 miles an hour to stay alive. Along the roads of New York, 761 small furry animals were found in a year by two New York State Department of Transportation employees who took time to count and lament the mangled corpses; included were squirrels, skunks, muskrats, raccoons and rabbits; their only offense was that nature had not taught them to dodge cars.

Some of the most noble animals, by man's standards, are in the most demand and therefore the most imperiled. More than 28 million live wild animals were imported by American pet businesses, laboratories, and zoos in 1967 alone, according to Daniel McKinley, a biologist at the State University of New York. "If only the rich were guilty," he laments. "But the Sears Christmas catalogue offered a leopard-skin 'trophy' for $700."

Many of these animals are protected in their native land. But poachers shoot and smuggle them out for shipment to New York.

New York State has made an admirable effort to stop the practice by passing legislation (The Mason Act) that would ban the
sale in the state of alligator, caiman, and crocodile hides and of
furs from the tiger, leopard, cheetah, and other spotted cats,
vicunas, red wolves, polar bears, mountain lions, or cougars. By
1971 the law covered ocelots, jaguars, and margays. This would
strike what *The New York Times* calls a "vital blow" against the
fashion and fur trades centered in New York City. Several industries using the hides challenged the act in court as unconstitu-
tional. A sympathetic State Supreme Court justice upheld the
industry. He was overruled by the State Court of Appeals. That
enlightened tribunal said the animals covered by the law were
"necessary not only for their natural beauty and for the purpose
of biological study but for the key role that they play in the
maintenance of the life cycle." Such understanding is the beginning of an environmental wisdom desperately needed in high
places.

The Interior Department, in February, 1972, proposed to add
eight species of spotted cats to its list of endangered species to
stop the "heavy exploitation by the skin trade." If enacted into
law, the measure would ban the importation into the United
States of any parts or products of the cheetah, leopard, jaguar,
margay, ocelot, snow leopard, tiger and tiger cat.

The department observed that "organized poaching rings
flaunt the laws of the countries where these cats originate and
send a flow of furs to feed the fashion salons of the United
States."

The Committee for Humane Legislation, Inc., stated that between 1968 and 1972 "U.S. furriers imported 18,458 leopard
skins from Africa as well as 31,105 jaguar pelts and 249,680
ocelot skins from South America."

According to the Interior Department, reports *The New York
Times,* "fewer than 2,000 tigers exist in India today, a reduction
of over 90 percent in the last 25 years, while probably no more
than 400 snow leopards survive in the entire Himalaya Mountains . . . probably no more than 2,000 chetahs exist in the
parks of Africa, and even the widespread and secretive leopard
is being drastically reduced in numbers." The jaguar, the ocelot,
the margay and the tiger cats were said to be facing "tremendous pressure" in Latin America.

Since 1970 most union fur workers in the United States have
refused to handle the skins of any endangered species, including

those of spotted cats. But "the United States and European markets remain open and poachers succeed in getting shipments through," according to the Interior Department. It joined conservation groups in expressing hope that the proposed ban would prompt European countries to take similar steps.

Another encouragement is the number of people who now see the killing of animals for their hides or for sport as immoral. They are putting increasing pressure on the less sensitive who wear these products.

The House of Representatives recently passed a bill that would make it a crime, for a private citizen, anywhere in the country, to shoot or fire at any animal from an airplane, unless he had a state permit to go after predators known to be damaging his property. The vote of 307 to 8 would indicate small disagreement, regardless of party, observes *The New York Times*. "But the incredible fact remains that the Administration is opposed, along with substantial numbers of senators. The federal government's own Fish and Wildlife Service sometimes uses this method to assist cattle and sheep raisers to eliminate predators."

Similar pressure is being exerted to outlaw the most vicious device ever invented to catch animals: the leg-hold trap. Its steel jaws (sometimes jagged) can tear off the leg of a struggling animal or hold it until the tortured creature starves, goes mad with pain, or is clubbed to death (shooting would damage the skin). Some creatures become so desperate that they chew or twist off their own limbs to escape. No record is kept of how many animals are caught this way. *Fur-Fish-Game* magazine states that roughly 30 million furbearers are trapped each year in this country. The figure is challenged by Mrs. Thomas S. Maxwell, director for Wildlife Preservation of the Humane Society of the United States, New Jersey Branch. In her opinion it's closer to 100 million furbearers and birds. The use of traps is uncontrolled, she points out: "Anyone can use them for any purpose. Children are big trappers, contributing to their early brutalization. Their target is wild animals but they are as likely to catch dogs, cats, birds, and other unintended prey that happen to pass by."

Hunters are also a menace to wildlife. Hunting interests are so powerful that in most states they are able to block almost all laws that would protect game. In New Jersey, for example, the State Fish and Game Council has total jurisdiction over wildlife, with the exception of migratory waterfowl, which are under federal control. The eleven-member council is made up of six men nominated by the New Jersey Federation of Sportsmen's Clubs. Three farmers are nominated by the Agricultural Convention (composed of farm organizations). The other two are commercial fishermen. These nominations are approved by the governor, as required by state law. No one speaks for the animals. The council vigorously resists and usually manages to block all local regulations in the state that would in any way curb hunting or trapping, says Mrs. Maxwell. "This means that animals belonging to all the people in the state are turned over to private groups, whose total membership represents probably less than 2 percent of the population of the state, for their personal pleasure and exploitation." The New Jersey council has its counterpart in practically every state in the union.

Many hunters kill with maniacal fury. They kill out of season, and often their victims are rare or protected species. A biology student counted 21 dead hawks and 16 golden eagles that had been shot by hunters beside telephone poles along a 10-mile stretch of dirt road in Utah. He surmised the killers were "frustrated city dwellers out for a weekend of plinking." In California maniacs with guns riddle land and sea animals with bullets and leave them dead or dying. They shoot creatures in the national forests, domestic animals on farms, animals on offshore islands. Anything that moves or flies is a target—seals, porpoises, sea lions, goats, sheep, birds. It is a bad situation, authorities say, and getting worse. Patrols now go out day and night to try to protect the animals from these madmen who are legally permitted to have guns.

For wildlife there is almost no escape. Civilization keeps pushing in on all sides. In addition to guns, poisons, traps, and poachers, technology is destroying and shrinking whatever remains of natural environment. In the Everglades the hard-pressed alligators are beset by poachers, restricted flows of waters, oil pollution from new wells being drilled, and a nearby runway left in operation from an aborted jet airport, a project that may only be dormant rather than dead. Elsewhere there is the crescendo of machines and motors—automobiles, trucks, snowmobiles that invade wilderness habitats, bulldozers, backhoes, chainsaws hacking down forests for new roads. In the intermountain West, there is a conflict of interest in the use of scarce water. The animals, as usual, will be the victims. "Several important wildlife areas are being jeopardized because the available water is being diverted to irrigation," says Roland Clement, vice president of the Audubon Society. "Not because we need the added production, but because increasing numbers of our own species need to be kept at work." Animals must even pay for the boondoggle.

Within 25 years between 75 and 80 percent of all species of living animals will be extinct, according to Dr. S. Dillon Ripley, secretary of the Smithsonian Institution. This view is supported by the biological past, and the rate at which the environment is becoming progressively hostile to life. Biologists warn that 99 percent of all life forms that existed on earth are now extinct, despite their efforts to survive.

Since 1699, 359 animal species have become extinct, half of them due to the effects of man, and 817 species are currently endangered, according to *The Red Data Book*, the bible of wild animal conservationists, published by the International Union for Conservation of Nature and Resources. Daniel McKinley, assistant professor of science at the State University of New York, says this includes 1 percent of some 4,000 species of mammals in the world that have become extinct, and 3 percent are now endangered. Of nearly 8,700 species of birds, almost 2 percent have become extinct in the same period, and another 1 percent are now endangered, he adds: "If we include many differentiated and interesting island forms, so instructive for the study of evolution in action, we come up with about 322 rare and endangered forms of birds. Some 200 of these live on islands where every airfield, every phosphate mining operation, every radar tower, and every increase of human population imposes new threats."

Experts estimate that the African wildlife population has been reduced to a tenth that of fifty years ago, according to Allan C. Fisher, Jr., writing in the *National Geographic*, February, 1972. He notes that a listing of endangered fauna, maintained by the International Union of Conservation of Nature and Natural Resources, "currently contains 656 mammals and birds, including 123 in Africa. Many more, though not on the endangered list, have dwindled alarmingly in population." Fisher asks if Africa's wildlife will be able to survive. "Or will they be reduced to small zoolike remnants, difficult to find and see, pitiful curiosities, a reproach to all mankind?"

Each chemical unleashed on the environment and every engineering project makes its small contribution to speeding up the process of evolution. Each change adds to the discomfort or danger of existing life forms that have adapted to existing patterns. No single threat will ruin the world, nor is it even likely that it can be scientifically proved to cause significant damage, says McKinley. "But the total of these invisible and apparently inconsequential insults is staggering."

Birds have been hard-hit by man's depredations. Some of the most magnificent winged creatures are gone altogether or on their way out. Their situation is so desperate that biologists are now trying to save a few specimens of wild birds through artificial insemination and breeding them in captivity under simulated natural conditions—a pitiful living museum in a dying world.

The list of victims is long and growing. The osprey is endangered, as is the brown pelican. The California condor is disappearing, along with the Florida Everglades kite and the Bermuda petrel. There are believed to be now at most only 800 nesting pairs of bald and golden eagles left in the United States (for 36 years a bounty was paid on the birds in Alaska because they were considered "damaging" to the salmon industry, and in that period more than 100,000 were killed for payment).

The whooping crane, almost extinct, and then carefully built up from only 14 left in 1937 to 56 now, is in trouble again. This time they are threatened by dredges hacking away at the cranes' feeding grounds off the Texas coast. The dredging operations make waste of the whole area, complained an Audubon Society official. "When one bay is stripped, they just move to another." The dredging (shell mining) rips up the bottom, silting the water and destroying food sources so the habitat no longer can support life.

In Denmark the national bird, the stork, will soon be gone. Last summer just 70 pairs crossed the border from West Germany and settled down in the old nests atop houses, farm buildings, or spires, reports United Press International. "Zoological experts predict that the birds will die out in the near future and that nothing can be done about it." A century ago 10,000 storks arrived each year from the wintering grounds in Egypt 8,000 miles away, and were greeted by delighted children's cries that "Father Stork" had arrived. The decline is blamed on lost marshes and swamps and the consequent loss in food for the storks, changing weather, and migratory perils that include "electric wires, trigger-happy farmers, and poison put out against the vast flocks of grasshoppers that sweep across Egypt."

It's all up with the peregrine falcon, that spectacular sky performer who thrilled many with his bold and villainous thrusts, raiding other birds in flight, seizing them, or striking with such force as to kill at once or send the victims plunging to the ground to their death where they would be leisurely plucked and devoured. It's a man's bird, a strong, silent, solitary raptor," explained Arthur Cleveland Bent, a naturalist, in *The New York Times Sunday Magazine*. His tribute is now an obituary note, a museum identification for a proud bird stuffed and with unseeing glass eyes, perched self-consciously on a plastic limb behind a glass pane. The peregrine is a documented victim of DDT poisoning; he was unable to reproduce because of calcium deficiency in his shells.

Must this always be? asks Hugh H. Iltis, botanist at the University of Wisconsin.

> Shall man come always first at the expense of other life? And is this really first? This may be expedient but in the long run impossible. . . .
> Butterfly and wild flower, mountain lion and antelope, blue whale and pelican, coral reef and prairie land—my grandchild may need to know them, to see and smell them, to hear and feel them, to be alive, bright and happy! Yet hardly anyone speaks for wild nature—for *Morpho* butterflies in a Peruvian valley, for timber wolves chasing caribou in Alaska.

What happens to man in a world without other creatures? asks another naturalist, Joseph Wood Krutch.

> What will he then have become? Will he not have become a creature whose whole being ceased to resemble *Homo sapiens* as we in our history have known him? He will have ceased to be consciously a part of that nature from which he sprang. . . .
> He will no longer have, as he now does, the companionship of other creatures who share with him the mysterious privilege of being alive. The emotions which have inspired a large part of all our literature, music, and art will no longer be meaningful to him. No flower will suggest thoughts too deep for tears. No bird song will remind him of the kind of joy he no longer knows. Will the human race have then become men-like-gods, or only men-like-ants?

Often we forget the importance of other creatures in our narrow view of life. We forget that the world rests on a biological base and will collapse if it is destroyed. In natural communities, still existing today, each species plays a role that is often narrow, specialized, and nonoverlapping, explains Daniel McKinley in *University Review*, published by the State University of New York:

> Fraser Darling lists the roles of big game mammals in Zambia—elephants, hippo, rhino, buffalo, zebra, giraffe, two kinds of pigs and 21 species of antelopes. . . .
>
> Each species seems to play a particular role in the ecosystem. Some make paths for others; some fell trees; others disperse seeds; some keep rivers open and fertilize waters; some graze in wetlands; others graze in special places; some browsers and grazers specialize in eating certain plants; some like woody stuff, others fruits, flowers, buds, leaves. And so on.

From animals we have learned about our own behavior, we are reminded by McKinley, a biologist. The smallest forms beneath view do essential jobs quietly and without dramatics or applause. We overlook the vital role of many of these forgotten creatures. "A nest of red ants in one summer destroys 3 to 5 million insects," according to Russian literature. Artificial sowing of ant nests successfully checks outbreaks of sawflies, webworms, and silkworms. One Italian worker found that ants on a million hectares of forest each year consumed 24,000 tons of food, including perhaps as much as 14,000 tons of pests:

> In some areas of the world, major parts of the soil profile are the work of ants. Termites are essential parts of the ecosystem in African Tropical Forests. In this tropical landscape nutrient minerals are lodged precariously in the biosphere. The cutting of forests not only kills trees, but kills termites, and lets minerals flush from the land. . . .
>
> We often assume the right to remove species that get in our way, especially when they seem useless. But unless we love the ecological array of mammals, termites and ants, caterpillars, and the gentle rain, we will inherit a land where much of the vegetative matter produced does not get eaten, where cattle eat by preference the young trees, where much of what remains is not properly trampled and comminuted for return to organic matter, where accumu-

lating litter ends up producing very hot fires that kill forest reproduction and where ashed nutrient is likely to be flushed permanently through chemically inert soils or lost due to surface erosion.

It is this warning to which we must pay the closest attention: the risk we take in altering and reducing the diversity of life. Nature's first principle is variety. Each plant and organism performing its own specialty is essential in the complex web that sustains us. In nature a species dies off or is eliminated when its job is no longer necessary—it becomes superfluous or the altered habitat will no longer support it. Technology is now taking over nature's role as the censor of life. It does not eliminate nature's needs but merely the species that fulfilled those needs. The need itself remains unattended. With each loss of an essential life form the fabric of life is damaged in some degree. The damage may appear infinitesimal, it may be undetectable, and yet touch off profound repercussions. Such effects may be beyond our capacity to perceive or understand until we see some drastic end result that is likely to be attributed to an incorrect first cause.

In the conquest of nature we have been blind to our own peril. By destroying and reducing the habitat of wildlife we have injured our own habitat. By wiping out other life we diminish our own. For billions of years evolution moved toward present conditions. It purified the air, cleansed the water, built soil, grew forests, created weather patterns that sustained the earth and interlocked its parts in a cosmic unity. Now man is abruptly altering these patterns and harmonies. He is, in effect, putting into motion a process of reverse evolution. Poisons long stored in the earth are being returned to the air. Soil is being destroyed. Atmosphere is polluted. Water is contaminated. Forests are drastically reduced. The sea has become a sink rather than a source of renewal. Weather patterns are disrupted. The environment is becoming increasingly harsh and less diversified; monotony precedes sterility. The whole system is breaking down under the impact of our witless meddling and mindless greed.

The warning is obvious if we have the sense to see it. The vanishing species tell us what is taking place. Their total numbers may remain the same, but the composition is changing. Refined creatures are being wiped out and replaced by the coarse, which are better adapted to a toxic atmosphere and will be the last to go. In place of the bass and salmon will come the carp, which already abounds in dying Lake Erie. The ocean bottom no longer will be vibrant with crabs, oysters, clams, shrimp and mussels; in their place will be bloodworms, sludgeworms, sow-bugs, and bloodsuckers that can live on the most vile wastes, almost without oxygen. The peregrine falcon is replaced by the sparrow hawk, the osprey by the pigeon, the whooping crane by the sparrow. Others will follow.

The same fate will ultimately claim man. The most suscep-tible and sensitive will be the first to go. Those who cannot measure up to the toxic standards will be eliminated. The sur-vivors will live in a joyless, poisoned, barren world, gasping for air. The end will be neither a whimper nor a bang, but gradual, progressive decay. The only inhabitants will be those with leather lungs, iron livers, and plastic souls. It will be the time of the sparrow, when the earth no longer can flower and only its poorest creatures survive.

INDUSTRY AND GOVERNMENT: POLLUTERS AND POLITICS

In 1967 President Lyndon Johnson gave the American people a chilling choice. "Either we stop poisoning our air—or we be-come a nation in gas masks, groping our way through dying cities—and a wilderness of ghost towns." Congress responded by passing the Air Quality Act of 1967, superseding the Clean Air Act of 1965. Both of these bills, so optimistically named, were hailed by their sponsors as landmark legislation that would clean up the nation's dirty air. Yet five years after the first and three after the second, the air was more polluted than ever and getting steadily worse. Why? The reasons are many and com-plex.

The primary problem is the alliance between industry and government, and government's failure to provide the people with the protection they are entitled to and think they have. Our pro-tection is almost wholly illusory. This unhappy situation has been touched upon briefly. Now it must be pursued in greater depth. It is not a story of villains and good guys. Rather, it is the result of government breaking down in its function under pres-sure from industry, which is largely a victim of its own short-sighted goals and failure to understand the consequences of its own acts. There are elements of Greek tragedy in it. We move toward a remorseless almost inevitable conclusion, as if the gods themselves were powerless to stop a system based on self-destruction.

There is no conspiracy between government and industry. But when a large number of people are linked by a common objective the results may be the same. In this case the common objective is to promote economic growth and preserve the status quo. Government takes care of business. Business takes care of government. Together they take care of the economy. The public gets what is left—an abundance of goods paid for by a ravished environment.

Throughout my researches I tried to determine how close we are to the environmental brink. The answers reflected the professional orientation of the person interviewed. Every engineer I talked to was optimistic. Every biologist was pessimistic. The engineer deals with things, primarily, not people or life. He sees the world as one vast mechanical operation. If things are a bit out of kilter they can easily be put to rights: shift a river here, cut down a mountain there, dig a harbor, design a new coastline, build another dam, rechannel a canal, add more NPK to the soil, whack down a few more trees, pump in more DDT, add a "few bugs to the water to pick up the BOD (biological oxygen demand)," * build a taller smokestack, a longer effluent pipe—and everything will be all right.

The engineering mentality developed technology as a perversion of science. It is based on some mysterious inner optimism that a destructive means will not lead to destructive ends. This same optimism has infected the American business community. It gave birth to Babbittism. It spread throughout the world during World War II when pastel military walls were plastered with that characteristically American slogan: "The difficult we do immediately. The impossible takes a little longer."

The biologists are a different breed. They deal with life forces. Almost to a man they predict disaster. The only real difference among them is when we will be struck down. They play a kind of numbers game among themselves—how long before catastrophe strikes and the form it will take. It seems almost a consensus among biologists that we have about thirty years at the most. George Woodwell, the chief ecologist at Brookhaven National Laboratory, gives us possibly ten years, if we continue as we are—and even now it may be too late to undo our transgressions

* BOD: the amount of oxygen required by bacteria to complete the decomposition of organic matter.

against nature. Almost every one of these scientists predicated their doom timetable on one thing: that we will keep going as we are now. Most expressed hope, if not confidence, that we could avoid the plunge into oblivion, or at least slow down the pace, if we changed our ways immediately. Not one biologist that I talked to believed this would happen. They felt we are too deeply committed to our way of life—and possibly death. A few felt that it is already too late to reverse the momentum of forces that are destroying us.

These forces, as noted before, are an interlocking of a laissez-faire attitude and exploitation of the earth's resources through technology. This exploitation is powered by an almost fanatic dedication to growth. Our gross national product, much of it at the cost of the environment, reached $1 trillion in 1970. It is expected to increase half again by 1980. This does not mean that things will be half again as good or bad in ten years as they are now. Rather we are caught in a process of doubling: in five years, for example, we will use twice as much pesticides as at present. Environmental decline will almost certainly follow a similar curve. At best, technology's destruction can be negated or reduced by a countercontrolling technology—recycling, for example. But little effort or money is being spent toward that end.

Industry's function is to produce goods and turn a profit. Government's basic obligation is public welfare. The founding fathers saw this as the protection of life, liberty, and the pursuit of happiness. Today it has deteriorated to, primarily, the protection of the economy rather than the people. Government bears overall responsibility for creating the milieu of policy, priority, and regulation within which industry operates, says Roderick A. Cameron, executive director of the Environmental Defense Fund, writing in *Look:*

> If industry has never been housebroken it is because government has failed, the more so because today's crisis was foreseeable long ago. Government has too often not governed at all. Thus it is by legislative indirection and privilege that we are faced with automobiles and roads strangling our endangered cities and mangling our degraded countryside; with a citizenry seduced by big-budget commercials into driving everywhere in inefficient cars that spew out polluting chemical gases; and with the lack of any attractive alternatives that we otherwise would have developed.

Many congressmen call for corrections, but they are caught in an intricate legislative web that makes corrective measures almost impossible. The seniority system enables powerful men to remain entrenched for many years. They become committee chairmen or control large areas of economic influence. Their decisions are usually made behind closed doors, completely subverting the democratic process. Through vote trading, exchanges of favors, other swaps and devices, a few individuals represent an enormous concentration of power in American life. Much of this power, in turn, is indirectly controlled by private interests to whom the politicians look for financial support. "Typically it is the steady support by an economic interest that permits a congressman to be reelected again and again," observes Cameron. "And it is that interest which calls the powerful man's tune."

Small groups of men in key legislative positions can block almost every effort at reform. A bill was recently introduced to provide for a Federal Consumer Agency to look out for the interests of the public. That such a bill was introduced is a commentary on government's present commitment. The measure seemed sure to pass both Houses. Instead it was killed by the House Rules Committee. *The New York Times* describes the committee's power:

> . . . the tyranny of the Rules Committee and its strong-willed chairman Representative Colmer of Mississippi[7] has converted it into a third chamber of Congress, with veto powers more absolute than those of the President. So magisterial is the authority of the committee chairman that he can doom bills by the simple expedient of refusing to let them come before the committee for a vote.

Industry is able to "buy" certain congressmen because it almost alone can supply the huge sums of money necessary to elect men to office today. The cost of the 1968 campaigns was $300 million, according to the authoritative Citizens' Research Foundation of Princeton, New Jersey. The cost of electing a President and Vice President shot up 67 percent over 1964, from $60 million to $100 million. Nixon's general election expenses were $24.9 million, and Humphrey's were $10.3 million. The CRF report shows that Nixon's general election campaign spent $9.02 million for television and radio time and newspaper advertising space. For the same purpose the Humphrey-Muskie campaign reportedly spent $4,229,000. Even this was not the full cost. A Federal Communications Commission study showed that the true cost was about 50 percent higher then the official totals of either the Nixon or Humphrey campaigns. The FCC study revealed that the networks and local stations billed Nixon $12.6 million, Humphrey billings were $6.1 million. The difference was spent by other supporters not under the candidates' direct control.

In New York, Representative Ottinger spent $1,840,000 just to get the senatorial nomination. The Rockefeller gubernatorial campaign in New York cost $6,794,627, his family contributing $4,124,500.*

The American practice of political campaigns subsidized by the rich has had some bizarre side effects, noted *The New York Times.*

> The United States is surely the only major power which practically auctions off ambassadorships to political contributors. . . .
>
> More significant is the undoubted influence which business executives and wealthy investors gain in particular spheres of public policy as a result of their political investments in winning candidates. These big givers can get the attention and often the support of congressmen and executive officials in matters which concern them such as oil import quotas, tax bills, airline routes, drug prices, and appointments to federal regulatory agencies.
>
> The unions, which usually support liberal Democrats, have their own selfish aims. . . .

* After Nixon vetoed a 1970 bill that would have curbed national election campaign spending, a similar measure became law in 1972. The new legislation has weaknesses but is seen as a first step in correcting election abuses, particularly the buying of elections with saturation advertising on television and radio. Traditionally there has been a wide gap between the amounts reported spent on campaigns and those actually spent. The new legislation imposes an all-media spending ceiling of 10 cents for each eligible voter with no more than 6 cents of that for broadcast advertising. Under the formula the Republican national ticket, for example, could spend only $8.4 million on radio and television during the 1972 post-convention campaign, compared to the $12.8 million reported spent for those items in 1968. The Democrats reported spending $6.1 million for radio and television in 1968. Under the act no Congressional candidate can be restricted to less than $50,000 for media spending no matter how small his constituency. Candidates are limited in the amount of their own money they can invest in their campaigns—$50,000 for President or Vice President, $5,000 for the Senate and $25,000 for the House. Overall media spending for each national ticket is limited to $13.9 million. The measure provides for a reporting system for Presidential and Congressional candidates, under which a list of all expenditures and contributions over $100 are to be filed every year.

Powerful men in political office who are supported by industry are often able to control the vital regulatory bodies and dominate their policies. They can influence appointments and appropriations or starve an agency by withholding funds. They can render it incapable of doing its job or cripple it into submissiveness. At the same time industry shuttles its own representatives back and forth among the agencies. The regulatory agency came in the wake of intolerable abuses in a laissez-faire economy that threatened to destroy itself by its excesses. The agencies were intended to preserve free enterprise while protecting the public and the environment from ruinous exploitation. How completely they have failed can be seen in the breakdown of public trust in government.*

The regulatory agencies have taken as their function not to protect the public but to protect industry. The laws may look good on the books, but more often than not they are not enforced or they are written with loopholes to accommodate special interests. Thus the entire system of public control over large segments of private industry breaks down. Public protection is more illusory than real. We are often left to the mercy of polluters and exploiters with little to protect us in the way of government action. Many of these failures have been documented in a series of reports by the dauntless Nader's Raiders. They point to inadequate laws, lax law enforcement, inadequate research funds and programs. Politics often dominate appointments; staff morale is poor. Employees frequently shuffle between regulatory agencies and the industries they regulate. In 1964 alone eighty-three FDA employees left their jobs to work for the food industry.

The FDA is one of the nation's most important and powerful regulatory bodies. It stands between the American public and a sea of dangerous chemicals that pollute foods; there were 1 billion pounds of additives consumed by Americans since 1958 (3 pounds per person per year). Many of these chemicals are powerful poisons. Others cause cancer in laboratory animals. Most are toxic in some degree.

The agency in charge of this death-dealing potential is in a state of "total collapse," according to a Nader report. The highly

* The University of Michigan's Center for Political Studies reported an opinion sampling that revealed a startling drop in the number of adult Americans who expressed a high degree of confidence in the Federal Government, reports *Time*. "In 1964, 62 percent of those polled expressed high confidence. In 1970 the figure was 37 percent."

competitive industry it is supposed to regulate cuts corners, often misrepresents its products, and refuses to cooperate with the government. The FDA is "continually neutralized" by the powerful forces arrayed against it—large well-funded efforts by Washington law firms, massive trade associations made up of the 50,000 food-manufacturing firms, "and a small group of industry-dependent 'food scientists' who more often than not routinely produce scientific studies that support the most recent industry marketing decisions." Says the Nader report:

> In the face of the $125 billion food industry, which is over six times as large as General Motors, America's largest industrial corporation, the FDA is unable to exert any meaningful influence on behalf of the food-consuming public. Impotence has characterized the FDA and its predecessor agencies since passage of the Pure Food Act of 1906. Due to the total collapse of the food protection effort of the FDA, which has allowed vicious, unchecked battles for profit to wrack the food industry, food restoration has become an important national goal for all Americans to work toward.

Of the 3,000 food additives it regulates, relatively few have been adequately tested for safety. Most are considered safe only in the absence of proper testing data. Industry spends much more selling its products than finding out if they are harmless or making them nutritious. But it is expert in packaging and selling its products. Some 18 percent of its annual $125 billion gross income goes to promotion.

The public has been subjected to incredible health risks over the years. Some are due to bureaucratic lapses. Some should be criminal offenses. It took FDA twenty years to partially ban the cyclamates after its own studies revealed that the artificial sweeteners cause cancer, mutagenic effects, and birth defects; then it acted only after public pressure. The pathologist who did the original studies in 1951 later told a colleague that his superior wrote the report, a watered-down version of the findings that he dutifully signed. "He didn't want to fight with people. That was his policy over the years," said his confidant. Even thalidomide was refused clearance only after a great deal of foot-dragging when it was shown to cause such dreadful deformities to children in Europe.*

* In December, 1971, the FDA belatedly warned against daily body bathing of babies and adults with antibacterial solutions containing 3 percent hexachlorophene. The agency revealed that hexachlorophene applied to the skin can enter the bloodstream in amounts that may reach levels that have

From within the agency have come the most damning admissions. For years FDA officials manipulated laboratory results to avoid conflict with established policies on food additives and pesticide safety, according to Dr. Howard L. Richardson, chief pathologist in the Bureau of Science. "Middle-level bureaucrats with limited scientific training have expunged conclusions and recommendations from reports on animal experiments because they cast doubt on FDA policies," he is quoted as saying in the *Times*. A few days after his disclosure, Richardson was demoted.

Some disclosures in recent years have shaken even professional apologists. The same official who set many of the present policies that regulate the use of antibiotics in drugs and animal foods was found to have picked up $278,000 by pushing the drugs in two private publications he edited for two pharmaceutical houses that produced antibiotics. When the scandal broke he was trying to get antibiotics in human food. The FDA generously permitted him to resign.

Pesticides in foods also cast a dark shadow. The FDA's most important proof that the public gets only a minimal and safe dose is the so-called Total Diet Test. Everything that an average nineteen-year-old boy would allegedly eat in a two-week period is chopped up and analyzed. An FDA adviser, Dr. William C. Purdy, found the tests unscientific, misleading, and worthless. In one district residue reports were made where there was no equipment to do the studies reported. "Foreigners who observed the tests—the competent analysts among them—can only leave in a state of shock when they find out about our Total Diet Program," Dr. Purdy concluded. "The tragedy about this entire program is that the data are presented in a manner that suggests that our food contains no harmful residues. In actuality . . . the citizens of this country are not being protected. Furthermore, they are being led to believe that there is no cause for worry." FDA's reply? The test data used by Dr. Purdy were confiscated and his contract "terminated."

---

produced brain damage in monkeys. Hexachlorophene is used in hundreds of products, ranging from soaps and shampoos to deodorants, and millions of Americans have been exposed and are being exposed to appreciable risks. A decade ago physicians reported hazards associated with the use of hexachlorophene. In 1967, scientists discovered that hexachlorophene can enter the body not only through wounds or other openings but also through the intact skin. By mid-1969, *The New York Times* reported, FDA scientists had already discovered basic evidence of brain damage to rats fed minute amounts of hexachlorophene. "Yet not until now, two and a half years later, has the FDA issued a public warning."

The food contamination scares that have shaken the public in recent months are but the tip of a much bigger problem involving inadequate government inspection of the products that Americans eat, *The New York Times* reported in December, 1971. "Authorities say that last summer's episodes of botulism in soups, mercury in swordfish and PCB chemicals in poultry were not rare exceptions of food contamination but merely spectacular examples of wide-spread and long-standing problems in federal efforts to insure the wholesomeness of food on grocery shelves."

The article noted the lack of inspectors. USDA's Meat and Poultry Inspection Program employs some 8,000 inspectors. Of the 8,000 inspectors, 1,500 are veterinarians who inspect animals for disease. In 1970 USDA inspectors examined over 118 million cattle, sheep, goats, swine, horses and mules. Of these over 400,000 were condemned as unfit for use in food. *The New York Times* article continues:

> By far the most common reason for condemnation before slaughter was that the animal was already dead in its pen. After slaughter many diseases were found that warranted condemnation. However, over 14 million animals found to be diseased or injured were deemed by USDA inspectors to be wholesome after the diseased parts were cut away. This included 2,000 cattle and 885,000 swine with tuberculosis and over 127,000 cattle with cancer.

When an FDA official was asked recently why its market basket tests had not found the mercury in tuna fish, he replied, "Well, we didn't think to look." How many other things is the FDA not "thinking" to look for? On the shelves of a good-size supermarket today there are some 10,000 different food items, most of them subjected to some kind of chemical treatment. The FDA spends only one third of its $75 million-a-year budget regulating food; the rest is spent on drugs, cosmetics, and medical devices that collectively represent a market only one sixth the size of food in terms of money.

A recent study of the FDA was ordered by its commissioner, Dr. Charles C. Edwards, who is credited with making many efforts to upgrade the agency. The study was conducted by a committee of five professors in leading medical schools. They expressed surprise and "dismay" over some of the things they found during the one-year survey, according to *The New York Times*. Their report said that some of the FDA's laboratories used advanced technology, had good morale, and gave every impression of being first-rate: "On the other hand," it said, "one can also find laboratories so poorly managed that scientists seem to be unable to describe their work coherently or to produce interpretable data books containing their findings."

The report noted that widely different estimates had been made of residue of drugs in meat and poultry reaching the market. "The committee said that in this situation 'there may exist a situation of possible danger to the public and potential embarrassment to the government.'"

The FDA, so tolerant of pesticide residues in foods, is not the only offender in failing to protect the public. It was supposed to advise the Department of Agriculture on the safety of pesticides proposed for registration, before control over registrations passed from USDA to the Environmental Protection Agency. Yet its recommendations were disregarded by Agriculture "virtually 100 percent of the time," according to a government study. In 1969 FDA recommended 5,052 label changes. Not one was adopted. The Department of Agriculture is dominated by the pesticide manufacturers. Both have their powerful protectors in Congress.

The FDA is subjected to unbelievable pressures, from both industry and congressmen representing special interests. When Herbert L. Ley, Jr., was kicked out as FDA commissioner in December, 1969, he told the *Times* he was "under constant, tremendous, sometimes unmerciful pressure" from the drug industry: "The thing that bugs me is that the people think the FDA is protecting them—it isn't. What the FDA is doing and what the public thinks it's doing is as different as night and day."

Frequently the agency does not even know what is happening in its domain. It is understaffed and underfunded for the massive job it is supposed to do. It depends primarily on industry's honesty and self-compliance with the law rather than enforcement. It has neither the money nor the stomach for strong enforcement. How many FDA employees are going to offend industry when they are looking to it for a haven when they leave government service? The failure of the voluntary compliance program is revealed in an official report quoted by James Turner in the *Chemical Feast*, a Nader report:

In 1965 a total of 711 firms suspected of producing harmful or contaminated consumer products refused to let FDA conduct inspections. Some 515 refused to furnish quality or quantity formulas to the Administration; 26 denied the Administration the opportunity to observe a manufacturing procedure. And 153 refused FDA personnel permission to review control records. Also, 111 would not permit the FDA to review complaint files, and 216 refused permission to review shipping records.

In Westchester, New York, in 1971, a lapse in safety standards caused a family tragedy. A man died and his wife became paralyzed as a result of eating tainted canned vichyssoise. Because the cans had been insufficiently heated, they contained botulin toxin. The manufacturer was the Bon Vivant soup plant in Newark, New Jersey. FDA inspectors, following the death, said inspections generally were made about once a year. Investigation revealed that the Bon Vivant plant had not been inspected by state or federal officials for nearly five years. Some plants, the FDA admitted, are not inspected for as long as ten years. In one case the agency discovered a New York plant that had been in business for twelve years, with an annual sales volume of $12 million, and it had never been inspected. "We just didn't know about it," said an FDA spokesman.

This is human pollution for commercial gain. If Congress does not actually approve, little or nothing is done to prevent bureaucratic breakdown or to insist upon enforcement of its own laws, particularly the "Delaney amendment" which bans the use of carcinogens in foods. The same legislative collapse is behind most environmental degradation: Congress' refusal or unwillingness to stop it. In many cases the federal government is the prime offender. Usually politics is the motivating force. Take the Corps of Army Engineers.

The Corps has almost unlimited power. Many of its projects are not only destructive; they cannot even be defended as necessary or desirable. The Corps is basically a vast pork-barrel operation to distribute patronage and favors. Projects are allocated throughout the country through the Corps' "highly developed sense of the relative political strengths within the Congress, and by making sure that each region of the country gets a little something each time," states Elizabeth Drew in The Atlantic. Not every section of the country needs an engineering project. But all are hungry for the funds that spill out. Legislators are eager to make political capital out of bringing this windfall to their district. It is always easy to invent some project, even if it damages the environment. Money is money. Votes are votes.

This becomes a virulent form of trading favors for mutual political gain. "There may have been a Corps project that was rejected on the floor of Congress, but no one can recall it," says Miss Drew. "Every two years—in election years—a rivers, harbors, and flood-control authorization bill is passed by Congress, and every year money is appropriated. . . . The most recent appropriation carried something for 48 states."

A few isolated congressmen have tried to battle the system. Former Senator Paul Douglas (and more recently Senator William Proxmire) occasionally spoke out against some particular project, or the " 'pork-barrel' technique of legislating Corps projects." But he was not taken seriously. Ultimately he gave up. After futilely battling rivers and harbors appropriations for years, Douglas was forced to throw in the towel. In a candid 1956 speech he said he thought it was "almost hopeless" for any senator to do what he tried to do when he first came to the Senate, "namely to consider these projects one by one":

> The bill is built up out of a whole system of mutual accommodations, in which the favors are widely distributed, with the implicit promise that no one will kick over the applecart; that if senators do not object to the bill as a whole they will "get theirs." It is a process, if I may use an inelegant expression, of mutual back scratching, and mutual logrolling. Any member who tries to buck the system is confronted with an impossible amount of work in trying to ascertain the relative merits of a given project.

Other federal policies are equally destructive. The government owns about 34 percent of the total U.S. land acreage; and federally supported highway, airport, reservoir, and flood-control projects transform an estimated 580,000 rural acres each year. Much of it goes under cement. Much of the environmental damage is a direct result of policies designed to aid special interests, promoted by their champions in Congress. Some of the environmental damage due to federal projects and policies was outlined recently by the President's Council on Environmental Quality: overgrazing has left much federal grassland in "desperate condition." The tendency toward monoculture in national forests "for the benefit of lumber and pulp" wiped out many native species. Federally financed predator-control programs have driven some animal species to the edge of extinction. The damage is varied, but it has a common focus that was noted by the *Wall Street Journal:* "The goals of most federal programs simply ignore the environment."

Congress has consistently rejected efforts to educate itself on the environmental effects of its various actions by refusing to pass a measure providing for a Joint Congressional Committee on the Environment. The act would give Congress a long-range committee concerned not with specific legislation but with a broad and continuing review of environmental matters. It would identify problems as they emerge, keep track of successes and failures in government approaches to the environment, hold hearings on the annual reports of the Council on Environmental Quality and, above all, as the *Times* noted, "enable the standing committees of Congress to act concertedly in this vital area, with a sense of coherence and priorities not always discernible in their legislative output."

There is no end of playing politics with the environment. Especially disturbing is the great concentration of power in the hands of some of those who have shown little regard, and in some cases disdain or hostility, toward environmental issues. One of the most powerful men in Congress is Representative Jamie L. Whitten, Democrat of Mississippi, chairman of the House Agriculture Subcommittee. Recently George H. Mahon of Texas, chairman of the House Appropriations Committee, gave Whitten control over the funding of all federal environmental and consumer protection programs. This put in Whitten's domain such powerful and sensitive agencies as the Environmental Protection Agency, the Council of Environmental Quality, the Food and Drug Administration, and the Federal Trade Commission.

Whitten, who has exercised personal authority over farm programs for two decades, now has the foremost congressional voice on how much money is spent by these key agencies and what it's spent for. All air and water pollution programs are under his jurisdiction, along with pesticide control.

Prospects seem especially bleak in the area of pesticide control. Whitten, it will be recalled, wrote a pro-pesticides book attacking *Silent Spring.* Is he going to feel charitable toward EPA if it cracks down on DDT and other pesticides? Will Ruckelshaus be able to make an unbiased determination on whether to ban pesticides while Whitten holds the purse-strings to his agency?

The League of Conservation Voters, which rates congressmen according to their votes on certain key issues, gave Whitten a strongly negative rating. He voted for funding the SST, for increased logging in the national forests, and against spending $1 billion extra to fight water pollution. A conservationist in Congress, noting Whitten's farm policies as head of the agriculture subcommittee, was quoted as saying that Whitten had "allowed the Agriculture Department to raise environmental havoc" in its soil, forest, and watershed programs.

Maneuverings and manipulations within Congress to benefit special interests are not new. Environmental plunder by private industry is not new. Government's support of private industry, often against public welfare, is not new. But with the recent focus on universal environmental damage, and the annihilating power of technology, the problem becomes more serious. Government not only has failed to stop plunder and pollution; it has often allied itself with the polluters against public interest. The Nixon Administration has taken a particularly schizophrenic position. It has continually talked conservation while it often actually aids polluters.

Ramparts states the case against the President harshly: "For Nixon, as for Johnson before him, conservation functions as the facade behind which environmental pollution is formalized and made part of a society which has come to depend on environmental pollution . . ."

In his 1970 State of the Union Message, the President made headlines with his speech warning against "the prospects of ecological disaster": "The great question of the seventies is shall we make our peace with nature and begin to make reparations for the damage we have done to our air, to our land and to our water? Clean air, clean water, open spaces—these should once again be the birthright of every American. If we act now, they can be."

A few months later the Council on Environmental Quality presented the bill to "make our peace with nature." The President, faced with dollar signs, seemed to have second thoughts about our environmental birthright. He spoke of "realism," of maintaining "a healthy economy while we seek a healthy environment," and warned against demanding "ecological perfection at the cost of bankrupting the very sources of funds for improvement."

The council said the cost would be high but "well within the capacity of the American economy to absorb." To achieve air and water standards now in effect, the cost between 1970 and 1975 would be $62 billion; solid waste control, now largely paid for locally, would cost another $43.5 billion—an overall price tag of $105.2 billion.

Of the $62 billion outlay for air and water controls, industry would have to pay $13.5 billion and the public $24.5 billion for water controls. For air, the private sector would have to pay $22.1 billion and the public the remainder. But industry was not willing to go even that high, according to figures gathered from business by McGraw-Hill and reported in *The New York Times*. Industry estimated the cost of meeting present air and water standards at $18.2 billion. But its total planned expenditure in 1971 was estimated at only $3.64 billion.

This speaks poorly for our prospects of correcting environmental pollution. If the public and industry invested the total $105.2 billion cost projected by the Council on Environmental Quality in the 1970–75 period, the annual cost by 1975 would be only 1.4 percent of the estimated gross national product for 1975. The lower figure of $62 billion to meet air and water controls, estimated by the council, would be only 0.8 percent of the estimated GNP. It hardly seems an extravagant sum to achieve clean air, clean water, open spaces—"the birthright of every American."

In his major appointments the President has also often shown a lack of sensitivity to the environmental problem. In 1969 Washington Columnist Jack Anderson wrote a biting column on Secretary of Commerce Maurice Stans: "While President Nixon was signing a bill to protect rare animals around the world, his Commerce Secretary, Maurice Stans, was oiling up his guns to kill two rare African animals that may well come under the act." Stans, it was pointed out, is known as a big game hunter.

Stans has almost consistently defended business and challenged environmental legislation. He told a businessmen's convention that "business is 99.44 percent pure"—a figure that hardly agrees with the Council on Environmental Quality's assessment of factory effluents, observed *New York Times* environmental writer Gladwin Hill. Stans told a chemical association that the Clean Air Act of 1970 showed "complete disregard for economic factors" and contained "unrealistic technological demands."

Before the National Petroleum Council he cautioned against plunging recklessly into environmental reform without first considering the cost (his audience presumably did not need much prompting). The Secretary warned against rushing into such things as banning DDT, stopping the use of phosphate detergents, killing the SST, or reducing noxious automobile emissions —on grounds that these things may not be as bad as they are thought to be, or other things may be worse, and the changes might, in any case, be expensive, commented a *Times* editorial. Stans called his speech "Wait a Minute," but he criticized those who would go slow in such industrial ventures as offshore oil well development or the trans-Alaskan pipeline: "Are we so afraid of what might happen to the Alaskan environment," he asked, "that we will sacrifice the enormous new sources of oil which we need for our homes and our cars and our jobs and our country?"

Stans is also credited with the precept that "a determination of the economic impact should be required before environmental actions are mandated." This seems odd from a Cabinet officer in an Administration that proclaims itself in favor of strong environmental protection. The *Times* made a pertinent comment about this:

> The making and saving of money, it seems, must not be made dependent on their effect on the planet, but rather the preservation of the planet must depend on what the process will cost, how many potential jobs might be lost and whether it will have a bad effect on business. . . . In short, the Stans doctrine goes, save the world if you want to—but only if the price is right.

Other Nixon appointees have made their mark. Consider a report on what the *Times* considered a "shabby statement" by Hendrik S. Houthhakker, a member of President Nixon's Council of Economic Advisers:

> Going down a list of environmental causes, Mr. Houthhakker found reasons to minimize them all: since there is no "satisfactory alternative" to the Alaskan oil pipeline, why worry about "a fraction of 1 percent of the total permafrost" in the state?; "extreme" conservationists are close to arguing that there should be no cutting in the national forests

at all; any air standard is only a "value judgment" which
should be left to local option, etc., etc.

One of the government's most important weapons in fighting
water pollution is the Federal Refuse Act of 1899. It prohibits
the pollution of navigable waters. It has been applied effectively
where other legal remedies failed. The law is clear and unequiv-
ocal. It specifically directs U.S. attorneys "to vigorously prosecute all offenders." The act is universally violated. In July, 1970,
a conservation group charged 214 manufacturers, businesses,
and municipalities with polluting Alabama streams in violation
of the 1899 act. Did the Justice Department step in and enforce
the law? It did not. Instead Attorney General John Mitchell,
himself, chose to ignore the law he had sworn to uphold.* He
ordered U.S. attorneys to act under the law only if the discharges
of pollutants were "accidental or infrequent." Mitchell said
chronic polluters would have to be dealt with by the Federal
Water Quality Administration. The effect was to give chronic
polluters license to keep polluting. The FWQA lacked the funds,
manpower and facilities to handle so vast a problem. Representative Ottinger said he would sue the Justice Department to enforce the law. Here we have the bizarre picture of the Justice
Department being sued by a member of Congress to force compliance with the law—the first time such a travesty had been
enacted.

The Justice Department has long shown a lack of enthusiasm
for curbing polluters. For fifteen years there was collusion
among the biggest auto manufacturers not to develop a smog-
control device that would give one a competitive advantage over
another and call attention to the polluting effects of cars. The
Johnson Administration did not know how to handle this hot
potato. It willed it to the Nixon Administration as a parting gift,
capriciously filing the suit just as Johnson was leaving office.
The Nixon Administration was equal to the challenge. The Justice Department worked out an agreement with industry so the
dirty linen would not get washed in public. The suit was dismissed with a legal maneuver known as a "consent decree"—a
model of "legalistic obfuscation," explains *The Vanishing Air*.
Shorn of its "tangle of legalese," it meant that the conspiracy
would end. A federal judge obligingly upheld the dismissal.

---

* Mitchell resigned as attorney general early in 1972 to direct Nixon's
campaign for reelection.

More recently the Internal Revenue Service threw its weight on the side of the polluters. It announced plans to revoke the tax-exempt status of public interest firms filing lawsuits against those who damage the environment. This struck at the heart of free government. The action violated fundamental constitutional rights, outraged critics charged. The ruling virtually closed the courts to citizens and provided indirect license to polluters. Only big tax-exempt conservation organizations could get together the huge funds necessary to sue the big polluting corporations. (A corporation going into court to defend itself in litigation can deduct such expenses from its gross income.) Newspapers reported that the action was prompted by industry pressure on the Nixon Administration. Conservationists charged that industry had direct connections with Bryce Harlow "that we don't have." Harlow, counsel for Nixon, is a former vice president and lobbyist for Procter & Gamble, a leading detergent manufacturer.

The IRS action indicated a "callous disregard" for the public interest, according to Senator Walter F. Mondale of Minnesota. He finds it ironic that the Administration, which stressed the need to work within the system, chose to attack public interest litigation: ". . . for this type of litigation symbolizes trust in 'the orderly process of government'—rather than reliance on demonstrations and rallies to stop pollution and obtain fair treatment for consumers." He noted that the action was not in a vacuum, but occurred in the context of "other Administration plans to shut off access to the legal system for those seeking to redress various grievances." Among the offenses cited was the Justice Department's refusal to enforce the Federal Refuse Act of 1899. "Each of these actions stems from a common desire—to reduce the legal pressure against powerful corporate interests and reluctant government agencies."

The IRS action stirred up an unexpected storm. Protests came from Congress, labor, conservationists, lawyers, consumer groups, newspapers, outraged individuals. Even members of the Nixon Administration opposed it. A Senate subcommittee announced it would hold hearings on the IRS decision. Just before the hearings were to begin the IRS backed down and cancelled the move, but announced new guidelines, including the denial of tax-exempt status to new organizations seeking it. At least the pressure would not become greater. The depressing part is that the ruling was made in the first place.

More recently the Nixon Administration moved to transfer major environmental lawsuits out of Washington. In one action the Justice Department petitioned to shift from Washington to Alaska a suit by conservationists to stop the trans-Alaska oil pipeline. The Justice Department claimed it was for "the convenience of the parties and in the interest of justice." Conservationists saw it as a new form of harassment; they contended there would be more local political pressure to gain favorable decisions for the government and polluters, and that it would work hardships for the conservation groups in added expenses for travel. They also held that the cases belonged in Washington because the public interest was national and not local in scope.

Each Administration move that aids polluters and brings new insults to the environment is accompanied by splendid rhetoric and pious motives. The Administration announced that it would realign federal agencies concerned with the environment under one department. This seems an excellent idea. It would, as the President announced, achieve better management. But the startling part was that the agency chosen for this delicate undertaking was the Department of Commerce, a department openly dedicated to aiding commerce and growth. Among its new responsibilities are the ocean and coastal waters—the new frontier just being opened for exploitation. The decision to place the Commerce Department in charge of most marine affairs is "somewhat akin to appointing the undertaker as lifeguard at the beach," commented the *Washington Star*.

In April, 1970, President Nixon formed the National Industrial Pollution Control Council to advise the government "on environmental programs affecting industry and on industry's proposals for dealing with the pollution it causes." The council was put under the Department of Commerce, headed by Secretary Stans, who, as mentioned, has shown more sympathy for industry than the environment. The council has 63 members and about 200 panelists or subcouncils that it can call on for advice or information.* Senator Lee Metcalf charged that the council is

* Some 400 panels and committees help the government make decisions in a wide variety of matters said to range from acoustics to the need for zinc in human nutrition. About 7,500 scientists, engineers and other specialists sit on these panels. In December, 1971, *The New York Times* reported that the National Academy of Sciences is now requiring its thousands of consultants to file statements of their biases and possible conflicts of interest if they are to serve on committees that serve the federal government. Dr. Philip Handler, president of the Academy, explained that bias might exist if committee members made positive recommendations in areas

composed of "the leaders of the industries which contribute most to environmental pollution." Certainly it is made up of top executives of some of the nation's biggest corporations, representing a tremendous concentration of corporate power focused on a business-oriented Administration. It functions at the top level of environmental policy-making and has direct access to the top policy-making levels of government. No conservation group has similar power or access to the President.

Decisions are made by Administration and industry officials behind closed doors—decisions that affect all of us. Usually we do not even know what those decisions are until they are in force. No minutes are kept. Environmentalists ask: Why do the President and the Department of Commerce prefer to be advised on pollution, in secret, by the people who do the polluting instead of those who protest it? Ralph Nader's Center for the Study of Responsive Law was among ten conservation and consumer groups denied permission to attend the first session. "Why must the public and press be kept out?" asked the Nader group. "Closed sessions," we are told, promote "frank and open discussions! But what have they to hide?" Henry Steck, a political scientist writing in *Environmental Action*, says the council may represent "an important step toward the creation of an industrial-environmental complex." Considering that some of the nation's biggest polluters sit on the council, says Steck, suggests "nothing so much as putting the town madams on a Vice Control Committee to control the town's bawdy houses."

Under the Nixon Administration there has been a subtle shift in the initiative for environmental policy-making from Congress to the executive branch. This power is largely concentrated in two agencies: The Council on Environmental Quality, which evaluates and makes recommendations, and the Environmental Protection Agency, which sets standards and is charged with enforcing them. The creation of these two agencies, along with passage of the 1970 National Environmental Policy Act, and the

---

involving companies in which they held stock or served as consultants or employees. He said that earlier this year the Agricultural Research Institute, which is linked with the Academy, sent a report on pesticides to the Secretary of Agriculture, who was then Clifford M. Hardin: "'Before forwarding the report, I included a letter with a caveat that noted that most of the membership of the committee which had drawn up the report was connected with industries involved in agriculture,' Dr. Handler said. All five officers of the 35-member committee are linked with chemical corporations, food processors or seed companies."

revival of the Refuse Act of 1899, establish what has been called an administrative base for environmental policy.

The Environmental Policy Act requires every executive agency, when recommending legislation and "other major federal actions significantly affecting the quality of the human environment" to file with the Council on Environmental Quality (also formed under the act) a statement of the environmental impact that a proposed action would have. The statement would include alternatives to the recommended action. It is required to be made public.

The purpose of the act was to give other affected agencies and the public a voice in decision-making so they could evaluate risk against benefit. But the act has been largely undermined in the Nixon administration. Several executive agencies have failed to file the required statements. Others made the big decisions and then issued statements merely as a formality, too late to change the action decided upon. Even if the statements were filed earlier it would not necessarily mean that "there has been any reality at all to consideration of these factors," says Russell Train, head of the Council on Environmental Quality. In one case, the Department of Interior released a statement assessing oil spill risks only upon announcement of its decision to proceed with a major auction of offshore oil leases.

The act is further undermined because the council does not have the staff to process the impact statements properly. It gives the council no power to veto inadequate statements, and the agency submitting the statement has the final authority for taking any action it proposes. The council lacks authority to require an adequate or, indeed, any statement at all from a recalcitrant agency, said Train. All the council can do, he said, is to ask the President to order the agency to satisfy the law's requirements. In other words, the law has no teeth.

There are disquieting signs that the Environmental Protection Agency's authority is being nibbled away. The *Times* notes what in Washington is recognized as " 'an offensive' by industry against the agency's implementation of the clean air and water laws." In passing the 1970 Clean Air Act, Congress purposely deleted a phrase requiring information of the economic impact of proposed actions in environmental statements. Human health was to be the only criterion. But the information is again being called for in committee reports.

"The environmental agency was also directed to require the

same information in reports submitted by industries when they apply for a permit to dump wastes into the nation's waters under the Refuse Act of 1899," reported the *Times*. "This 'requirement' was the work of Representative Jamie L. Whitten, Democrat of Mississippi, chairman of the subcommittee handling the appropriations.

"Language in a committee report is without legislative effect, but officials at the council and the agency are in a quandary nonetheless. A fact of life they have to live with is Mr. Whitten's power over their appropriations. But if they yield to Mr. Whitten, they say, the effectiveness of both the Environmental Policy Act of 1969 and the Refuse Act of 1899 will be gravely weakened."

The Council on Environmental Quality has been useful in defining many environmental problems, but it has been timid in recommending corrective action. Its first report was criticized as "hortatory and often vague." The *Times* spoke of the council's "uneasy relationship with the Administration":

> The report is anchored in specifics only when it is endorsing one of the Administration's legislative recommendations. Then, "should" changes to "must," and Congress is sternly told to get on with the job. President Nixon's name is invoked, if not on every page, at least scores of times as if he were the most inspiring leader in this field since Henry David Thoreau. . . .
>
> It is apparent that the council feels itself very much part of the Administration and is loath to go beyond the pace set by the White House. In short, the council is monitoring its own political fallout pretty carefully.

Item: The government wants to clean up the nation's waters. But first it must determine what is going into them. Of the 300,000 manufacturing businesses in the country, all but 25,000 discharge into municipal sewers. Sewage plants treating the wastes are incapable of removing the inorganic pollutants, such as poisonous metals and phosphorus from detergents. A questionnaire is sent to 250 industries asking them to define what goes into their wastes. But this inventory is voluntary rather than mandatory. Government claims it has no authority to demand compliance. Legislative authorities on Capitol Hill deny this. They insist the government has authority to require answers from industry under the 1899 Refuse Act that prohibits polluting the nation's waterways. The Administration has declined to use its powers under the law, they say.

The voluntary inventory is worse than nothing, according to Senator Lee Metcalf's office. It sets a premium on dishonesty. "The nonpolluters would reply and the polluters would ignore the request. That would leave the illusion . . . that something is being done to clean up the nation's waters when actually nothing is being done."

Possibly the criticism stung. Nixon is too adroit a politician to leave himself so open to attack. Late in 1970 he took two steps that could help curb pollution. First he formed a new Environmental Protection Agency (EPA) to consolidate several environmental functions under one roof. This allegedly would add efficiency and promote protection through greater law enforcement.

The second step followed the explosive mercury scandal. The President announced in December, 1970, that the 1899 Refuse Act would be enforced. The act requires permits from the Army Corps of Engineers for the discharge of wastes into the country's streams, giving it some control over what goes into the water. In 71 years since the act was passed the corps has issued only 415 permits. This despite the fact that some 40,000 of 300,000 industries in the nation seriously pollute the waterways (10,000 firms account for 90 percent of the water used by industry). Congressman Henry Reuss of Wisconsin calls the handful of permits issued "disgraceful." Under the act no permit would be issued without full disclosure of the nature of the effluent. State agencies would also certify that the waste met existing water-quality standards. In addition the federal government is preparing its own new water-quality standards that would have to be met. EPA would have final veto power over whether permits should be issued.*

It sounds encouraging. But so have many other moves that turned out to be window dressing or just another new name on the door with nothing really changed inside. Will the EPA be

* A federal district judge in Washington ruled in December, 1971, that all waste discharges into nonnavigable waterways are illegal and he barred the federal government from discharging them with discharge permits. In addition, *The New York Times* reported, Judge Aubrey E. Robinson, Jr., ruled that the government could not issue discharge permits even for navigable waters unless it first prepared a study on what impact each permit would have on the environment. Some 20,000 applications have been filed by industries, many of which already dump treated or untreated wastes into waterways, it was stated. Government officials said they did not know what effect the ruling would have, but one said the decision "appears to raise serious questions regarding the basis on which a federal permit program for pollution control can be conducted."

given the money and manpower to do a thorough job? asks the *Times,* which is also skeptical. Or will it have to fall back on sporadic spotchecking? After permits are issued, will there be regular monitoring to see that the holders abide by the original conditions? What penalties will there be for violators? Will they be enforced? Or will the permits turn out to be an official license to pollute?

There are two basic ways to attack pollution. One is at its source. This means not raping nature and not producing unnecessary goods beyond basic needs. The second is to continue to exploit the earth and to try to control what comes out of the smokestack into the air and the effluent pipe into the water, the wastes dumped onto the land and buried beneath its surface. The first method strikes at the heart of America's economic, political, and social system—a cultural ecosystem in itself. This would mean a reduction in growth and the GNP. It would mean loss of jobs and lowering the "standard of living." It would mean changes in values and upheaval in national life. Our society generally rejects that. It is terrifying to politicians, abhorrent to businessmen, frightening to those wedded to the comforts of the status quo. Only the young and the wild-eyed old call for such radical means of survival. The "prudent" tell us that reasonable men do not try to recapture the past. They do not go backward. "Progress" cannot be stopped or reversed, they intone. This leaves reasonable men no choice but to keep going forward on the suicidal path they are on.

The problem is simple enough. We must continue as we have been doing and yet convince ourselves that we have taken steps to prevent catastrophe, if not to correct the fundamental problem. We must pay lip service to reform at no sacrifice in comfort or growth. We must continue to exploit finite resources in a finite world as if both were infinite. We must create a new mythology to bolster the crumbling fantasy that has sustained us. This is the remarkable achievement of our pollution laws and the men who so skillfully engineered them, the players in the game of politics.

How has this mythology worked? We have now been actively aware of the environmental decline for a decade. We have passed numerous laws to clean up the air and water. Billions of dollars have been spent. Niagaras of rhetoric have been spilled. Where do we now stand? The fact is that we are worse off than when we started. We are plunging downhill at an increasing rate. "We are not only failing to end the pollution of American waters, but the situation is actively getting worse," according to Senator William Proxmire in a 1969 Congressional speech that was described as "detailed and shocking." Ralph Nader was equally blunt a short time later about the state of the atmosphere: "Not one particle of air pollution from smokestacks has been reduced" in two years under the Air Quality Act of 1967.

A third supporting appraisal comes from the "Environmental Quality Index" published by the National Wildlife Federation in September, 1970. It says we are losing ground on almost every front: "Our air is dirtier. Our water is more polluted. Land for food, wildlife, and living space is deteriorating. Certain minerals may soon be exhausted. Apathy is our biggest problem."

New laws are not the answer, states the "Index." We must enforce laws already on the books.

Item (*New York Times*): "Because 'administrative agencies often cannot be relied upon to enforce the laws,' Senator Hart asked the President today to support the bill that he and Senator George S. McGovern, Democrat of South Dakota, introduced to facilitate citizens' suits to protect the environment."

Two Presidents have toyed with our lives since it was recognized that an environmental threat exists. Both have played politics with the environment. A huge cast of congressmen, politicians, and industrialists have used the danger for their own benefit. It is a deception made up of part skulduggery, part fantasy. One reason is that the problem is too overwhelming to face honestly. The cost of correction is too fantastic to consider. The sacrifices are too great to comprehend. To handle a problem of such magnitude a set of legislative myths were provided to sustain the fantasy that something is being done about the problem and all will be well. The fantasy is made up of many components: tall smokestacks that reduce and disperse visible smoke, buried effluent pipes, dumping nerve gas into the ocean, swapping one pesticide for another, trying to cut automobile emissions without reducing the number of cars, appropriating pennies to cleanse the air while spending millions on the SST to foul it.

The Air Quality Act of 1967, touched upon briefly before, had almost every possible shortcoming written into it. It was written by Senator Muskie and his staff in collaboration with industry lobbyists. It was based primarily on the Manufacturing Chemists Association pamphlet ("A Rational Approach to Air Pollution Legislation") published fifteen years earlier. It provided for two basic points: The objective of air-pollution control "should be to limit the amount of foreign material in the air so that there will not be 'too much' but at the same time allow the atmosphere to function usefully to its fullest capacity." This meant pollute it just short of the point of death—an atmospheric extension of the "philosophy of poison."

The second point was to regulate the air only in specific locations. This meant where industry was concentrated. Other areas would not even be affected. The whole sordid account is outlined by John Esposito in *The Vanishing Air*. "Congress in 1967 found itself allied with the polluters' lobby," the Nader investigators charge.

In essence the act called for taking air pollutants one by one and finding the level at which they were harmful and then establishing standards for them. At best this is a simple-minded approach. Only in the laboratory can pollutants be investigated one by one. In the air we breathe, known and unknown pollutants are mixed together in complex and continually changing relationships, explains Dr. Raymond Slavin, a specialist in internal medicine at St. Louis University. The act did not require industry to prove that a substance was safe before it could be dumped into the air. The government had to prove it dangerous in order to remove it or set an emission standard. Such proof, in view of the complexity of the problem and the lack of research, and research funds, was virtually impossible.

The act provided for ambient air standards—the measurable mixture of known pollutants in the atmosphere. Individual polluters were protected by the overhanging pall so it was impossible to fix individual responsibility. The country was divided into "air-quality regions." Each region set its own standards based on local conditions and approved by the government. The result was a hodgepodge of different standards rather than a single standard. The country was divided to suit local conditions and the lowest possible standard became the norm—another extension of the "philosophy of poison." State officials were able to set up air-quality districts "so ridiculously gerrymandered to protect privileged industries as to defy rational explanations," complained Ottinger:

Two counties—Rockland and Westchester—were divided up in eleven—that's right, eleven—air-quality districts, ranging from the highest to the lowest standards. Some of the low standard districts were considerably less than fifteen miles square. Neither the federal law nor the state plan took account of the fact that no legislation can provide, nor bureaucracy decree, how the winds shall blow or when. . . .

Under this federal mandate New York State was able to call for substandard "air-quality districts," such as that for Buffalo which permitted standards of pollution 25 percent greater than the minimum standard set by public health officials as the worst air you can breathe without noticeable adverse health effects. Only after citizen outcry were the proposals altered. But the appalling fact is that it was even possible for state officials to suggest it under existing law.

The act was ill conceived; critics called it an outstanding example of "wretched draftsmanship," poorly executed, and almost unenforceable. A diligent polluter could delay enforcement for seven to seventeen years, according to Ottinger. It was so vague it could be interpreted almost at will, so formidably complex as to be almost beyond understanding, so cumbersome it would take years to put into effect, so technical that local officials called on the only experts available to explain the meaning— those from industry. The measure concerned itself only with pollution in areas of concentrated industry and not overall effects. By the spring of 1970 not a single criteria document on pollutants had been completed. When completed they did not have the force of law but served only as guidelines for state officials. Even all air-quality regions had not been designated. The law stopped no one from dumping as many contaminants into the air as he liked as long as they were not too visible as dark smoke. Outside the designated air regions not even this restriction applied, unless controlled by local statute—a rarity.

"The highly touted Air Quality Act of 1967 proved to be worse than ineffective," said Ottinger. "It is actually sanctioning continuing pollution of the air and may be protecting polluters from effective enforcement already available under law."

The Nader report hit even harder: "The act provides a virtual license to degrade the last remaining unspoiled areas or to perpetuate those numerous company fiefdoms—towns and counties all over the nation that are the special preserve of one company or one industry." From a polluter's eye view, the basic standards implementation plan was ideal:

It leaves relatively undisturbed the notion that there is an absolute right to pollute, absolute at least up to a certain level (i.e. until ambient-air standards are exceeded). The Air Quality Act of 1967 institutionalized the attitude described by Louis C. Green, former Chairman of the Missouri Air Conservation Commission, that it is socially desirable to permit anyone—specifically any industrial plant—to dump his wastes into the ambient air, and thence into the nostrils and lungs of you and me.

Why was such a fraudulent act passed? Because of politics. President Johnson wanted an air-control bill stronger than the meaningless Clean Air Act of 1965. He was taking a beating in Vietnam. He needed an impressive victory at home. His advisers wrote the toughest revisions bill possible, one that called for strict national standards. They had no illusions it would be passed, but saw it as a means of enhancing the President's image as a public defender. Image is the triumph of reflection over substance. Those who watered down the Johnson bill would be the villains. The tough Johnson bill caused panic in industry. Its lobbyists came swarming into Washington to find out if the President meant business. They were quickly assured there was no danger. Policy had already been agreed upon even before the public hearings were held by Muskie. Everyone knew the finished product would call for regional standards and the ambiguity, vagueness, and confusion later written into the bill. In the congressional hearings the charade was played out. Industry executives and its scientists appeared for public benefit, hammering away at national emissions-control standards, which had already been eliminated.

The final measure was conceived and guided through Congress by Senators Randolph and Muskie, the Nader report observes. Both came from states little affected by the regional approach. The law was written under the influence of the coal interests. Joe Moody, until recently President of the National Coal Policy Conference (composed of related coal interests) allegedly boasted that he wrote the entire act. Was it true? he was asked by a Nader investigator: "He smiled and responded in a tone of mixed pride and apprehension: 'Well, I have no recollection of that particular incident, but I might have said it. I worked very closely with the staff of the committee. There's no telling who wrote what.'"

On November 21, 1967, President Johnson proudly affixed his

signature to the Air Quality Act of 1967. Each faction had won something, except for the public. *The Vanishing Air* states:

> Lyndon Johnson got a law whose title he could claim as his own. Jennings Randolph had made a good showing for the coal industry. Edmund Muskie won the real battle to determine the direction of the federal air-pollution control program for the next several years, and he did so without creating enemies, especially within the ranks of industry. Industry was happy to have a law it could live with without changing its plans. Of course, the omnipresent Joseph Moody was content, for the new law protected the pocketbooks of the coal operators. That is, after all, what the man is paid to do.

For industry the new law provided many advantages. It gave an illusion that something was being done without actually curbing industry. The minimal standards could be met without outrageous expenditures. Under the regional concept standards would be adjusted to local conditions, which could generally be controlled. The bill relieved the threat of truly tough legislation later on if nothing was done. Most important, the right to pollute was now written into law with government a full partner in responsibility for cleaning up industry's mess. The public was the villain, pointing the finger of guilt at itself. This guilty public would now bear most of the cost of cleaning up the environment if it came to that.

There was also another odd benefit. The uncontrolled polluter would be curbed. He was giving all industry a bad name and cutting in on big corporate profits. In this sense the law protected industry against its own excesses, rather than the public. Henry Ford II said that effective government regulation of industry was essential: "Without it the company that spends nothing for pollution control gets rewarded by the lower costs and higher profits than its more conscientious competitors. It is futile to rely entirely on corporate good citizenship if the system encourages the poorest corporate citizen."

For industry the law worked beautifully. The conscionable could conform without ruinous cost—extolling their control measures through extensive public relations campaigns (tax deductible). For the unconscionable there were few real hardships. In the beginning the law provided for control through abatement procedures—confrontations, conferences, and hearings. But these soon proved to be unworkable. In the pollution world "abate" does not mean to remove but merely to reduce in amount. Enforced abatement procedures were distasteful to the NAPCA. Abatement was dropped. In its place was substituted "voluntary compliance" in the name of sweet reason. Such sweet reason was tried on the aviation industry.

It did not seem unreasonable to ask the industry to cut down on the 3 or 4 billion pounds of pollutants that jets spewed into the air every year. Emission controls would cost only an estimated 0.1 percent of 1968 revenues—a passenger cost of about 10 cents on a $100 ticket. The NAPCA invited the forty-three airlines to a conference. Twenty-four were represented by a spokesman, General Clifton F. von Kann, USA retired, the head of the Air Transport Association. Twelve airlines declined to discuss the subject. Seven did not bother to reply. At the hearing von Kann said, in effect, that no problem existed and no solution was therefore necessary. He not only embarrassed NAPCA; he did it on television. Then three states got tough with the airlines; New Jersey, Illinois, and California threatened regulations, legal actions, and stiff fines. When the federal government had tried to curb the airlines they insisted they were under jurisdiction of the states. Now that the states went after them they insisted they were under federal control. The airlines got the government to bail them out. The accommodating secretaries of Health, Education and Welfare and the Department of Transportation worked out an agreement for the airlines to install smoke-reduction devices over a two-year period. The moral was lost on the government: one ounce of firm action is worth pounds of sweet reason.

A short time later the Federal Aviation Administration recognized that the noise of jets should be toned down. Did it tell industry to "get busy and find a way to turn it off, boys"? It did not. Instead it issued a tepid notice asking for industry comment on the effect federal subsidies might have in reducing jet noise, and how they felt about a compliance deadline if a requirement were imposed—an invitation to prescribe their own placebo. Aircraft noise is "the single greatest impediment to airport development in the United States at the present time," proclaimed an accompanying statement from FAA Administrator John Shaffer. "Prompt remedial action is required if we are to meet the projected growth of aviation in the 1970s and beyond." No mention was made of the harmful effects of noise on people, just concern for the industry's health and a possible corrective schedule that would not inconvenience it.

In 1971 the Federal Aviation Administration proposed federal standards for measuring airport noise. Following industry protest, the FAA withdrew the proposed measure and announced that it would permit the air transport industry to help rewrite the regulations.

The noise of jets at Los Angeles Airport became so unbearable that three nearby residential neighborhoods covering 400 acres were razed of nearly 2,000 homes at a cost exceeding $200 million. But no longer is the jet noise assault confined to those living near airports. It assails anyone under or near a flight pattern within dozens of miles of a jetport. In metropolitan areas this includes millions who cannot escape. It is impossible to carry on conversation when the jets thunder overhead. They shatter sleep. They addle the innocent brain.

The Federal Aviation Administration, reacting to complaints, has issued a preliminary notice that says it may order airlines to install noise-reduction insulation. But the airline industry is fighting even this mild remedy, arguing that modifications would cost it collectively about $1 billion for marginal noise improvement. Little mention is made of research to find means to tone down the jets. California reportedly plans to impose the nation's first state noise standards on jetports, but industry lawyers are confident the rule will be declared invalid by the courts. The aviation industry is still convinced that there is a better solution to the noise problem than installing expensive abatement equipment—moving the people at public cost.

Industry, with some exceptions, has vigorously resisted environmental reform, although the technological know-how exists to eliminate most pollution. Why hasn't this technology been used? "Because it costs money to control pollution," points out Representative John D. Dingell, "and many corporations will fight anything that interferes with profits, regardless of who suffers."

Industry's resistance takes many forms. It ranges from meaningless pretenses at compliance to outright obstruction and defiance. Corporations avoid compliance with laws by getting friendly officials to grant variances, postponements and exemptions. Delays on abatement schedules are stretched into years. Promises are disregarded, if they were ever intended to be kept. Many industries not only have evaded the laws but often managed to capture their control. In most states representatives of the biggest polluters sit on boards that are supposed to administer antipollution laws. A 1970 *New York Times* survey revealed that the membership of air and water pollution boards in thirty-five states is dotted with industrial, agricultural, municipal and county representatives "whose own organizations or spheres of activity are in many cases in the forefront of pollution." It was almost impossible to find a board with a person representing the public rather than industry.

"The roster of big corporations with employees on such boards reads like an abbreviated blue book of American industry, particularly the most pollution-troubled segments of industry," it was asserted. "The state boards—statutory part-time citizen panels of gubernatorial appointees and state officials—are in most states the entities that set policies and standards for pollution enforcement and that then oversee the enforcement." In other words, the polluters set their own antipollution regulations and then "police" themselves. Like this, according to the article:

> One Colorado state hearing on stream pollution by a brewery was presided over by the pollution control director of the brewery. For years a board member dealing with pollution of Los Angeles Harbor has been an executive of an oil company that was a major harbor polluter. The Governor of Indiana recently had to dismiss a state pollution board member because both he and his company were indicted as water polluters.

The takeover of state boards by polluters is documented extensively. Typical: "Ohio—where four of the five members on the Air Pollution Control Board have ties with the pollution sector (with industry represented by Procter & Gamble, a soap company with an acknowledged pollution record)—has the smokiest city in the country, Steubenville." Since May, 1967, the Connecticut Water Resources commission has issued 863 orders regarding pollution abatement; compliance has been obtained in less than half the cases, according to official records. "Louisiana's air and stream control commissions—composed of state officials and representatives of such groups as the Louisiana Manufacturers Association and the Louisiana Municipal Association—have never imposed a fine on anyone, and the air commission has brought only one corporate polluter into court in five years."

A confidential vignette of one board's activities was provided recently by an official in a Midwestern state where pollution problems are conspicuous:

. . . "the chief problem," he said, "was a general atmosphere of timidity [on the board] due to a hostile, lobby-ridden legislature and an apathetic Governor.

"We had money troubles constantly, so we didn't get much done. Some members would knuckle under if industry seemed to be getting to the Governor. The Governor had some tie-ins with the power industry, which restrained us from adopting tough emission restrictions." *

News Item: "In four years . . . air pollution has increased from an annual total of 142 million tons of contaminants to well over 200 million tons."

News Item: "More than three years after the statutory deadline . . . only 18 states have adopted water quality standards satisfactory to the federal government."

News Item: ". . . Federal officials have information indicating that thirty-two states have extended various abatement deadlines without the approval of the Secretary of the Interior—technically a violation of federal law."

On a community level the whole system of pollution control breaks down. The breakdown is self-aggravating and eventually reaches a point where the political and economic controls go in reverse and actually promote pollution, says Dr. George G. Berg, a biophysicist at the University of Rochester and past president of the Rochester Committee for Scientific Information:

> Zoning, licensing and inspections come to protect the polluter from legal liabilities and government controls without curbing the damage he causes. Public health and sanitation procedures come to protect the polluter from public censure by certifying that he is not responsible for anyone's sickness. The marketplace favors the polluter—his credit is better and his product is cheaper. Finally, the political process deadlocks reform at three levels. The town official who enforces clean water rules is fired; if he is not, then the whole administration of the town is voted out of office on the issue of taxes; and if the voters back the administration, business moves out of town to a more friendly political climate. . . .
> This is the situation at its worst and the worst is by no means rare.

* A year later, in December, 1971, the situation was reportedly improving. A *New York Times* survey showed that a score of states have taken or plan to take steps to eliminate or reduce the representation of polluting interests on state boards that regulate pollution. "The number of states with air or water pollution boards reflecting pollution interests dropped from 35 to 32 during the year."

The small and mighty alike have defied the pollution laws. Take Harold Polin, who would be a folk hero in simpler times. Harold operated a chicken rendering plant in Bishop, Maryland, that made the air so hideous with odors that his neighbors blanched and retched. For fifteen years he managed "to outsmart and outmaneuver three governments and two courts," according to *The Vanishing Air*. Even Washington itself was defeated by Polin and his stink. His victims included the NAPCA, the Clean Air Act of 1965, and the Air Quality Act of 1967, plus a larger cast of would-be enforcers and regulators. Finally, Polin's company was ordered to cease all manufacturing and processing operations by February, 1970, and the ruling was affirmed by court decision.

Some of the biggest corporations use their great economic power to defeat the laws. They can threaten to shut down a plant or move if controls are imposed or enforced. For many communities such action would be economically ruinous; it brings officials quickly to heel. U.S. Steel's plant in Duluth threatened to close down if local standards were enforced, insisting the operation would no longer be profitable. At stake were 2,000 jobs and a $25 million payroll. An abatement schedule agreeable to the company was worked out.

Union Carbide has defied the government for five years. In its Marietta, Ohio, plant, company officials refused to supply data or admit government investigators to resolve a dispute over emissions, according to *The Vanishing Air*. A nearby Carbide plant at Alloy, West Virginia, doses 220 people in the area with 28,000 tons of smoke, grit, and dust particles annually—"one of the largest single sources of particulate pollution in the nation," according to Ralph Nader. The company reportedly has repeatedly missed abatement deadlines, and local officials did little or nothing. Frustrated residents of Anmoore, West Virginia, finally brought a $100,000 damage suit against Union Carbide and federal and state air-pollution control officials. Newspapers reported that the company eventually bowed to federal pressure and agreed to clean up air-pollution emissions caused by its Marietta, Ohio, and Alloy, West Virginia, plants. Three months later, in January, 1971, Union Carbide reportedly said that to comply with the federal antipollution order would require laying off 625 workers. Nader called this an "act of economic and environmental blackmail." William D. Ruckelshaus, head of the Environmental Protection Agency, reportedly said he was shocked by the threatened layoffs: "How the company can decide in January, 1971, that it will be unable to find lower-sulfur coal by April of 1972 is beyond me."

Seduction also has its adherents. Nashville, one of the nation's most polluted cities, is the home of a du Pont plant. The city was about to establish regulations based on Public Health Services research. Some 123 air samplers provided 200,000 readings (samples) of gross contamination in 1958 and 1959. One sampler, however, was clean during the years 1961 through 1968. A du Pont public relations man was able to convince the city fathers that they should put their faith in the one clean sampler, according to *The Vanishing Air*.

In at least one case a polluter was even honored by local officials. The president of a polluting firm was given a plaque for his contributions to restoring the environment. He allegedly accepted the tribute with tears of joy.

The 1970 amendment to the Clean Air Act did not get off to a promising start despite President Nixon's pronouncement: "America pays its debts to the past by reclaiming the purity of its air, its waters and our living environment. It is literally now or never."

Two bills were submitted: one from President Nixon, the other from Presidential aspirant Muskie. The Nixon bill was preceded by tough talk of implementing strict national standards and a $10,000 a day fine for violators. But the actual proposed measure turned out to be worse than its predecessor. "The President's advisers have taken a bad idea, twisted it out of context and made it worse," complained *The Vanishing Air*. "It would be hard to imagine a more ineffective measure than the 1967 act, but the Nixon Administration has managed to draft one." The law was riddled with loopholes and lacking in basic enforcement mechanisms.

The Nader group was unable to write this off charitably as shoddiness or inexperience. It looked at a number of preliminary drafts that were circulated within the Administration prior to the bill's release.

> Although all of them contained serious flaws, most were at least respectable attempts at constructive legislation. In the journey up the echelons, every meaningful provision of the earlier drafts was systematically pruned. Private rights of action to enforce violations of standards, power to sub-

poena information, and expedited enforcement provisions
were only a few of the Nixon proposals that fell unceremo-
niously to the cutting room floor at the White House.

Senator Muskie, author of the 1967 debacle, after declaring it
a legislative triumph, offered a new version that included strict
national standards and big fines for offenders. The relatively
strong bill that emerged was worked out in legislative infighting.

In this brief report of progress under the environmental laws,
we cannot overlook the state of the nation's waters. They too are
falling steadily behind. A General Accounting Office study,
quoted by Tom Wicker in *The New York Times*, suggests how
serious the decline is:

> On a stretch of unidentified interstate river, where $7.7
> million had been spent on municipal sewage disposal plants
> since 1957, these facilities had reduced total pollution of
> the river by 3 percent while the amount of industrial wastes
> dumped into the same river in the same years had increased
> by 350 percent. Everywhere the GAO looked it found the
> same grim pattern.

How could this be, under our various laws destined to clean
up the water? Water, like air, has become a victim of politics.
"Things were bad enough under the reign of conservationist
Udall," observes *Ramparts*. "But under Hickel and Nixon water-
pollution control has been further weakened." Water programs
are far behind schedule. The sewer program has largely broken
down for lack of funds, setting back the water-purification
schedule.

At the time federal water standards were set, many were
lower than those already established in some states. This was
understandably embarrassing to the government. It feared that
its levels would become maximum rather than minimum. It
asked the states to sign a "nondegradation" pledge that they
would not lower existing water quality. Less than half the states
signed. The clause is meaningless, like many other "tough" conservation measures of the Interior Department, *Ramparts* as-
serts: "It is riddled with loopholes. A measure of its potential
meaninglessness is seen in the declaration that existing standards cannot be lowered unless 'such change is justifiable as a
result of necessary economic and social development.' "

Social and economic development are not terribly hard to prove today—not in the new Orwellian use of language.

Proposed new laws for the waterways virtually legislate them into open sewers. Indelicate expressions are not used, of course. Instead we have such euphemisms as "assimilative capacity" (the maximum amount of filth it can carry) and "beneficial use of water" (use it to carry filth). Is there an unbeneficial use? And who decides how much filth the river or lake can tolerate? This will be done by state and local agencies and the polluters themselves. Nixon's proposed law calls for industries and municipalities failing to meet water quality standards or implementation schedules to be "made subject to court-imposed fines of up to $10,000 per day." It sounds good. But the fine is left up to the discretion of the judge. It is a rare court that clamps down on polluters to such an extent. Or to any extent.

Many industries have discovered that it is cheaper to pollute public waterways than to clean up their own wastes. Minimal fines cost less than control equipment. Some of the larger firms, concerned with their corporate image, spend millions showing the public a sparkling stream they preserved, a squirrel happily perched on the stump of a vanquished tree, an oil well camouflaged among frolicking animals, only to be hauled into court on a pollution charge. But most polluters are less sensitive. They will sacrifice image to profit. A list was compiled recently to show who was being charged with polluting the nation's waterways. It included some of the biggest corporations. All were charged under the 1899 law. Conviction carries a maximum fine of $2,500 or one year imprisonment for each offense. (No one is known to have gone to jail in the United States for polluting.) Had all the fines been collected they would have amounted to a total of $40,000, hardly enough to discourage the giants on the list.

The root obstacle to cleaning up the mess is always the same: the cost is formidable. A continuing annual outlay of 2 percent of gross national product would be required just to ensure that environmental deterioration is gradual rather than rapid, according to experts meeting in Washington recently. This would amount to just under $20 billion in the United States, the *Washington Post* explains. Holding the line, on the other hand, "would cost about 4 percent of GNP, and actively cleaning up the past— and preventing future—pollution could cost three to four times as much. That means $160 billion."

In 1969 Secretary of the Interior Udall estimated the cost of basic national water pollution control needs through 1973 as totaling nearly $30 billion.

What are we actually spending? In 1968 President Johnson cut a $700 million congressional authorization to $214 million. The following year President Nixon cut a $1 billion authorization to $214 million. Congress, almost contemptuously, overrode him and appropriated $800 million.

The President's budget request for environmental control in fiscal 1973 was almost exactly the same as the previous year's despite the continuing environmental decline and inflation. In January, 1972, as in the previous two budgets, Nixon requested "far less money for pollution control than Congress authorized in the basic legislation on air and solid waste and less than it will authorize in the pending bill on water cleanup," *The New York Times* reported. The President is expected to spend little, if any, of the appropriated funds that exceed his request—the so-called appropriations gap.

The President requested for the Environmental Protection Agency, with its vast functions and responsibilities, only $2.246 billion. Of this amount, $2 billion—the same as requested for fiscal 1972—would be for the federal share of sewage facilities; the remainder is for the agency's research facilities and its abatement and enforcement activities. For the pending water pollution control bill, Congress is expected to appropriate at least $3 billion—$1 billion above the President's request, which he "as certainly will refuse to spend."

To implement the 1970 Clean Air Act the Administration estimated that $320 million would be needed in fiscal 1973, and the act finally authorized $300 million for that year. Nixon asked for only $171.5 million—"$148.5 million less than his own estimate of need."

For the fiscal year 1973 Congress authorized $238 million for the Solid Waste Program. Nixon requested $23.3 million, "or $11.3 million less than last year."

In the area of research and development, there was a cut of $5 million for development of improved waste water technology. The government said private industry would be relied on to do the basic research. But the government has also complained that industry isn't doing it, which seems to mean that it will not be done by anyone.

Against the 1973 budget calling for environmental-control expenditures of $2.246 billion, the President requested for defense spending $83.4 billion—a $6.3 billion increase over the previous year.

A brief look at our economics gives insight into our real environmental intentions and values. In the past dozen years, since pollution reared its ugly head, federal, state, and local governments have spent a total of $5.4 billion on corrective measures. To clean up Lake Erie alone would cost an estimated $10 billion; the Hudson River, $7 to $10 billion; New York Harbor, $4 billion. At the same time we have been spending $25 billion a year for our involvement in Vietnam. We have money for death but not for life. The total $5.4 billion spent on the environment is less than the current estimated cost of developing a Poseidon-submarine-launched missile, says Senator Clifford Case of New Jersey. "The difference between the original and current estimate of the cost of developing the Walleye II TV-guided glide bomb exceeds the amount spent at the federal levels in the past twelve years to subsidize construction of municipal waste plants." We spend billions for an electronic Maginot Line to protect us from foreign nuclear attack while poisoning ourselves at home.

The President's budget bears little relationship to his rhetoric on environmental concern. Of each dollar spent by the federal government, 52 cents goes for the armed services and the space program. Air and water pollution control get four tenths of one cent.

Little is being spent for the technology to find a way out of the bind. General Motors, for example, with gross annual sales of $24 billion, reputedly budgets only about 0.17 percent of that amount for pollution research—approximately $40 million a year, one sixth of its annual advertising budget of $240 million. Even the most enlightened companies rarely spend more than a fraction of 1 percent of their profits for pollution control.

The hard truth is that the necessary technology to clean up the environment does not exist at this time. There has been, as noted previously, no real widespread advance in the treatment of sewage waste in forty years, although some new methods are in the experimental stage. The secondary treatment the nation is now striving for does not even perform the limited task set for it. It merely removes some solids and converts organic wastes into inorganic materials that are returned to the water to nourish algae blooms which reconvert inorganic forms back to the original organic, as Dr. Commoner explains. Frequently the solids removed are dumped into the ocean, merely bypassing the estuaries and coastal lands but returning on the tides and currents. We pretend that by hiding, burning, burying or diluting, the hazard no longer exists. We protect one area at the expense of another. Self-deception is implicit in most of our present pollution laws.

No one must give away the game. That is one of the unwritten rules. And woe to the foolish and reckless who break it. One did: R. P. Clinton, president of the Clinton Oil Company of Wichita, Kansas. His company plans to build a $90 million refinery at Brunswick, Georgia, with satellite plants at Savannah and Charleston. Clinton said waste effluent would be piped into the ocean. "We are going to pollute," he said bluntly and honestly. "It's only a question of how much. But I think with proper marketing and construction we're not going to pollute this area. What we're going to do is contribute to the pollution of the world."

Was this candor appreciated? It infuriated Interior Secretary Hickel. He "exploded in anger . . . at what he called the 'arrogant' attitude of an oil company president toward the pollution of coastal waters," states the *Times*. " 'That was an unwise thing for Clinton to say.' Then his voice rising, he said, 'It was an arrogant statement industry does not need. If they want to challenge me, I'll find authority [to deal with such pollution]. They ought to be ostracized for such a statement.' "

Clinton's sin was not arrogance. It was shattering the myth that we have real protection against pollution. He merely stated a factual limitation of present technology and the consequences of additional industrial growth. We have not gone far beyond changing one form of pollution for another in most areas. Pollution control is said to be a potential $25 billion business, but at this time there are still conflicting claims on whether there is a proven way to remove sulfur dioxide from smokestacks.

There is rich irony here. Some of the biggest polluters are the most active in the booming antipollution field. Especially the big chemical companies. Monsanto Chemical Company, one of the leaders in providing control equipment and chemicals, boasts that it has a device to remove sulfur dioxide from smokestacks. Unfortunately, it has not been put to the test. Monsanto has complained publicly that it is unable to sell its new process. It would seem that Monsanto could install one of its own devices in one of its own plants. The company has a serious problem with sulfur emissions. But no, Monsanto believes that would be a wicked waste of money. *The Vanishing Air* states:

> Monsanto has made the calculation that it is cheaper to continue to pollute than to expend money for control. The "calculation" has been aptly described by economist Kenneth Boulding as the "famous freeloading" problem. The individual interest is to go on polluting as long as the rest of the society picks up the tab.

Industry's primary excuse for polluting is the lack of effective technology to prevent it. It points out that only primitive equipment is now available. Scrubbers and electronic precipitators remove particulates and leave most other contaminants untouched. Why? The basic reason is that almost no research funds have been allocated—a failure not altogether accidental, according to Dr. Woodwell at Brookhaven National Laboratory: "The exploiters of the environment have much to gain by subtly opposing further knowledge on subjects that might inhibit their rights as exploiters. Thus they create doubt and force compromise by questioning the accuracy, integrity and objectivity of science when it fails to support them."

Nor is science without blame in this failure, he suggests in *Ramparts*. It has, in general, failed to address itself "clearly and effectively" to the problems of environment. It was busy elsewhere during the "devastation of the earth" and the grass-roots protests that made ecology fashionable. Scientists were not leaders in these protests. Now they find themselves "conspicuously unprepared for the environmental crisis and often even antagonistic."

He blames growth for this continuing decline of the environment. "While a growth of 15 percent is attractive in a portfolio of stocks, it is very difficult to maintain a highly integrated, complex social system that dumps poison at that rate into its own environment." These poisons and pollutants have transformed the ecosystem of the planet from the complex communities that have built the biosphere and supported oceanic fisheries and man to the simplified biota of cesspools like Lake Erie. "The losses are not simply robins, bluebirds, eagles, mackerel or trees, but the potential of the earth to support life, including man. And through all of this the scientist seems to have been more of a curse than a cure."

Science has taken two paths that brought it to its "current disarray and helplessness." One was supported by industry, the other by public funds. The overriding objectives of industrially supported science have always been stated candidly: greater profits for industry. Woodwell states: *

> Profits are increased by spreading the costs of waste treatment to as wide a group as possible, so that no individual notices a significant effect and the costs need not be borne by industry itself. Obviously such a commitment leads much more rapidly to a deteriorating environment than to spontaneous solutions for that deterioration. Despite the increasing number of advertisements by industry on the efficacy of their own research into pollution, profits continue to come from spreading the costs of waste treatment widely. It is fatuous to hope that any cures to the environmental problem will arise in an industry.

The second branch of science is different, Woodwell explains. This publicly supported arm potentially serves the broader public interest. Its objectives have been set, for the most part, by Congress. It reflects the country's official sense of what is important:

> Research money has long been available for war—including chemical and biological warfare and bomb-related physics —and for cancer-related chemistry and biology. And thus the tax-supported segment of science has developed an elaborate technology and a corps of brilliant specialists in biochemistry and physics.

These are the men who won Nobel Prizes and other awards. They were specialists who followed the "reductionist" approach to science. They reduced phenomena so they could learn more and more about less and less. They were trying to understand the whole by reducing the minute to its finite point. This, in essence, is the antithesis of ecology, which concentrates on the whole.

Throughout their triumphs, the scientific aristocracy generally scorned environment. Now, after decades of neglect, the environment is in crisis, Woodwell notes:

> When wise and massive action in science and government is demanded, the American man of science is stumbling. He

* *Ramparts*, May, 1970.

is stumbling in part because he was led by the availability
of money to participate in building into society a large intel-
lectual deficit—a deficit of research, of knowledge, of public
understanding of the structure and function of environment, and a deficit of administrators experienced with environmental problems.

This man of science, like the environment he neglected so
long, is in danger. He fears he is obsolete, too specialized, too
vulnerable to the changing winds of the politics of science as
determined by the needs of industry and government. "In uncertainty, he is rushing to re-examine himself, the problem and his
tools; but there is danger that he will simply try to rename his
work 'environmental' and set to work applying corn plasters to
the cancer."

ADVERTISING: SELLING THE IMAGE

Few things are necessary to sustain life. A few more are necessary for creature comfort. Beyond that, possessions represent a
progressive scale of luxury that stands as an invisible wall between the possessor and a "natural life." The more goods the
more unnatural the life. In our system of values, possessions are
directly related to the standard of living. The standard of living
is the measure of "success." The more goods one owns the higher
his standard of living, and thus the greater his success. These
values, like the products they reflect, are forged in the marketplace. But this takes no account of the consequences of ownership—the marketplace is not interested in consequences, only in
selling goods. Ultimately the consumer no longer is served by his
possessions. He becomes their victim, indentured by created desires rather than real needs. To fulfill these artificial desires
means destruction of the environment. Thus, the individual is
unwittingly caught in a lethal trap: in fulfilling his life goals he
is inadvertently working toward his own destruction. Nor does
the irony end there. He is so engrossed in the pursuit of goods
that he is oblivious to his own plight.

The primary motivating force in this drama of destruction is
advertising. It first creates desire and then becomes the medium
of fulfilling it. It does not sell goods directly; indeed it tells precious little about the objects for sale. Rather, it sells image.

We cannot be satisfied with what we have. The unattainable
dream beckons. We have a reduced workweek but cannot rest.

We work nights at a second job, or harder at the first in order to earn more to buy more.

Image is bolstered by public relations—now an accepted operation of practically every large corporation and public and private agency. It can be useful and informative. More often it deceives or helps create false values which promote the sale of goods and a climate of opinion favorable to industry. From most public relations departments flows a steady stream of propaganda, self-serving information, misinformation, distortions and outright lies. Truth becomes as polluted as the environment. This broadside of packaged fantasy is aimed primarily at the mass media that usually accept it without question. More often than not the commercial message is skillfully buried in what appears to be the innocent feature, "news" article, or television documentary. It is designed to shape public attitudes and values. This becomes increasingly important in manipulating public response.[8] In 1959 the Pentagon spent an acknowledged $2,755,-000 for public relations. Between then and 1970 overall defense spending almost doubled from $43 billion to $77 billion. But military public relations expenditures shot up fifteenfold to a declared $40,447,000, and it was undoubtedly several times that amount.* It costs a great deal of money to sell an unpopular war, to make slaughter palatable and environmental devastation acceptable.

Industry underwrites its message with an estimated annual outlay of $20.4 billion in advertising and another $3 billion for public relations.† Its direct and disguised propaganda helps dictate public attitudes in every area of life, from the length of skirts to the myth of DDT's safety perpetrated for twenty-five years while the public was being slowly poisoned. It sounds im-

---

* An unpublished report by the 20th Century Fund suggests that the true total of Pentagon spending on "public affairs" on the home front may be $190 million a year, according to the CBS documentary "The Selling of the Pentagon" (February 23, 1971). The money is spent to condition the public to acceptance of war and the expenditure of vast sums for arms. The documentary disclosed that the Army spends $6.5 million a year in making its own films. A total of 12,000 radio and television tapes are mailed annually to 2,700 radio stations and 546 television stations. More than 2 million releases, reporting medals, promotions and reassignments, are mailed to 6,500 weekly and daily newspapers.

† The commercial conditioning of a child begins early and continues throughout his life. By the time a child is eighteen he will have seen 350,000 TV commercials, according to one study. Another states that in an average American home the television set is on more than six hours a day; between the average male's second and 65th year he will watch over 3,000 full 24-hour days of television—almost nine full years of his life.

pressive when the petroleum industry boasts that it is spending a million dollars a day on pollution control. But consider the size of the industry. Total sales come to some $60 billion a year— more than $200 million a day. The $1 million spent on pollution , control amounts to only 0.5 percent.

The paper mill proclaims that it removes 85 percent of the pollutants from its effluent. This too sounds impressive. But on examination we learn that ordinary sewage is 99 percent pure water. It's the last 1 percent that causes the mischief.

As Gladwin Hill points out in *The New York Times,* the difference between 96 percent control of soot from a large coal-burning power plant and 98 percent control, which sounds like a trivial amount, is the difference between surrounding residents being exposed to one ton of soot per hour or only half a ton. Thus we have not only air pollution but mind pollution.

Education and commercial propaganda take us away from our feelings of place in nature.[9] Our training has made us so bemused with our own gadgets and powers that we have no time for anything else, even in recreation, according to Dr. David J. Wilson, Professor of Chemistry at Vanderbilt, and acting chairman of the Nashville Committee for Scientific Information:[*]

> We have been trained to admire and enjoy consumption, the dissipation of power, convulsive growth rather than dynamic balance. Instead of canoeing on the river, we blast down it with a 60-horsepower stinkpot, going so fast that we have no chance to appreciate what the river has to offer, and making such a racket and spewing out so much oil that we destroy anyone else's chances of appreciating it too.

For industry, generally, such problems do not exist. It still sees pollution and environment as more of a public relations problem than anything basic in the way business is conducted. The environment becomes a vast image problem. For example in *Time* magazine, November 7, 1969, we see an ad by Potlatch Forests, Inc., fourth largest manufacturer of lumber in the nation. Pictured is a beautiful swift-flowing river, banked by thick forests, and overhead a clear sky. Potlatch has a message: "It cost us a bundle but the Clearwater River still runs clear." The ad proclaims Potlatch's "total commitment to pollution control":

> On the Clearwater River in Idaho we have spent over $3.25 million to assure clear water. Our latest effort is a

[*] Personal correspondence (answer to survey query) June 23, 1970.

The fact that ads often are misleading or even lying, in many cases, is not the worst of it, Mander observes. Their greater sin is that they divert the reader from a more central understanding about what's really going on. "I fear that all the recent government rhetoric, magnified by industrial assurances, might have the net effect of encouraging a society already dazzled by technology to be further assured that technology is solving the problem—people want so much to be assured—and so it's back to the television set."

7

The Choice:
Cataclysm
or Reform?

By 1980 we may be living in a dying world in which the last bell has tolled. By then the grisly shadows of famine and war may have darkened the lands of the underdeveloped nations. Pesticide drainage from the continents into the oceans may have snuffed out all marine life and brought famine to Asian and other nations whose very life is taken from the sea. Those living along the coastlines may have been forced inland by the foul slimes and noxious gases generated in the corpse of the world's ocean. All of us in the United States may be living with the certain knowledge that we are doomed to premature corruption and death, as the long-term corrosive effects of poison gases in the lung and pesticide residues in the liver finally become clear. The centers of major American cities may finally have exploded into continuous guerrilla warfare, as the environment of the ghetto deteriorates below the point of human tolerance.
— Former Congressman Richard L. Ottinger

THE NEED FOR CHANGE

The American dream is badly tarnished. Technology and science have let us down. The earth's bounty turned out not to be inexhaustible after all. The abundance of goods on which we set such store has brought us neither happiness nor peace. The social structure is in chaos. Our cities are in turmoil and decay. The suburbs are on their way to becoming cities. Rural areas are turning into suburbs. Industry follows the people. Crowding and pollution increase. The new roads are so choked with traffic that the powerful new cars can only creep.

The Bomb may one day no longer seem a threat, but a release, says Dr. Rene Dubos. The real danger is not that we will be wiped out by pollution and degradation of the earth but that we will adapt.

> Children can readily learn to accept treeless avenues, starless nights, tasteless food, a monotonous succession of holidays that are spiritless and meaningless because they are no longer holy days, life without flowers or birds. Loss of these amenities of life may have no obvious detrimental effect on their physical well-being, but it will almost certainly be unfavorable to the development of their mental and emotional potentialities. . . .
> The real spectre that pollution casts over men's future is not, perhaps, the extinction of *Homo Sapiens* but his mutation into some human equivalent of the carp now lurking in Lake Erie's fetid depths, living off poison.

Is there no alternative to these dismal prospects? Surely there must be an out. In preparing this book I sent letters to some 100 people prominent in science or environmental concern, asking for any thoughts they might care to offer on the environmental problems and what could be done about them. Most did not reply. Some sent polite notes that they could not spare the time such a request entailed. A few were generous beyond belief. Three who responded (among the most gracious replies) were Nobel laureates.

The responses, generally, are not encouraging. Or possibly only pessimists take time to answer such requests. In any event a dark threat runs through the replies. The denominator is that our enveloping tragedy is not necessary but is almost inevitable because we refuse to change our ways. We are victims not of fate but of our own narrow view of life. My correspondents join a public chorus that calls for change if we are to save ourselves.

Albert Szent-Gyorgyi, a biologist at Woods Hole Marine Laboratory, and a Nobel Prize winner, recalls his book, *The Crazy Ape,* published in the spring of 1970. It outlined the madness with which we are destroying the world and ourselves. Now he observes that when he wrote "my little book I gave a 50 percent chance that mankind would survive for a few more decades, but the behavior of man and political evidence makes this estimate too optimistic." He sees our prospects as "very bleak." We are under the "terrible strain of idiots who govern the world and moving inexorably and insanely toward ultimate calamity," he told a *New York Times* correspondent. We are a death-oriented society, he says. "If you watch and if you read the newspapers, a great part of it is taken up by war, by killing, by murder, atomic bombs, MIRV's, gases, bacterial agents, napalm, defoliants, asphyxiating agents and we have war. All our ideas are death-oriented."

Szent-Gyorgyi is seventy-six. He watches the pageant unfold with "quiet desperation and disgust." He works in his laboratory, putting his hope in youth, expecting nothing from "the present leading class"; he swims, fishes, sails and shakes his head over our common fate. "What sort of world is this? Tell me. Shakespeare said life is the dream of an idiot. I say it is the nightmare of an idiot."

Another Nobel Prize–winning biochemist, Dr. George Wald, professor of biology at Harvard, sees the same breakdown of American ideals, the emphasis on death.

> How many of you can sing about "the rockets' red glare, the bombs bursting in air" without thinking, "Those are *our* bombs and *our* rockets, bursting over South Vietnamese villages?" When those words were written we were a people struggling for freedom against oppression. Now we are supporting open or thinly disguised military dictatorships all over the world, helping them to control and repress peoples struggling for their freedom. . . . The only point of government is to safeguard and foster life. Our government has become preoccupied with death, with the business of killing and being killed. So-called defense now absorbs 60 percent of the national budget, and about 12 percent of the gross national product.

He says we must get rid of atomic weapons, here and everywhere. We cannot live with them.

> I think we've reached a point of great decision, not just for our nation, not only for all humanity, but for life upon the earth. . . . The thought that we're in competition with the Russians or with Chinese is all a mistake, and trivial. We are one species with a world to win. There's life all over this universe, but the only life in the solar system is on earth, and in the whole universe we are the only men.
> Our business is with life, not death. Our challenge is to give what account we can of what becomes of life in the solar system, this corner of the universe that is our home;

and, most of all, what becomes of men—all men, of all nations, colors, and creeds. This has become one world, a world for all men. It is only such a world that can now offer us life, and the chance to go on.

The politician is hopeful. He places his faith not in human nature but in corrective legislation, although he has watched most other corrective legislation fail in Congress or in the regulatory agencies. It failed because it was the wrong kind. Or there was not enough of it. The answer is more legislation. We need a new environmental agency with concentrated powers to protect the environment. We must underwrite vast ecological studies. We must form a conservation bank to collect all the data gathered for a systems approach that will put nature on a mathematical formula. We must evaluate the environmental impact of all new projects and initiate corrective measures. We must appropriate more funds to put the unemployed to work rebuilding our ravaged and decaying cities. We must have new and bigger family-planning programs.

We must have a new ethic. A new environmental Bill of Rights. An equitable land-use policy. Large appropriations for research and reform—yes, above all, larger appropriations. But money alone does not guarantee solution. Nor does law alone guarantee success. We now have many good laws but they are not always administered by good men. The best laws are worthless if not enforced. Bad laws, as we have seen, are worse than no laws when they provide the polluter a screen to hide behind. A new name is printed on the door in the State House in New York: Department of Environmental Conservation. But what is inside? Representative Ottinger tells us: "When we open that door we find the same tired faces, the same paralyzed bureaucracy, the same indifference to our predicament, and the same passion for erecting monuments at the expense of a ravaged environment."

Outrage is not lacking. Some are still capable of it. The students are angry over the kind of world being handed to them. They demand change—real change, not just a new slogan or a new name on the environmental door. Dennis Hayes demands such change. He was the national coordinator of Earth Day and now heads the politically oriented Environmental Action Inc., based in Washington, which publishes *Environmental Action*. Hayes recognizes that Earth Day changed nothing. He protests the unnecessary production of goods, the unnecessary waste and expenditure of "insanely large sums on military hardware instead of eliminating hunger and poverty. We squander resources on moondust while people live in wretched housing. We have made Vietnam an ecological catastrophe. We have turned a largely self-sustained country into a dependent fiefdom."

Hayes takes Congress to task in testimony before one of the endless committees investigating the need for more new environmental legislation. He protests the "piecemeal programs and insipid rhetoric" that we have had so far. He does not think that most of the politicians and businessmen who are jumping on the environmental bandwagon have the slightest idea what they are getting into.

> I don't think they realize that we are going to need fundamental changes in the values of this country. I don't think they realize that the students see this as a long and serious fight for profound changes in what the country is all about. . . .
>
> They are talking about emission control devices for automobiles; we are talking about bans on automobiles. They are bursting with pride over plans for municipal waste treatment plants; we are challenging the ethics of a society that, with only seven percent of the world's population, accounts for more than half of the world's annual consumption of raw materials. This country is robbing the rest of the world and future generations of their natural resources. We have to stop.

From others comes the same call to conscience. Dr. Karl Menninger, head of the Menninger Clinic in Topeka, Kansas, is concerned about what is happening to the world and its people and creatures. He looks out the window of his office, at smog overhanging the Kansas flatlands, as he talks: "We need an ecological consciousness that recognizes man as a member, not master, of the community of living things sharing his environment."

He does not think that the deer and the air and the rabbits and the earth belong to us. He does not think that we have any right to allow people in Alaska to sell the oil that's there.

> What right do they have to sell it? Who bought it in the first place? Do some people just get a free handout of natural resources which they can sell to other people? Are we going to agree that that's a good policy?
>
> We have agreed that that's a good policy. The guys that get there first get the grabs. Now it seems to me that that's

a dangerous weakness. I don't believe the Communistic system will work; I believe in the capitalistic system, but I think its weakness lies in the fact that we allow people to grab something irreplaceable, something that they don't possess and sell it to others. We are a wasteful society in America. We're wasteful, we're untidy, we have a low aesthetic sense and we underestimate our national intelligence.

From the scholar comes a call for dramatic change if we are to survive. "We shall be very fortunate if our society can survive the remaining days of the twentieth century without an irrevocable collapse," writes Dr. Everett M. Hafner, Dean of the School of Natural Science and Mathematics, Hampshire College.

Dr. Linus Pauling, the Nobel Prize–winning chemist who was one of the powerful voices in helping stop the atmospheric nuclear testing program, visited India and was shaken by the poverty and human degradation he saw. It set him to brooding about population problems, the obligations of government and men's relations with one another. "Clearly there are too many people already," he decided. After much thought he decided we should have 160 million people in the United States—not more:

> I believe that it is now becoming a duty of government all over the world to think about the question: "How many people should there be in our country?" I think that the principal duty of the government is to improve the well-being of the people of the country, rather than just to preserve the integrity of the borders. The time has come when the people of the world are recognizing that they are joined together by the bonds of brotherhood, and that nationalism must be abandoned. We must all work together for the welfare of human beings everywhere, of humanity as a whole.

Others call for change purely on practical grounds. The renowned physicist Dr. Jean Coulomb, of the University of Paris, speaks of the "basic fragility of our world." We must impose a "speed limit" on our civilization, he says. We must limit scientific and technological advances and the consumption of resources or the earth will not endure.

Limits will have to be put on the country's growth in the next decade to avoid destruction to the environment, says Stewart Udall, no longer hampered by the inhibitions of public office. Such thinking runs counter to the current American idea of continuous expansion, he agrees. It might be considered by some to be an "alien if not un-American concept, . . . but it will have to be done because the country is running out of frontier area."

Even the business executive is uneasy. J. Paul Austin, president of Coca-Cola, tells his company's annual meeting that he fears the earth could become a barren planet with undrinkable water and unbreathable air. "Unless all of us begin immediately to reverse the process of impending self-destruction which we have set in motion, this green land of ours will become a graveyard."

More and more people realize that growth no longer means the better life. We have reached the point of diminishing returns. More goods imposes greater strains on the environment and more pollution. "Personal convenience and comfort are actually going to decrease as everyone gets more cars, more gadgets," points out columnist Anthony Lewis in *The New York Times*. Even the economists are beginning to doubt the traditional wisdom of growth. A letter in the *Economist* of London from Kenneth Hunter of the University of Pennsylvania makes the point:

> Will economists never learn? This economist did over the summer in Los Angeles, where he found it impossible to take a deep breath without ripping his lungs apart. There comes a time when extra growth creates more problems than opportunities. It is just possible that Britain is close to that point right now. If the Japanese have oxygen vending machines in Tokyo at the moment, one can't help wondering how much longer it will be before their "economic miracle" makes life intolerable.

The trouble with such a view is that it would be so formidable to sell, politically, points out Lewis. "Who is going to campaign on the slogan 'Less is more'? We may come to that someday— and sooner than we think—but right now it is not likely to get past the stage, the necessary stage of economic and philosophical discussion."

How to achieve the massive changes required to halt the present destruction of the environment? First we must carefully define the problem. We must evaluate exactly where we are, where we want to go and how to get there. This is something our society has not done for some 300 years. We have lost all sense of direction. We have moved so fast that we have outpaced all of the props that once stabilized and supported us. The old rules do not apply to this new world. We are futilely trying to apply outmoded tools to a new situation, blindly clinging to the past as we hurtle toward an unknown destination.

Several thoughtful men offer possible solutions. Dr. Glenn T. Seaborg, a Nobel Prize–winner and former chairman of the Atomic Energy Commission, suggests a great conclave of thinkers to come together to consider our environmental problems:

> Nowhere do we have any body of thinkers whose duty it is to take all those things—the air, the water, the land, population, the environment—and put them all together. We must do this because, for the first time, the planet is one. For the first time man has looked on the earth from the outside and he has seen how small and fragile it is.

A similar convention of minds is proposed by Dr. John Platt, a biophysicist who is associate director of the Mental Health Research Institute at the University of Michigan: natural scientists, social scientists, doctors, engineers, teachers, lawyers, and many other trained and inventive minds,

> . . . who can put together our stores of knowledge and powerful new ideas into improved technical methods, organizational designs, or "social inventions" that have a chance of being adopted soon enough and widely enough to be effective. Even a great mobilization of scientists may not be enough. There is no guarantee that these problems can be solved, or solved in time, no matter what we do. But for problems of this scale and urgency, this kind of focusing of our brains and knowledge may be the only chance we have.

Platt reminds us of how fast we have traveled and how it has separated the generations:

> Within the last 25 years, the western world has moved into an age of jet planes, missiles and satellites, nuclear power and nuclear terror. We have acquired computers and automation, a service and leisure economy, superhighways, superagriculture, supermedicine, mass higher education, universal TV, oral contraceptives, environmental pollution, and urban crisis. . . . It is hardly surprising that young people under 30, who have grown up familiar with these things from childhood, have developed very different expectations and concerns from the older generation that grew up in another world.

We must go even farther than seeking a nationalistic solution, says Colonel Charles Lindbergh, who has devoted his life in recent years to conservation. He sees a need for a cosmic approach to our difficulties. The worldwide threat is so great that the environment cannot be maintained in this technological era through commercial organizations acting independently: "Contemporary pressure lobbies together with the often ruthless exploitation and spoliation of our country stand witness to the need for quick and firm governmental action. We need a policy and plan that covers our entire planet and extends to the utmost of human capability into space and time."

Such forums must take place in public. The people must have a role in determining what they want to be and the kind of world they want to live in. Much of our present plight is due to the way people have been excluded from having a say in their destiny, suggests Senator Gaylord Nelson:

> Up to now, the decisions that have destroyed our environment have been made in the board rooms of giant corporations, in the thousands of government-agency offices protected from public scrutiny by layer on layer of bureaucracy, and even in the frequently closed committee rooms of Congress, all by the consent of a lethargic public. Now the matter must be brought out and fought. We need action. The cost of not acting will be far greater than anything we have yet imagined.

This must be a brave new approach, free of what is politically expedient, proclaims Dr. William A. Niering, a botanist at Connecticut College. "A further challenge will require a revaluation of what the GNP means in terms of a healthy and productive society," he says. Neiring continues:

> In a nation that spends over $200 million annually for tranquilizers, and finds one out of every ten of its citizens suffering from some serious mental or emotional disturbance, there is a dire need to reevaluate our goals in the light of our past uncritical progress. As expressed by anthropologist Ashley Montagu, "We are now at one of those watersheds which will demand a monumental change in thought and outlook."

And so we find ourselves at this crossroads, where we pause to look at ourselves. What are we? What are our expectations? What compromises have we accepted? John Kenneth Galbraith helps fill in the portrait. He speaks of

. . . the family which passes through cities that are badly paved, made hideous by litter, blighted buildings, billboards, and posts for wires that should long since have been put underground—they picnic on exquisitely packaged foods from a portable icebox at a polluted stream and go on to spend the night at a park which is a menace to public health and morals. Just before dozing off on an air mattress, beneath a nylon tent, amid the stench of decaying refuse, they may reflect vaguely on the curious unevenness of their blessings. Is this indeed the American genius?

The dimension deepens under the hand of Stewart Udall: "We have accepted noise, foul air, dirty rivers as inevitable consequences of industrialization. This has been our psychology. But now we are changing our basic assumptions. We have been a filthy generation. What will become of our grandchildren if we don't? [change these assumptions?]"

The question is, "How did we get to such an unhappy stage?" The historian Lynn White, Jr., traces the path: We are dominated by an implicit faith in perpetual progress unknown to Greco-Roman antiquity or to the Orient. It is rooted in the Judeo-Christian concept of religion. God created the world and gave man dominance over it and its creatures. "No item in the physical creation has a purpose save to serve man's purpose. And, although man's body is made of clay, he is not simply part of nature; he is made in God's image."

Unlike modern man, ancient man—primitive that he was— had a religious view of nature. For him a spirit resided in every tree, every spring, every stream, every hill. The world was a mysterious and hallowed place. All nature was treated with reverence. Christianity demolished this animism, even cutting down trees to eliminate the abode of the resident god. "By destroying pagan animism, Christianity made it possible to exploit nature in a mood of indifference to the feelings of natural objects."

Now, what we do about ecology depends on our ideas of the man-nature relationship, says White. "More science and more technology are not going to get us out of the present ecological crisis until we find a new religion, or rethink our old ones."

This means change of attitude. Attitude always determines a civilization's fate. Attitudes shape the world we live in, and at the same time the world shapes us, so there is always an interaction between the world house and its occupants. Eventually it is impossible to say which is cause and which is effect. This is "ecology" in its larger sense—a word that, in the original Greek, means "house" or "home." The world is our home; as we destroy it, we are in turn destroyed.

The need for putting our house in order, in its physical and deeper cultural-spiritual sense, has been written about extensively by Dr. Paul Sears. Replying to my query he points out the basic truth that the good life can only be enjoyed if human beings are alive to enjoy it. Therefore the first problem is to meet the physical imperatives necessary to survival.

> We shall have to develop cultural values and sanctions that will make this achievement possible. Once we have them, the technical means should fall into place. . . .
>
> This leaves us with the major challenge—how bring about the kind of change that is required? Information is needed, of course, but without inspiration it is not enough. Perhaps I am wrong in estimating the potential of the creative artist here, but unless goals can be soundly dramatized nothing much is likely to happen. We have a gift for learning the hard way.

There is a devastating basis of logic that underwrites our irrational environmental destructiveness. Collective ruination may be the inevitable result of many individuals acting "reasonably." This is implicit in what Dr. Garrett Hardin, professor of Human Ecology at the University of California at Santa Barbara, calls "the tragedy of the commons." He takes the classic definition of tragedy—its essence is not unhappiness but it resides in the solemnity of the remorseless working of things: "The inevitableness of destiny can only be illustrated in terms of human life by incidents which in fact involve unhappiness. For it is only by them that the futility of escape can be made evident in the drama."

The scene of tragedy is the commons—a pasture open to all. Each herdsman sees it to his individual advantage to keep adding to his herd. The more animals he has, the greater his profit. If his animals don't eat the grass, someone else's will, he tells himself. Each individual herdsman has the same reasoning. The rational herdsman concludes that "the only sensible course for him to pursue is to add another animal to his herd. And another and another . . . But this is the conclusion reached by each and every rational herdsman sharing the commons." At some point the carrying capacity of the pasture is exceeded by over-grazing. This means disaster for all.

Therein is the essence of the environmental problem. We are all rational herdsmen. Each of us pursues his individual best interest toward the ultimate destruction of the group—so many actors destroying the stage on which they perform. This perversity of human nature is expounded on by Representative Ottinger:

> If, by a selfish act, a man can benefit himself while harming the community, he is likely to perform that act. Population control is for all those other people. An electric dishwasher is a marvelous convenience, pollution of a river is a distant abstraction. Those who insist on greater support for mass transit often see it as a way to relieve highway congestion so that they can drive their cars in freedom and comfort.

The rational herdsman fills in his part of the estuary and points out that "My little bit doesn't hurt." He is the industrialist killing Lake Erie who says, "My effluent doesn't smell, it's those damned fish." It is the nuclear scientist who writes, "Plutonium is my favorite element. . . . The reputation of plutonium as a toxic material perhaps has contributed more than any other thing to my being supported in the modest but comfortable manner to which I have grown accustomed." It is the mining engineer who gouges wounds in the earth and talks of his love of the land, the big-game killer telling about how he loves the animals he slaughters. It is the public health official who fails to close the contaminated beach for political reasons. It is the politician who protects the polluter. It's the polluter who buys the politician. It is former Governor Dan Samuelson of Idaho who sees his state faced with the choice of whether Sun Valley is to become a national park or the beautiful scenic wilderness torn up by roads and a mammoth gaping open-pit mine (700 feet long and 700 feet deep) producing molybdenum that would not be needed for decades to come, piously mouthing the rallying cry of despoilers everywhere: "The good Lord never intended us to lock up our resources." It is the refusal of the nations to stop killing the ocean and its life. Is the SST, and all those automobiles choking the roads and poisoning the air, the highway fund, oil depletion allowances that encourage the rape of the earth, phony bans on pesticides, standards that legislate waterways into sewers, courts that make mockery of law by supporting pollution. It is, above all, public officials calling for growth and more growth, paying lip service to environmental protection and in every way possible encouraging the destruction of the environment.

Curtailing or halting growth implies horrendous problems, resolving inequities between rich and poor, nations that have and those that have not. But what if we fail to address ourselves to these problems? What if we continue as we are? The answer comes from 33 of Britain's leading scientists, including the biologist Julian Huxley, who, in January, 1972, issued a document called "Blueprint for Survival." It declares that we are heading toward catastrophe:

" 'If current trends are allowed to persist, the breakdown of society and the irreversible disruption of the life-support systems on this planet, possibly by the end of the century, certainly within the lifetime of our children—are inevitable.' "

It calls upon the British people to stop building roads, to tax the use of power and raw materials and eventually cut the country's population in half. *The New York Times* account continued:

" 'The key factor in the present situation,' the blueprint said, was man's 'deeply rooted beliefs in continuous growth,' a notion that had to end in wars, famines and social crises. . . .' "

" 'Governments are either refusing to face the relevant facts or are briefing their scientists in such a way that the seriousness is played down,' the report states. As a result, 'we may muddle our way to extinction.' "

Is this overdramatic? A chorus of scholarly Jeremiahs singing a death song? Then listen to the prestigious Club of Rome, made up of what *Time* called "70 eminently respectable members," including Aurelio Pecci, the Italian economist and former Olivetti head, industrialists, and scientists.

The club wanted to study the most basic issue of all—survival. Toward that end it turned to an international team of scientists led by MIT computer expert Dennis Meadows. The Meadows group, using a special computer that could simulate the major forces at work in the world today, fed it "data ranging from expert opinion to hard empirical facts—the world's known resources, population growth rates, the incidence of pollution connected with nuclear power plants," according to *Time*. The question Meadows had to answer was: How long can population and industrialization continue to grow in this finite planet?

Thousands of variables were fed into the computer, and the results were published in a study, "The Limits to Growth." Meadows is quoted as summing up his conclusions tensely: "All growth projections end in collapse."

The computer merely confirms what is already mathematically certain: it is not possible to keep drawing indefinitely from a limited supply. Every advance in technology, says the computer, consumes scarce natural resources, throws off more pollutants and often has unwanted social side effects. Because of the Club of Rome's membership, states *Time,* "it is as if David Rockefeller, Henry Ford and Buckminster Fuller suddenly came out against commerce and technology."

The Meadows team offers a possible cure for man's dilemma, reports *Time:*

> . . . an all-out effort to end exponential growth, starting by 1975. Population should be stabilized by equalizing the birth and death rates. To halt industrial growth, investment in new, nonpolluting plants must not exceed the retirement of old facilities. A series of fundamental shifts in behaviorial patterns must take place. Instead of yearning for material goods, people must learn to prefer services, like education or recreation. All possible resources must be recycled, including the composting of organic garbage. Products like automobiles and TV sets must be designed to last long and to be repaired easily.
>
> As the report presents it, the result is a sort of utopia—not the stagnation of civilization. "A society released from struggling with the many problems imposed by growth," the report says, "may have more energy and ingenuity available for solving other problems." Research, the arts, athletics might well flourish in a no-growth world. Nor would developing nations necessarily be frozen into everlasting poverty. Without the burden of increasing population they might provide citizens more amenities.
>
> The report makes one thing abundantly clear: "There is a limit to everything," says Japan's Yoicha Kaya, a club member and systems analyst. . . . "There is no use wringing hands. We can and must try to do what is humanly possible and we must act soon."

That the world is imperiled is a tremendously big notion to absorb. The enormity of the idea eludes us, and is part of our problem; certainly it must have been easier for the ancients to accept the concept that the earth was "round" rather than flat

—a concept that intelligent men resisted for generations. Time presses, and all we have done to date is immunize ourselves to the environmental problem without fully comprehending its threat.

We have our warning: change or perish. It is that clear. We know the problem and the solution. But is the price too high? Are we ready to make sacrifices, or do we preserve our comforts and march resolutely toward destruction? The computer confronts us with its imperative logic of doom. As rational herdsmen we must rise or fall together. Is ours to be an era of reform or the Age of Extinction?

THE LAW: BY WHAT RIGHT?

People come to challenge the principles their society lives by only as they learn to see themselves in a new light. Usually this means a new cultural self-vision. It involves some of the biggest questions a person can ask: What is a human being? What are his obligations? What are his rights? Our ways change as vision clears or clouds. Often the insights we live by come in odd and unexpected ways from strange teachers. Dennis Puleston, the technical information officer at Brookhaven National Laboratory and an amateur ornithologist, became aware of one of his rights through the spring peepers, the small marsh frogs that add to the chorus of spring in the wake of retreating winter. Puleston lives on Long Island. Every spring it was his custom to take his son to a marsh nearby to hear the peepers announce the end of winter with their piping mating cry. Each year the sound became fainter. One spring it was stilled.

Puleston knew, or suspected, that the peepers had been killed off by DDT sprayings for mosquito control. He was also sure that DDT was to blame for the disappearance of the ospreys he had studied for so many years on Gardiner's Island. The loss of the peepers and the ospreys had diminished his world. Did some bureaucrat who ordered the spraying have the right to intrude on his pleasure in this way? Why did an aesthetic value not have as much legal standing as property value? Property has untold advocates—but who was to speak for the spring peepers? Who was to explain or interpret their delicate and mysterious small role in the great symphony of life? Do spring peepers have rights—or must they and all other creatures live by the tolerance of men?

Carol Yannacone came to the great question of human rights

by a similar path. In April, 1966, she read in a local Long Island
paper that the Suffolk County Mosquito Commission would
undertake another summer of massive DDT spraying. The year
before she had seen a massive fish kill in her childhood swimming pond attributed to the Mosquito Commission. It upset her.
What right did the county have to kill the fish that belonged to
the people and endanger the health of children swimming in the
pond? Was there nothing a private citizen could do, she asked
her husband Victor, a thirty-one-year-old lawyer whose office is
in Patchogue. He explained that she had not suffered immediate
personal damage through the spray programs and under current
law she had no standing in court. "But," she said, "you always
say that equity will permit no wrong to be without a remedy."
Yannacone began to brood over this legal anomaly. He started to
ask himself some big questions about human rights. Later that
day he prepared a complaint and tossed it to his wife. "Here," he
said, "read this and sign it."

This led to a landmark legal action. It was a "class action"
filed on behalf of all the people in the county, asserting that their
common environment had been impaired by the use of the broad-
spectrum, persistent biocide DDT. It did not seek money damages but, rather, injunctive relief to halt the proposed spraying.
The suit received wide notice. Several scientists who had been
chafing about environmental damage from DDT volunteered to
testify. Among them were scientists from the State University of
New York at Stony Brook, including Dr. Charles Wurster, a biologist and DDT specialist; Dr. George Woodwell, the ecologist at
Brookhaven National Laboratory, offered his services. Dennis
Puleston joined in. So did Anthony Taormina, regional director
of the New York State Conservation Department, and others.

The case was heard by Judge Jack Stanislaw in Suffolk
County Court. Judge Stanislaw, a farmer who used DDT himself,
told Yannacone he had two hours to present his case. Six days
later he was still listening. This marked what one writer called
"the first great marshaling of responsible environmental scientists in a courtroom" to present the case against DDT. Technically the suit failed. Judge Stanislaw, after issuing a temporary
injunction, finally ruled more than sixteen months later that the
legislature should decide whether DDT was to be regulated or
eliminated. But that temporary injunction marked a victory: for
the first time the public learned facts about DDT that had been
successfully suppressed for almost twenty-five years. This was

an amazing reversal over a related case in Federal Court in Brooklyn ten years earlier when a group of conservationists, led by Dr. Robert Cushman Murphy, tried to get an injunction against a mass DDT aerial spraying that the government contended would "eradicate" the gypsy moth. The conservationists were virtually yawned out of court by a bored judge who had to ask the plaintiff's lawyer the meaning of "ecology." He had no idea what was at stake, and he was not interested in finding out. In the end he automatically ruled in favor of the government spray program that was carried out and, predictably, failed to eradicate the gypsy moth.

But now, for the first time, an injunction had been issued restraining the use of DDT. During the proceedings Judge Stanislaw said he would never use DDT again and lamented his own role in killing the crabs in Smithtown Bay ("All of my beautiful blueclaw crabs . . ."). To Yannacone the case had broad implications. It proved that a layman could understand complicated facts of biology. In his opinion, the judge had ruled (despite his legal decision on jurisdiction) that DDT adversely affected the environment, that DDT was "potentially harmful and inherently dangerous," that it reacted negatively on good and destructive insects alike, and that other chemicals were available to control harmful insects. During the injunctive period the county switched from DDT to a less persistent pesticide and has not since returned to DDT.

Nine months later Yannacone and Woodwell formed the Environmental Defense Fund. Its original ten members included several who testified at the Long Island action. Puleston was named chairman of the board ("I'm the only one with gray hair"). EDF was primarily to be made up of scientists who would use the court to protect and restore the environment. Yannacone recalls:

> EDF was organized for the express purpose of presenting to the public scientific evidence on controversial matters relating to environmental quality, utilizing the courtroom as forum and adversary litigation as the method. We felt that the courtroom and the litigation process had certain inherent safeguards and a kind of built-in resistance to the conventional political, economic, and social pressures. My mission was to present the scientific evidence that had been denied a hearing in other forums, and test the opinions of

the establishment scientists in what we in the trial bar euphemistically call the "crucible of cross examination."

In short, he says, "You separate sense from nonsense."

Yannacone believed the courts were the alternative to revolution. Before the United States District Court in Detroit he said
that "when you close the door of the courthouse to the people you
open the door to the streets." He argued that the conservation
movement rarely influenced Congress because legislators are
often no longer responsible to the people but to special interest
groups whose concerns only occasionally coincide with those of
the general public. Appeals to administrative agencies are
equally fruitless, he said. They had been created to act as judge,
jury, and executioner of their own actions without regard to the
will of the electorate. Yannacone's Law is that "civilization declines directly in relation to the increase in bureaucracy."

More than 200 scientists offered their services to EDF, without charge. Many of them had been bruised and disillusioned in
fighting for environmental reform. They had learned the futility
of calling for corrective legislation, enforcement of laws that did
exist, voluntary compliance by industry, petitions, and letters to
the editor. They had learned through harsh experience that the
only answer short of violence was going to the courts in a show
of strength. The ivory tower as a way of life was dead. Now the
environmental scientist, to be effective, had to be an activist.
Possibly he had learned from his restive students.

Many of the scientists were further disillusioned by what they
had previously seen in the courts. Lawyers were more interested
in winning cases than in presenting truth. Most lawyers are part
of the environmental problem and not part of its solution, said
Dr. Clarence C. Gordon, a botanist at the University of Montana,
paraphrasing Eldridge Cleaver. He observed that any time a corporation or government agency is threatened "by persons concerned about the ecological state of affairs, large assemblages of
lawyers can be bought, who, for money, are more than willing to
defend the rights of the despoilers." He said:

> Lawyers write the bills for the legislators, and the legisla-
> tors pass or reject the bills depending on their beliefs, their
> party affiliation, or the lobbyists' influence. Although lawyers
> may be competent when it comes to knowing the laws of
> man, they are for the most part completely incompetent

when dealing with biological laws. They expound at length on the rights of the individual man and free enterprise and yet have little to say about the rights to existence of the other millions of species which inhabit this earth. In his daily reading the lawyer rarely encounters articles explaining the dependence of man on his ecosystem or the increase in carbon dioxide from one hundred years ago.

He told about lawyers who talked out of both sides of their mouth about the way they compartmentalized their lives. One spoke of his "private role as a 'conservationist' and his public role as a defender of DDT: 'For $700 a day, I would defend the devil.' " Gordon said that, as with scientists, the brightest young graduate law students are sought and usually hired by the more prominent law firms or industries. "Obviously these firms became affluent not by defending the poor or oppressed but by securing local industries or railroads as their clients. Unless the plaintiffs have all the evidence and happen to have bright lawyers fighting for them, this difference in talent and financing presents a horrible imbalance in any court case."

Well-paid corporate lawyers have at their call many scientific "experts" to testify on behalf of their clients. Gordon observed that these "experts" come from universities and research centers and are highly paid consultants who have a vested interest due to industry-financed research grants or the consulting fee itself. Social conscience, scientific honesty, and man-to-land ethics offer little reward compared to a large research grant and/or a consulting fee (of $150–$200 a day, plus expenses, compared to $3 or $5 an hour paid by local and state governments for air pollution "expertise").

Gordon revealed his own frustration and disillusionment in a speech before the 1969 convention of the American Trial Lawyers Association when he urged its members to join in the fight to preserve the biosphere from further disintegration: "For a long time I made speeches. I thought that was the answer. Now I am convinced that the answer is in the courtroom. The nation just hasn't time to wait to correct environmental insults."

Such insults belonged in a new category of "corporate crime," according to consumer crusader Ralph Nader, who urged the same assemblage of lawyers to treat this crime as one that utterly dwarfs into insignificance crime on the streets—in terms of people injured and killed because of dangerous machinery, pollution, and unsafe household products. Yannacone, another speaker, told the lawyers that those companies are "pushing mankind to the brink of doomsday. We must knock at the door of courthouses throughout the nation and seek the protection and quality of our environment. . . ."

Yannacone, not given to euphemism or subtlety, at the same time gave EDF its rallying cry: "Sue the bastards!"

He believes implicitly that the courts are the last hope to save the environment. In a 1967 speech before the National Audubon Society convention, a speech that possibly launched environmental law as a separate discipline, he said:

> Sad experience has shown that at this time in American history, litigation seems to be the only way . . . to focus the attention of our legislators on the basic problems of human existence, short of bloody revolution. . . . All of the major social changes which have made America a finer place to live have their basis in fundamental constitutional litigation. Somebody had to sue somebody before the legislature . . . took long-overdue action.

Yannacone is representative of a new group of emerging environmental lawyers. Most are young, shrewd and impatient of traditional law and environmental abuses underwritten by the courts. They ask new questions and demand new answers. They are stirring up a new public morality, challenging special privileges in the name of human rights. The idea they expound is simple but profound in its implications: Who owns the land and its natural resources? This reflects the land ethic of the American Indian—that land belongs to all, regardless of whose name is affixed to the property deed. This idea of real property ownership goes far beyond the usual narrow political-economic context of commercial or private land ownership. This new morality based on the renewal of the philosophical view of man in his environment recognizes that the world is finite. No part of it can be damaged or degraded without detracting from the whole. No man has the right to do this to other men, for all are equal shareholders in this fragile biosphere that supports all life.

This new morality is now being tested in the courts. It demands a new kind of law that is responsive to human needs rather than to economic interests. The traditional law that underwrites the "American way of life" is that a man has a right to do as he pleases with land. This is based on the right of private property inherent in the private enterprise system. Under this traditional concept an individual can seek redress in court only if he has been damaged. Damage is equated to dollar loss, which implies that everything has its price. If there is no monetary loss then there is no damage. But often in environmental cases damage is impossible to prove. How can I prove that I have been damaged if an estuary is destroyed? Or if a species of wildlife is wiped out? Or if a tanker fouls the ocean with oil wastes? Traditional law does not recognize such damage to an individual. Thus the individual is denied standing in court.

Theoretically, my interests are protected by the government, primarily through the various regulatory agencies. But the regulators are usually on the side of industry. So for me there is no real law. I am, in effect, excluded from the courts and without real protection from the destroyers and polluters of the environment. (For many people, especially the poor and destitute, the law does not exist outside a theoretical availability.) I am further discriminated against because damage must first occur before it can be challenged; such damage, even if provable, may not be possible to undo—the ravished estuary or poisoned wildlife, for example. Nor does the polluter have to prove that what he is doing is "safe." I must prove that it is harmful, and this is often impossible. So the law has often been on the side of the polluters rather than the polluted.

The courts have been cool to the new morality when it has been presented to them. Law is almost invariably designed to protect the status quo. Anything that departs from immediate precedent is regarded suspiciously by the legal mind, even with hostility. "The courts protect the rights of the individual to create of the earth a wasteland as long as they will pay damages to those injured by degradations," explains a professor of environmental law. "So damages and not preservation of the environment are the measure of justice."

Traditionally anyone pleading the cause of conservation has found little sympathy in the courts—a condition that, fortunately, is now changing. Many judges are still better schooled in property rights than human rights. Many are knowledgeable about the law but remain incredibly ignorant of nature. Many judges, as younger lawyers, served as counsel for wealthy corporations and were partners or associates of big law firms that are little different from the corporations they serve. Many judges are more interested in their own portfolios than in the environment (they too are rational herdsmen), and this bias is shared by many lawyers. They are more interested in the fee than in the cause; the big fees, as Professor Gordon noted, come from those who destroy nature and not those who try to protect it. Most legislators who write environmental statutes are lawyers with this professional outlook.

In conservation cases there is occasionally a question of conflicting interests. A big law firm may serve as counsel, without charge, for a conservation group, and the same law firm may represent the very industry the conservation group is suing. Or there may be overlapping interests. The same law firm, for example, represented the Scenic Hudson Preservation Conference and the Chris Craft Corporation, which owns half the Montrose Chemical Company, the nation's biggest manufacturer of DDT. The problem becomes increasingly sticky because of today's interlocking corporate structures. With the rise of conglomerates, nobody knows who owns what and where, and conflicting interests may surreptitiously overlap.

The big law firms representing both sides insist there is no conflict of interest because different lawyers in the firm handle each matter. This may be true. But the attorney assigned to the conservation group as a public service is usually a young lawyer in training, while his "opponent" representing the well-paying corporation may be a partner and senior member of the firm. How far will the ambitious young lawyer dare go in presenting his case if it means offending his employer or someone who could affect his career?

Conservation groups are usually made up of conservative people. They are represented by conservative lawyers who practice traditional law and they have had limited success in the courts. They are hampered by lack of funds. They are largely dependent on contributions from private donors or foundations, and such donations can have invisible strings on them. Those the conservation group must look to for support are often the very ones that it dares not offend. The conservation organization is thus deprived of one of its most important weapons. If it engages in political activity it will lose its tax-exempt status as did the Sierra Club three years ago when it ran a series of advertisements on the Redwoods National Park bill and other environmental issues then before Congress. The effect of the IRS ruling —"an outrage," according to a *New York Times* editorial—has crippled many organizations in the conservation and consumer protection fields. "They dare not be nearly so active and aggressive as they would like in furthering the causes which they were founded to defend because they are afraid of jeopardizing their tax status."

Corporations and businesses engaged in lobbying do, as a matter of fact, possess certain tax advantages that are expressly denied to nonprofit organizations. As a *New York Times* editorial observed:

> . . . every major corporation has a Washington office or belongs to a trade association which lobbies in Congress in its behalf. . . . it is surely as reasonable to grant comparably favorable tax treatment to non-profit organizations that lobby in behalf of the unorganized public interest.

The new environmental lawyer is usually without corporate ties. He is generally more interested in the cause than in the fee. He disputes the narrow traditional interpretation of the law that is more concerned with property rights than human rights. He demands a new dimension from the law. In effect he says that the world's resources are a unity that belongs to all. An individual should not have to prove monetary damage to have standing in court, he maintains. Some things have an intrinsic value rather than a retail price tag, and therefore they must be protected just because they are priceless and irreplaceable. This is true of air, water, soil, wilderness, wildlife, and the earth's dwindling mineral resources. If the earth that supports mankind is injured in any way, whether the injury touches a particular individual directly and momentarily or not, it affects individual welfare. He therefore has a right to go into court and seek an end to the abuse. He does not stand only on traditional legal rights. He proclaims a god-given right as a human being and a citizen to a decent environment as his birthright. That right would be guaranteed to all through a Constitutional amendment proposed by a few congressmen.

One of the new breed of environmental lawyers is Joseph L. Sax, a thirty-five-year-old professor of law at the University of Michigan. He is author of a revolutionary environmental law adopted by Michigan. It permits any citizen or group of citizens to go into court against any public agency, at any level of government, or private industry that they think is damaging the environment. Then the burden of proof is not entirely on the public; it would be up to the defendant to prove that there is "no feasible or prudent alternative" to the activity complained of, and that the activity is consistent with public health, safety, and welfare.

In many states individual citizens have been unable to start actions. They have no standing in court because they have not suffered immediate personal damage. "If a plant is fouling up a stream and you own a house on the shoreline the suit will probably be allowed," says Sax. "If you live far away but go there to fish, the court will throw it out."

The courts also throw out suits on grounds that actions should be brought by local, state, or federal agencies in charge of protecting the environment. The citizens of Santa Barbara were unable to file suit against the Department of the Interior to prevent it from granting leases for offshore drilling before the disastrous oil spill there—despite common knowledge that it was a hazardous earthquake fault zone. The courts held that the interests of the citizens were already being properly protected by the Interior Department. This is the same department that grants the right to drilling offshore and acts as broker to sell off the nation's resources to private interests.

The Michigan bill goes even further in overcoming longstanding abuses. It also gives a citizen the right to bring suit even if he is not directly affected if he believes a threat exists to the overall environment. A Detroit resident, for example, can bring suit against a state agency that is spraying DDT in a remote forest preserve. A citizen can ask the courts to shut down a company for contaminating a waterway, challenge regulations of state agencies as too lenient toward industry, or sue a neighbor for fouling the air with a trash burner. Circuit courts, in addition to having the power to grant injunctions and impose conditions to stop pollution, may even direct government units to upgrade standards they find deficient.

The Sax bill became the model for proposed federal legislation that would assure citizens an almost unlimited right to file suits in federal courts for the protection of the environment. The measure was promptly opposed by the Nixon Administration. A spokesman argued that the bill would leave to the judgment of the courts what constituted "unreasonable impairment" of the environment, and it claimed that the courts are not equipped to handle such technical questions. The Administration said such questions should be left to the administrative agencies. Proponents of the bill noted that it was proposed precisely because the administrative agencies have failed to fulfill their obligations.

They noted further that other Administration bills for environmental control did not flinch from having the courts rule on technical issues. Senator Philip A. Hart of Michigan, who introduced the bill in Congress, observed that the water-pollution bill compelled a judge to consider, in enforcing an order, whether it was practicable. "This permits a court to say that an administrator [of the antipollution law] is being too tough. We want the court [through citizens' legislation] to be able to say that the administrator is being too lenient."

Gradually the federal government has been opening the door —once completely closed—to citizen protests. The trend is to accord full hearings to citizens' groups protesting federally sanctioned developments such as highways and power plants. But there have been exceptions. The Department of Justice bluntly opposed citizens' groups challenging the decisions of federal agencies, as *The New York Times* pointed out: "This opposition has extended, some lawyers assert, to outright 'harassment' by the Internal Revenue Service through tax investigations of litigant organizations or of the personal finances of some aggressive attorneys." The attempt by the IRS to eliminate the tax-exempt status of conservation groups suing the government in conservation cases virtually cut the ground out from under citizen protest.

Many environmental lawyers question how wide open the courthouse door really is in a political sense. And like Yannacone they fear for the consequences if it is not. Among the apprehensive is Marvin During, a Seattle environmental lawyer, who is quoted in *The New York Times:*

> To put it in half-joking terms, the bar has been saying we've been through the legislative phase in environment and found that laws aren't necessarily enforced. Now we're coming to the judicial phase, and if this doesn't work there's nothing left but "Green Panthers"—a suggestion that the only recourse will be a militancy such as the Black Panthers have brought to other facets of the social scene.

Not only are there citizen demands to open the courthouse door for class action suits directed at the government; the new environmental lawyer demands new laws to meet the environmental threat and to reexamine the premises on which present law is based. "The statutes now available don't suffice," says Yannacone. "It requires a new approach. It must not be rooted in conventional statutory law but in equity. I say that what we must do is constantly probe the fundamentals of human rights and obligations."

He traces the development of American law over the past century, how it proceeded apace and led to building a mighty industrial empire and went astray. "In frantic reaction to excesses of laissez-faire capitalism, the people demanded government control and out of that the bureaus and the alphabet agencies were born. Unfortunately, one form of callous disregard for the individual was replaced by another."

In the early days of the nation the individual's rights were protected by two simple equitable maxims "enshrined in law and drawn from the wisdom of the Talmud":

1. Let each man so use his property so as not to injure another.

2. No wrong shall exist without a remedy.

These basic maxims regulated American life until the late nineteenth century. Then the Supreme Court made its famous "slaughterhouse"* decisions. These marked a legal and social milestone and an ecological gravestone. For the first time the corporation was legally given human rights. "Even though the corporation as a person is a figment of legal imagination, the Supreme Court clothed it with all the rights, privileges and immunities of a human being. The corporation was indeed a person in the eyes of the law but it had no soul to save or ass to kick." As a nation we were set off on a new road that is still being traveled—corporate privileges without the necessary human restraints. He continues:

> Since the slaughterhouse cases, the corporations, and their legal counsel, many of whom became Supreme Court Justices, further expanded corporate rights without ever establishing corresponding corporation responsibilities, relying always on a belated legislature to limit the corporate rights.

This meant exploitation underwritten by law. In the past the nation had depended on equity law with its two basic guiding principles, the law that had built Western civilization. But now it

---

* A landmark series of legal decisions in the late 1800s that began with a meat-packing firm in New Orleans and culminated with the U.S. Supreme Court extending to corporations the privileges and immunities previously granted to individuals under the 14th amendment.

was largely replaced by inflexible statutory law. This is the fundamental difference between the two:

Traditional damage law rooted in statutes is effective after an act or wrong has been committed. The injured person then must sue for monetary damages. In equity an individual can try to stop an act from taking place *before* the damage is done. Only in equity is it possible to get an injunction to prevent the damage.

Conventional common law is concerned only with the statutes and precedents representing judicial interpretation of those statutes, Yannacone explains:

> It calls for finding a statute that does what you want it to do: protects what you want to protect, prohibits what you want to prohibit, or mandates what you want done. Then you look for an agency that does what you want done—enforces the law you want enforced. The conventional conservation lawyer follows this approach. He finds a law that says you should not build a dam on a particular river, or he petitions the legislature to pass such a law. But that may take years, during which time the resource is destroyed.

Unlike statutory law, equity does not depend on finding an existing statute that can be applied to a specific fact situation. The equity side of the court is involved with adjudication of fundamental rights and obligations of man and his institutions. It recognizes no legislation other than the natural law—or as one court said: "the work of the Great Legislator of the universe."

Over the years lawyers have turned away from this "natural law" of equity and put their faith in the works of the legislatures. Yannacone says:

> Equity has gone out of fashion. It died about 100 years ago. The country lawyer was an equity lawyer. He went to work and really dug for the justice of issues and the good of the country. Abe Lincoln was an equity lawyer. In the last 100 years America has lost her independence. The big corporations took over. The corporate law firms were made in the image and likeness of their corporate clients, and the morals of the market place replaced the ethics of the legal profession. The small independent barrister who went to seek justice was starved out, in the same way as the family doctor. It's a thankless job. No one wants to pay for justice.

Now lawyers must return to equity if they are to deal with the problems of environment, according to Yannacone and some other environmental lawyers. "Equity is a kind of creative law," he says. "It has the most force. Many lawyers are afraid of it. It would, for example, demand that anyone using a public asset for private profit owed the public the cost of replacing or renewing that asset for the next generation. The corporate lawyer rejects this view. All of his corporate clients would be vulnerable."

Equity law is especially geared to meet the new problems because it is unfettered by statute, he says. "You go to equity to establish a fundamental right." Then why is it not used more?

It demands justice. A lot of people don't want justice. I tell the conventional conservationist, who is usually a pre-servationist, that in equity you must justify wilderness to the people in the tenements of Watts, Bedford-Stuyvesant and San Antone. The preservationist recoils in horror. I say, "You must justify the protection of the Bermuda petrel against the potential of thousands of deaths from malaria in the Third World if we stop using DDT." The preservationist says, "No, we don't." I said, in 1969, "If you want to protect the Florissant fossils, I must say they belong to all the people. The fossils cannot be preserved just for the paleontologist." The preservationist says, "If we let everyone into the wilderness it won't be wilderness." I say, "If you want to justify limiting the number of people in wilderness areas, you must not discriminate against the poor, black, the Indian and the inarticulate." The preservationist says he'll try another way than equity—which demands justice for all. The equity petitioner must go into court with clean hands. Many petitioners today must wear gloves. . . .

The problems of equity are compounded by the bias of the courts and the ponderous inefficiency of the legislative process to meet the needs of the people. The urgent need for new laws to protect the environment was dramatized by the threat to the Florissant Fossil Beds, one of the great natural resources of the world. They cover 6,000 acres 35 miles west of Colorado Springs, where seeds, leaves, insects, and plants from 34 million years ago are "remarkably preserved in paper-thin layers of shale." The fossils are studied by scientists from all over the world, as the richest of their kind anywhere on earth; 144 different plant species and more than 60,000 insect fossil specimens have already been found.

A bill to establish the Florissant Fossil Beds as a National Monument was passed by the U.S. Senate in 1969. While Congress was deliberating, four speculators bought the land. They said they intended to begin bulldozer excavation on roads immediately. The Defenders of Florissant, made up of scientists and citizens, turned desperately to the courts; they filed suit "on behalf of the people of this generation and those generations yet unborn who might be entitled to the full benefits, use and enjoyment of that unique national treasure, the Florissant Fossil Beds."

The next few weeks saw a running drama. The Defenders, represented by Yannacone, shuttled between courts, utilizing the cumbersome and inadequate legal procedures available. They asked the U.S. District Court for Colorado to hold back the bulldozers until Congress could complete its deliberations. The court refused to interfere with the absolute rights of private ownership. An appeal was made to the U.S. Court of Appeals: "This court must not countenance destruction of a 34 million year record, a record, some would say, written by the hand of God."

The court questioned its own power to grant a temporary restraining order, asking for precedent: "What statute does this excavation violate?" The Defenders had to admit that there was no direct statutory protection for fossils. "So what right have we to control the use of private land?" asked the court. Yannacone held up a fossil palm leaf from the site, pleading:

> The Florissant fossil beds are to geology, paleontology, paleobotony, palynology, and evolution what the Rosetta stone was to Egyptology. To sacrifice 34 million years of geologic history for 30-year mortgages and the basements of the A-frame ghettos of the '70s is like wrapping fish with the Dead Sea Scrolls.

The Court of Appeals responded to this dramatic appeal. In a remarkable precedent-setting ruling, it restrained the speculators from disturbing the fossil beds until July 29, 1969. On that date the District Court heard testimony and argument for a preliminary injunction. During that time Congress continued its deliberations but without final resolution. Again the District Court held that nothing in the Constitution prevented a landowner from making whatever use of his property he chooses.

Back to the Court of Appeals. Now it was critically late. The bulldozers were at the border of the monument, ready to start gouging out the beds. The speculators offered consolation, however. They said they only intended to scrape off the top four feet

of fossils and that would leave as much as 20 feet of shale remaining. Yannacone argued: "You would just as well say scrap-
ing the paint off *Mona Lisa* would cause no irreparable damage
because there's still more canvas underneath."

Again the 34-million-year-old fossils were spared by court
order. Congress finally passed the necessary legislation. The Florissant fossils have now been spared as a National Monument
rather than lost to a recreational housing development. But most
of the time there is no happy ending. No kindly judicial hand
stays the onrushing bulldozer from the insatiable appetite of
progress and the sanctity of private property rights. There is no
committed advocate at hand to plead the case. There is no
money. There is no will. There is no one to raise the question: by
what right . . . ?

The Environmental Defense Fund was something new in conservation groups. It did not go into the courts, empty bowl in
hand, figuratively begging, "More porridge please." It became the
most militant of the country's 150 conservation organizations. It
did not even like being referred to as a conservation organization. It was not defending nature as much as demanding human
rights.

"We're hawks," cries Dr. Wurster. "But we operate entirely
within the sociological structure. We don't block traffic. We don't
sit in. We don't riot." He found traditional conservation groups
"legally weak, scientifically naive and politically impotent." The
old groups played by the old rules, seeking traditional relief with
traditional law. EDF called for solid law, solid science, and injunctive relief. The academic scientist had burst out of his ivory
tower and was ready to fight for a principle. He was taking on
the industry-hired scientist whose commitment was to money
rather than what was befalling the environment. EDF, through
its scientists, was asking the revolutionary question: by what
right?

EDF was colorful and impudent. Wurster drew a deadly bead
when he said that "using DDT to control mosquitoes was like
using atomic weapons against criminals in New York City."
Yannacone conducted a trial like a staged drama. It was made
up of scenes and scenarios. Like most good trial lawyers, he
spoke in the language of the theater: "If you can't make the trial
swing, you'll get yawned out of court, especially when the case is
technical." He had a sense of drama, an amazing grasp of diffi-

cult scientific concepts (he had trained as a biophysicist and kept up with current medical and scientific literature). He demanded that his scientist witnesses do their homework; if they didn't know something crucial to a case they had to go to the laboratory and find it out. Significant research was done as the cases progressed.

The organization had only limited funds—the curse of all conservation groups—and had to pick its targets carefully. It went into Michigan to challenge the state's right to dump 9,600 pounds of deadly dieldrin over a 100-square-mile area for an alleged Japanese beetle infestation. Under cross-examination it turned out that the infestation consisted of approximately 300 beetles. How many beetles did it take to make an infestation? Yannacone demanded. He received an astonishing reply from the state's witness: "One."

EDF started an action against a kraft paper plant in Missoula, Montana. The town is in a beautiful valley but smoke from the plant blots out the sun at midday; residents must turn on their headlights to navigate the streets; the smoke particles smell like skunk. It is a blot on the American conscience that people must sue to establish their right not to have to endure such subhuman conditions and an outside organization must come in to protest. How were people ever lulled and seduced into such a low opinion of their worth and their rights?

EDF waged its most prolonged and epic battle in Madison, Wisconsin, in a massive offensive to ban DDT in the state. The trial lasted over six months and more than thirty scientists testified. More than 250 offered their services. Under cross-examination some startling facts never before revealed were put on the record. Dr. Harry W. Hays, director of the USDA Pesticides Registration Division, testified: "If the data appears to us . . . to be adequate . . . the product is registered. We look at the data, but we don't do it analytically. We don't check it by the laboratory method."

Yannacone made Dr. Hays repeat his statement again and again, as if he could not believe it. Later he said, "At last Americans were told that the Department of Agriculture relies entirely on data furnished by pesticide manufacturers and does not conduct any tests of its own." The incredible lack of concern for the safety of the American people became apparent on further cross-examination.

Dr. Hays made many startling admissions. If a pesticide was

checked at all, it was checked by an entomologist only for its effectiveness against the target insect and not for its effects on beneficial insects and wildlife. "We don't assume that the intended use will cause any damage," Hays explained, then admitted that although he has personal knowledge of scientific studies showing damage to fish and wildlife from DDT, the USDA is not doing anything about possible environmental hazards from the pesticide. (Dr. Hays proudly stated, however, that the Department of Agriculture was solely responsible for the registration of pesticides and for determining whether they may be shipped in interstate commerce.) He reluctantly admitted that the public had no access to USDA records of pesticide registration.

"Only in adversary legal proceedings was it finally demonstrated that the United States Department of Agriculture is really serving the agri-chemical industry and not the American people," said Yannacone.

Unfortunately, the courts often were not ready for the challenges Yannacone and EDF posed to the judicial system and the corporate structure. But during the early development of environmental law, its most important function was in using the courts as an educational medium for the American public, exposing environmental skeletons in corporate and governmental closets. Some newspapers carried stories about the revelations developed in court. A few judges and lawyers learned some environmental science. Hopefully it moved them toward a more enlightened view of the need for environmental reform whatever the cost to the corporate sacred cow. There are indications that more courts are recognizing this necessity and deferring to it.

Yannacone, Wurster, and others in EDF often felt frustrated. They knew they had barely scratched the surface. In a 1967 speech before the Audubon Society, Yannacone sounded what was to become EDF's rallying cry:

> What can you do when a municipality decides that the highest and best use of the Mighty Missouri River is an open sewer?
> What can you do when government agencies seriously consider drowning the Grand Canyon or much of central Alaska, or when a combination of government agencies and private speculators act in concert to destroy the delicate ecological balance of the entire state of Florida?
> What can you do when the United States Department of

Agriculture publicly states that it does not consider the possible adverse effects of chlorinated hydrocarbon pesticides
such as DDT on non-target organisms, but permits them to
be sold and used even when their adverse effects become
generally known?

What can you do when timber and paper companies cut
down entire forests of redwoods and other exotic species in
order to "reforest" the area with faster growing pulpwood
trees?

What can you do when real estate speculators insist on
dredging estuaries in order to fill salt marshes or strip the
topsoil from irreplaceable prime agricultural land in order
to plant houses?

Just what can you do?

I know of only one answer: Sue the bastards!

And so he did. A modern David in legal loincloth taking on
corporate Goliaths. In 1970 he outdid himself, filing a class-
action suit against the eight major manufacturers of DDT, seeking $30 billion in reparations for the environmental degradation
caused by DDT, on behalf of his wife Carol and "all the people of
the United States not only of this generation, but of those generations yet unborn, who are the equitable owners of the environment and natural resources of the United States. . . ."

After filing the suit Yannacone and EDF parted company.
There were rumors. Some said EDF was embarrassed about the
$30 billion suit. Some said Yannacone could not be controlled.
Some said the organization wanted to pursue a more conservative route, use more conventional law. The truth was probably
that there were too many egos knocking together. But following
Yannacone's departure EDF became less militant and there were
fewer headlines without the colorful attorney.

During my research I spent a day with Yannacone. He is, as
his critics say, brash and flamboyant. He is impudent and not
easily impressed or intimidated. He keeps asking the vital question and demanding an answer: by what right?

He works closely with the most unlikely of allies, Martin J.
Merta, proprietor of Martin's Camera Center in Stony Brook,
Long Island. Merta, curiously, used to work for the USDA in the
Pest Control Division of its Agricultural Research Service. For
ten years he spent much of his time in the Golden Nematode
Project, with one season of spraying DDT for the gypsy moths
that refused to be eradicated, except in USDA press releases.

Merta recalls a day when he came out of a farmer's potato

field and saw a spray rig approaching. "I looked up and there was a bird. It was dead. A robin. It drank the chemical and its feet were still clutched to the bar where it was taking the drink of water." Later Merta developed symptoms of lead poisoning—"probably from all the lead arsenate I was exposed to." He left USDA and opened the camera shop, returning to commercial photography. He began to reproduce material for EDF, met Yannacone, and got involved in environmental litigation. More and more he was printing legal documents for Yannacone's environmental battles. Eventually he had to sacrifice much of his commercial work. Now his subterranean workroom, a former bowling alley, is stacked with legal papers. Merta says he has practically gone broke, "along with Yannacone," for conservation. "This is my penance."

Merta is devoted to Yannacone's point of view. He listens as Yannacone talks. Yannacone gets very excited. His hands wave. He observed that EDF, following his departure, petitioned the Secretary of Agriculture to ban the use of DDT. "Petitioning effectively the same Harry Hays," he said incredulously:

> EDF is petitioning him to please do something about DDT after all the evidence. I said by no means petition. I say challenge the right of a bumbling bureaucrat like Harry Hays to have sole charge of the environment of a large percentage of the world with no checks or balances. He is virtually a dictator. I say challenge the right of this bureaucrat. Challenge him on the grounds that nowhere does the Constitution of the United States give the director of the Pesticides Registration Division of the USDA the right to poison people. . . .

This is a footnote in history. Hays is now retired. The Pesticides Registration Division has been transferred from the Department of Agriculture to the Environmental Protection Administration. Yannacone and EDF have gone their separate ways. But Yannacone's remarks are relevant because they illustrate the attitude he brought to EDF, and possibly this marked the turning point when environmentalists ceased to ask for favors and began to demand human rights.

In his office at the Brookhaven National Laboratory nearby, Puleston broods about the fading environment. It is the future generations that will lose the most by this, he muses. "They will never know many of the things of this earth. They will see ani-

mals only between bars. They will never stand in a forest surrounded by trees, and if they do they will be terrified because it
is unfamiliar." He looks out the window and poses a question, as
if speaking to himself: "Why do we sell out so cheap?"

POLLUTION FIGHTERS: THE PROFESSIONALS

Scott of Illinois: A unique environmental experiment is being
carried out in Illinois. The outcome may determine whether the
nation, and possibly the world, can survive the environmental
crisis. The man carrying out the experiment is Attorney General
William J. Scott. He is a former Illinois state treasurer and crime-
fighting federal attorney. He was elected to his present office in
November, 1968, on a platform devoid of any ecological pledges,
and even little indication that the environment was a particular
concern of his.

Soon after Scott took office he learned a startling fact—that
Lake Michigan, a few blocks east of his Chicago headquarters,
would die within ten years, and in the same period Chicago
could expect a catastrophic air inversion that could kill 20,000 to
30,000 people. Scott started examining the statutes to find out
what could be done about this. He was shocked to learn that,
other than three ineffectual administrative boards, there was no
vehicle to control this pollution. In other words, Illinois' officials
and the people were ready to accept environmental cataclysm by
default.

Today Illinois has what is possibly the toughest air and water
pollution law in the nation. The Illinois pollution control act
was signed by Governor Richard B. Ogilvie on June 29, 1970.
It consolidates the pollution-control authority of several state
boards and commissions in a single Environmental Protection
Agency. It also created a State Pollution Control Board to set
pollution standards and hear cases brought by the enforcing
agency. The act abolished special exemptions granted to polluters by localities. It provided authority for the environmental
control board to initiate complaints against polluters and for the
state attorney general to fight for board rulings through a system
of direct appeals to state appellate courts. It can force a polluting
corporation to post a performance bond to ensure compliance
with ordered improvements, and it gave the Environmental Protection Agency power to seal up offending equipment during pollution emergencies.

In Illinois, to pollute is to commit a crime. Scott has hauled into court some of the nation's most powerful corporations and utilities. He is initiating suits against cities in the adjoining states of Wisconsin and Indiana that are polluting Lake Michigan. Some of the actions, such as airline and automobile suits, extend far beyond the borders of Illinois.

The *Sierra Bulletin* poses an interesting question: of all the states in the Union, why is Illinois the first to defy the powerful industrial polluters of its airsheds and water resources? "Why not California, where vast forests of ponderosa pine are turning brown and dying in an artificial and permanent autumn brought on by Los Angeles smog? Why not New York, whose western border is made up of a dead lake from its surface waves to its bottom mud?"

Illinois, it is pointed out, left as a level prairie when the shallow seas of the Carboniferous Period withdrew, has no sculptured Yosemite, no tall, proud redwoods, no primeval everglades. "Nevertheless, here in the heartland of America an ecological conscience is awakening."

Scott is credited with being primarily responsible for this awakening. And herein lies a moral for all of us. If salvation is to come, it will not be due to political rhetoric, pompous pronouncements, or appeals to good citizenship. It will be due to the acts of a few strong individuals who believe in themselves and what they are doing. They are the real heroes of the pollution wars. They are the ones who are trying to roll back the relentless tide of environmental destruction and calling on their fellow citizens to join the battle.

Almost daily Scott sounds his creed and rallying cry to his fellow Illinoisians: "In America we will get the kind of government and the kind of environment we demand. The message must be loud and clear—the American people will no longer tolerate the poisoning of their air and water and the continued destruction of our most priceless heritage, our environment."

Scott's actions go far beyond the borders of Illinois. He is testing whether the American legal system can meet the environmental crisis. So far it appears promising. Certainly it has shaken the polluters. In May the Illinois Senate began considering even broader environmental legislation to increase Scott's enforcement authority over air, water, land, solid waste, noise, and nuclear pollution. "Hundreds of lobbyists representing every major polluter in the country made their way to the prairie capital," reports the *Sierra Bulletin.* "Yes, they are worried . . ."

The legal concepts Scott has evolved and defended in Illinois courts can be used in all state and federal courts, explains the *Bulletin.* "And the National Association of Attorneys General, which recently named Scott to head its new Committee on Environmental Control, has passed a resolution urging member states to adopt antipollution legislation similar to that passed last year in Illinois."

Even without these powers Scott initiated 225 antipollution suits under an Illinois common-law principle. Many states do not have even this authority under existing laws. Scott told a meeting of state attorneys general that the federal government usually has been passing the buck back to the states, asking them to prosecute violations: "But in many states the attorney general is powerless to move unless the pollution control boards are willing to take on the big polluters. Unfortunately, a lot of state and local boards don't seem to have any interest in doing so."

How reluctant many states are to bring big polluters to heel was suggested by other attorneys general at the meeting. Wisconsin Attorney General Robert Warren said that only four pollution cases had been referred to his office for prosecution by Wisconsin pollution control agencies in the past eighteen months, according to a *New York Times* report. Scott offered a suggestion: that state attorneys general without specific authority to initiate pollution control suits could claim authority to sue an offender on the ground that they were acting as lawyers for citizens who were being damaged by pollution.

The response to his suggestion revealed how state officials are crippled by the lack of strong laws. Attorney General Robert M. Robson of Idaho said he was considering such action against the Federal Atomic Energy Commission, which, he said, was "using the state of Idaho as a garbage can for atomic waste. But to do so I have to prove to the court that the whole state is affected and is suffering economic damage as a result, and you have to be a real gambler to go into court on that basis."

The test of any system of law is whether it can—under emergency situations—respond to the needs of the people, says Scott. "We are in a crisis situation, fighting to save our planet. We have built the finest system of legal justice in the world. Now we must prove that working within it we can meet the greatest challenge ever faced by any generation—the battle for survival."

Scott has used the new antipollution legislation as only one of his tools. Enforcing it is the use of "every incentive—legal, economic and public relations—to make it unprofitable for polluters to continue to pollute." He began to educate his state by setting up weekly seminars with experts in ecology and related fields, points out the *Sierra Club Bulletin:*

> Lawyers toughened by careers in criminal prosecution would learn to write complaints in defense of herons and egrets whose nesting grounds were threatened. The consumer would learn from publicity generated by the Attorney General's lawsuits who the polluters were and what they were doing to the environment of Illinois. Prior to Scott's lawsuit few people knew, for example, that Commonwealth Edison was responsible for 60 percent of the sulfur dioxide in Chicago's air.

At the same time that Scott was reorganizing his own department and planning a statewide antipollution strategy, he began the battle to save dying Lake Michigan, reports the *Bulletin:*

> In the summer of 1969 Lake County, north of Chicago, closed all its beaches. Scott predicts that by 1972 every Illinois beach on the Lake Michigan shoreline could be closed. "We will not stand by and watch it become a dead lake in nine or ten years," Scott announced, as he brought suit against the North Shore Sanitary District—the first filed under the new legislation.
>
> Scott assigned Henry Caldwell, Assistant Attorney General, to head the Lake Michigan pollution probe. After one investigator had become nauseated and another fainted while taking water samples from the lake, Caldwell reported back that U.S. Steel and Republic Steel were pouring 800 million gallons of polluted water per day into the lake; Abbott Laboratories was dumping large amounts of chemicals and biological wastes; and wealthy, suburban Lake Forest was discharging raw sewage into a ravine that emptied into the lake. All soon found themselves the objects of lawsuits.

Other suits followed. Neighboring cities—Milwaukee, Wisconsin, and the Indiana cities of East Chicago, Hammond, and Gary—were charged in federal court with dumping wastes into the lake. Major corporations were charged with air and water violations.

One week after Scott had filed suit against Republic Steel for water pollution, the giant steel firm found itself back in the Circuit Court of Cook County on charges of air pollution. U.S. Steel fared little better—they had a month from the time Scott first sued them for water pollution to the time he yanked them back into court for air pollution.

Scott also ran into money problems. Months after the antipollution legislation was adopted the legislature provided no funds. Then it appropriated only $79,000. "When you're suing U.S. Steel that's a drop in the bucket compared to their resources," Scott says, laughing ruefully. He turned to what he calls "creative law enforcement." Antitrust action was taken against the auto manufacturers to take some of the load off the pollution division. The automotive giants were charged with violating federal antitrust laws by conspiring not to install antipollution devices on cars. ("This is the opposite of the free enterprise system, where the first guy that gets the device sells to others.") He started his action one day after the federal government let the auto manufacturers off the hook. Next he went after the airlines and helped bring about the agreement for them to install antipollution devices on jets by 1972.

He went after Commonwealth Edison for air pollution, hitting the utility where it hurt. Commonwealth Edison was seeking a $45 million rate increase. Scott opposed it, arguing that public utilities have a greater obligation to serve the public interest than any other business, and that included compliance with air-pollution laws, the *Sierra Bulletin* states: "For the first time in United States history, an Attorney General had intervened in a rate proceeding to argue that a utility seeking an increase 'must come to the Commission and courts with clean hands as well as proof of need for an increase.' "

Commonwealth Edison insisted that its forty-year-old furnaces required high-sulfur-content coal; low-sulfur coal was not available, and pollution-control devices were not available for such old furnaces.

After Scott intervened in the rate case, the utility discovered millions of tons of low-sulfur coal were available, and, by golly, it would burn in their old furnaces. To cap this run of luck, they had found an inventor who had developed a sulfur removal system that they could install at their generating stations.

Scott has filed air, water, solid-waste, and nuclear pollution suits the length and breadth of the state. But perhaps one among them is his favorite—the fight to save a small, 12,000-year-old tamarack bog. The Volo Bog of northern Illinois is threatened by a $94 million commercial and recreational project. The bog is at a higher elevation than adjacent land being drained for the project; it will die if its water table is lowered. "We want a restraining order and a permanent injunction against any development that threatens the ecological integrity of the bog," Scott said.

If he wins, he will have developed a new legal concept. "Like drugs that must be tested before the federal government will let them go on the market, we're saying to the developers that before you start building, you must prove to us you're not going to ruin the environment," he explains. This is in the face of the tradition that the burden of proof rests on those challenging the environmental impact of proposed commercial and recreational developments.

Sullivan of New Jersey: Richard J. Sullivan is director of the new Environmental Protection Division of New Jersey, one of the most highly respected pollution-control offices in the nation.

New Jersey might be considered a microcosm of the environmental bind that is plaguing many states as populations increase, industry multiplies, and pollution becomes more severe. Almost every problem the nation faces can be found in New Jersey, often in concentrated form. If Jersey can be cleaned up, any place can. If it cannot, God help us all. The "Garden State" is small in size but a giant among polluters. It lies adjacent to New York, the master polluter of them all, sharing the polluted Hudson River. Jersey has the heaviest population density in the nation—1,000 people per square mile, a concentration greater than the Netherlands, India, or mainland China. Most are congregated in the northern part of the state adjacent to New York. It is the most urbanized state in the nation, and has one of the biggest industrial complexes (number one in chemical plants and a leader in refineries and other industries); it is among the leaders in car ownership and highway usage. It is laced by the New Jersey Turnpike, a carbon-monoxide corridor linking New York and Washington, and the heavily traveled Garden State Parkway leading to the Jersey shore. Consequently, the state has some of the nation's dirtiest skies (which it shares with New York).

Over Port Newark hangs a perpetual grayish-brown shroud,

bearing the toxic kiss of emphysema, which joins the chemical miasma from New York; in the late afternoon it spreads out to the suburbs, causing damage to vegetation and crops in every county in the state and doing God knows what to people. Jersey's water is as foul as its air. The Passaic River, a drinking water source for hundreds of thousands, as noted earlier, is classified by the federal government as one of America's ten most polluted streams. The Hackensack River, which cuts through the dumping ground in the meadows, is a lifeless industrial sewer. The Arthur Kill, leading into the refineries at Port Newark, is a notorious oil highway whose poisoned spoils are periodically dredged and carted away to the ocean-dumping ground off nearby Sandy Hook. New Jersey's natural drainage is Newark Bay, one of the most contaminated bodies of water in the United States.

The state has a tradition of laissez-faire toward industry; it provides a corporate climate more hospitable than that of more squeamish states, wrote a historian. It is a stamping ground for organized crime and political corruption, reactionary organizations and conservative thought. It has no state income tax, depending on a sales tax and property taxes, and it is in a constant fiscal crisis. It is a bedroom community for thousands of commuters with no real sense of commitment. Pollution, like corruption, is accepted as a way of life.

It is against this background that Sullivan must work. Since 1967 he has been in charge of enforcing Jersey's strong antipollution laws. During that time he has won 5,000 court convictions, fines ranging from $50 to $13,000. The results show the weakness of even these "strong laws." The fines have not been enough to deter hard-core polluters. Many look upon small fines as cheaper than control equipment, in effect a license to keep polluting.

Unlike many enforcement officials, Sullivan has not made a token show by going after "small cheese factories." He has concentrated on many of the giants, hauling them into court after they ignored his pleas and warnings to reform. His victims include city and small-town politicians, big coal companies, major railroads, labor unions, builders, powerful corporations, and nine major airlines using Newark Airport and dumping 1,000 tons of pollutants in Jersey skies annually. One of his most important tools, short of jail sentences, has been the injunction. In a bold thrust he was able to get a court order banning all building in nine communities along the Rockaway River until their sewage disposal plants were upgraded. Jersey City has been forced to upgrade its almost worthless treatment plant at its reservoir near Boonton.

Sullivan has nothing of the zealot, reformer, or wild-eyed enforcer about him. He is a professional, forty-five, a graduate engineer from Stevens Institute of Technology; he has an MA in English and a Master of Public Health Administration from Columbia. He is a "law school dropout," which helps him understand the legal problems. He is low-key, calm, easy-going, and expresses a refreshing human concern for the problems of both the polluter and the polluted. He sees himself and his role in an unusual context for a political appointee:

"I am a professional bureaucrat. My job, despite the impediments in the bureaucratic system, is to make things happen, to make some of the pollution go away; maybe in the process to come up with some creative new measures to prevent and control pollution." (The state, with federal aid, has developed a new device to test auto emissions and is now trying to put it into use—a politically touchy business in a politically sensitive state such as New Jersey.)

He believes that any real solution to pollution must come through the courts. "You need strong laws and you must take court action and treat every chimney the same." You begin with good laws. Then you enforce them. There can be no exceptions.

He believes it is important to go to court and to win your cases:

> In that way we establish a tone for the community. The chimney and outflow owners know that the state must be taken seriously. We make our activity visible. One of our policies is to name names. We don't put out speeches or that kind of hogwash. We put out a list of violators. We want the public to know that we are active, that we are doing what the law says we must do.

Sullivan's department is representative of the way several states are trying to attack the explosive environmental issue— setting up a single department of professionals rather than citizen boards that are so easily dominated by industry. New York has a similar department.

No one, however, has yet gone to jail in New Jersey or anywhere else in the United States. It is still more respectable to kill millions of people slowly than one quickly. It is all right to poison the breath of life but not to damage personal property. Many thoughtful people now question whether anything short of prison sentences for the executives of polluting companies will resolve the problem. Among those raising the question are Henry Steele Commager, the Amherst historian, and M. Ethan Katsch, a law professor at the University of Massachusetts "A very short prison term imposed on the responsible executives would be a much more effective deterrent than all but extraordinarily high fines," says Katsch.

The tone for enforcement is set by the top official, says Sullivan. He must have conviction on the issue of environmental protection or no progress is made, no matter how effective the remedial statutes. But there is a whole second language among enforcement officials. If he tells a public hearing that "we must control air pollution, but we must take great care that we don't recklessly move ahead and draft regulations that would drive industry from our borders, that's code language for 'lay off, don't make industry angry.' "

Despite heroic efforts by Sullivan, pollution has not gone away in New Jersey. It's worse than ever, as it is throughout the nation. The reasons are complex, as pollution itself is complex. A state's problems are enormously difficult. There must be a balancing of social, economic, and political judgments, self-interests and prejudices. A simple problem like sewage disposal in a single community can become almost overwhelming.

New Jersey has 750 sewage treatment plants serving 90 percent of the people. Every county with sewers has a treatment plant at the end of the pipe. It is probably the only state in the country that has no county putting wastes directly into a stream. Still the water is polluted and the Passaic River is one of the ten most polluted rivers in the nation. Most of the sewage plants offer only primary treatment. Plans are now in effect to consolidate them into a network of 150 regional plants. It will cost a billion dollars to bring them up to the "state of the art"—the best present methods of treatment. The next billion must be spent for new methods of treatment. Many New Jersey cities now have single sewer systems; in rainy periods the sewage bypasses the treatment plants and goes directly into the streams. To replace the sewers in old cities like Newark and Hackensack would mean tearing up every street. "It would cost more than the cities are worth," explains Sullivan. Most Jersey industries now dump their effluents into municipal sewer pipes, often knocking treatment plants out of operation. The public now must pay for the treatment of these wastes—a form of indirect business subsidy. "But the money has to be spent or facilities will not be constructed, and the water will get dirtier. It is as simple as that," says Sullivan.

A single local problem shows how difficult is the job of correction. Princeton, New Jersey, the home of Princeton University, one of the nation's most enlightened and beautiful communities, has a treatment plant that was built in 1932 and has long been inadequate; designed to handle 3.5 million gallons per day, it now is deluged by more than 7 million gallons per day during storms. This means untreated sewage pouring into the Millstone River, a tributary of the Raritan, which is an important source of drinking water for thousands. In 1964 the need for a new treatment plant was recognized. A regional plan was conceived involving twenty surrounding communities, most of them rural and served by septic tanks. But the plant's estimated cost was $78 million (based on a $200,000 preliminary engineering study). All but seven of the communities dropped out because of the cost. The remaining seven now form the Princeton Sewer Operating Committee, headed by Foster Jacobs, a Princeton official. The cost of the new plant is estimated at $34 million, about $75 per family per year. To proceed requires a formal engineering study that would cost $1,340,000. Some of the communities are not sure if they want to continue at this price, especially since they are faced with the added cost of putting in sewers for existing homes as well as areas that may not be developed for many years. Financing would require a bond issue. The state would pay 25 percent of the plant's cost and the federal government 55 percent, theoretically. In reality, Jacobs explains, the government rarely pays more than 15 to 30 percent "and you have to sue to get that." Plans must be approved by the state and all of the partners must agree to stay in, or the whole thing must start over again. The formal engineering study would take a year. Building the plant would take at least two more years. The soonest the plant could be in operation is 1973; more likely it will be longer. Recently the problem was complicated anew because the state upgraded its requirements, says Jacobs. The state now demands removal of phosphates and nitrates to a level that present technology does not permit. "This means that the plant can't be built until the necessary technology is achieved, or the state must reduce its standards." This must be worked out in a series of meetings. Meanwhile, inflation keeps nibbling away, raising costs about 1 percent per month: $340,000 each month. In brief, seven years after the plan was conceived the plant has not even been started. The plans are still incomplete—and this in a community that wants to comply with the law and can afford the necessary funds.

Almost every community has a similar problem. All of them fall into Sullivan's lap. He is refreshingly candid about them. He believes that the lack of perfect control equipment is no excuse not to install the best that is available and upgrade as new methods are developed. "Go with the best you have at any given moment—the highest state of the art." A new industrial plant wanted to install smokestacks with scrubbing equipment of 60 percent efficiency. He held that the state of the art allowed for 90 percent efficiency and demanded that. State law imposes on any new polluting source the condition that it meets the "highest state of the art."

He is not satisfied with what he has done in New Jersey. He considers current levels of pollution intolerable to the public. He feels that many public officials have made a mistake by not being more forthright with the people, and this has led to a breakdown in public confidence.

> If public officials and teachers admitted our ignorance instead of deluding themselves and the public about the miracles of modern science, those at hand and those to come, there would be a greater sense of reality among the people. They would not be so easily misled into thinking there is a solution for every problem. For many of our problems there is no excuse.

To clean up a state like New Jersey will cost big money. Where is it to come from? In 1969 the people voted a bond issue for $271 million to clean up the state's waters—a big step in a state that starves higher education and almost every social and cultural endeavor.

There will also have to be stronger enforcement at the local level. Sullivan is outspoken in criticizing the failure of local governments to enforce state laws "against even the most crass and obvious sources of pollution"—a shortcoming found in most states. He wishes, almost wistfully, that federal regulations were not so confusing. He is almost heretical in believing that the federal government should play a "stronger role" in moving to protect the public interest in the area of the environment. "If the states don't move, the federal government should have the power to intervene. . . . I see too many cases where the states have simply failed to exercise the constitutional power they have, to provide the protection for the public that it deserves."

He is critical of the way some industries have taken advantage of confusion arising from differing levels of responsibility to avoid regulation. He cites the automobile and aviation industries. "The federal level is where such regulation ought to be in the first place." He would like to step up deadlines to reduce automobile emissions. He recognizes that this has political repercussions. It treats the voter as the cause and not the victim of air pollution. "This is a very risky undertaking for a government agency, but if we don't do it, I am convinced we won't make the progress we need."

Sullivan does not consider himself a "calamity merchant." With the exception of the state's continuing growth and the unrestricted use of land, "I generally take a hopeful view. A lot can be done now. It's my job to see that it gets done."

THE CITIZEN STRIKES BACK

Many people have found a way to do something about what they feel is an intolerable and worsening environmental situation. Some try to help save the world from its depredations. Others settle for trying to preserve their own sanity in a society that threatens to engulf and suffocate them. Some drop out—many of them educated people with advanced degrees or professional background; they may join communes, take to the road or drugs, or move to isolated areas to live in abandoned, run-down dwellings and farm worn-out lands, avoiding chemicals, raising families, trying to live more harmoniously with nature and themselves, finding reverence for life and spiritual relevance in the soil.

Some become political activists. Some form or join conservation organizations. Others try to practice conservation in their own lives; they sort out their wastes, have compost heaps, do not burn leaves, avoid detergents, choose foods without chemicals, sign petitions, hold protest meetings, picket polluters, collect no-return bottles and cans and sell or give them to the manufacturers for recycling or dumping. Concerned citizens have thrown themselves dramatically in front of onrushing bulldozers (none lost so far), locked arms around threatened trees, ripped down fences from "progress"-imperiled parks, cleansed streams of litter, led Scout troops in planting trees or collecting used papers for recycling. They have formed boycotts, demonstrated, given speeches, and attended countless meetings. Colleges and high schools give courses and seminars in environment and ecology. There is a burgeoning of books, films, and other educational material. Environmental missionaries lecture their neighbors, scold offenders, bore their friends, and often distress their unconverted and unrepentant mates. One frustrated woman conservationist in Alabama slapped a water-pollution board member.

At least one other has taken to violence. "The Fox" is a minor environmental legend around Kane County, Illinois. He strikes back at polluting companies, a sort of anonymous environmental guerrilla, or an antipollution "Zorro," as the Chicago *Sun-Times* describes him: "Whenever he blocks a company's drainage system, tries to seal off its chimney, puts a dead skunk on the porch of an executive, or dumps dead fish in a lobby, he leaves a note telling why, and always signs it 'The Fox.' " The Fox has had many adventures and left a trail of nausea, screaming secretaries, and panicky executives rushing for air after he dumps cans of rotting fish and sewage in a polluter's lobby.

> During his adventures, "The Fox" has been chased by plant guards, has crawled through drainage pipes, climbed roofs in the dead of night, and a shot has been fired in his direction. It is rumored that at least one manufacturing firm has posted a reward—cash or a job promotion—for his identification.

A one-man crusade has been carried on by Martin Schneider, a top photographer in New York. Schneider has had epic experiences on behalf of conservation. He teaches a course at Cooper Union: "Polluted Environment."

"At the first session he brought in a bag of canned goods, hauled them out one by one and told his students that this one could cause brain tumors, that one cancer, the next blood poisoning," relates *The New York Times*.

Schneider was assigned by *Life* to do a major story on environmental destruction. He was particularly interested in the phosphate industry in Florida, near Tampa: "the combination of sulfuric and hydrofluoric acid which blows out of the stacks of phosphate plants strips the paint off cars, causes emphysema and ulceration of the throat, has killed off many thousands of head of cattle and is wiping out the citrus growers in the area."

To dramatize the situation—"which Florida officials have whitewashed"—Schneider came up with a typically outrageous idea which he thought would make an excellent cover for the special issue, *The New York Times* states:

> He took some carcasses of autopsied cows—so he could *prove* exactly what they'd died from—out to a stretch of public land in front of a Mobil-owned phosphate plant. Then he turned his fluoroscope on them. Naturally, the accumulations from the absorbed hydrofluoric acid showed up crystal clear, outlining the cows' skulls in grisly perfection.

By then it was getting dark, so he picked a particularly photogenic skull, rigged up a homemade projecting device, and projected the image up above the plant onto a low-lying bank of clouds:

> One hell of an effective picture, you must admit: the factories outlined by the light from their own stacks, those deadly fumes pouring out, and above it all, a million times bigger than life, that hovering skull of one of the industry's victims. Schneider was busily photographing away, when out of the bushes popped a flock of plant guards, waving guns and cameras and claiming to be sheriffs (which happens to come under the heading of impersonating an officer of the law . . .)

The story gets wilder. Schneider fogging the guards' cameras with his fluoroscope, a chase through the dark, his car's lights off, overturning the car in a ditch, the escape, mailing the pictures, which then mysteriously disappeared from the Air Express office, the sky shot taken again and once again it vanished "this time in the hands of a *Life* editor who claimed that he himself—not a messenger, mind you, therefore no receipt—personally slipped a three-inch-thick package containing this photograph under Schneider's door." *Life* "sat on the story for a year." When it appeared Schneider tore his hair and threatened to give up photography.

> The magazine had run all his shots of the polluted atmosphere, but not a single picture of pollution's murderous effect on people and animals, thus undercutting completely the impact of Schneider's visual essay. Furthermore they

had replaced his text with their own, which toned down the most scarifying aspects of his carefully documented research.

Others carry on the campaign less dramatically. Pete Seeger sails up and down the Hudson in the sloop *Clearwater*, singing his plaintive folk songs and calling for restoration of the humiliated Hudson. Seeger, a gentle man, sits on the vessel's rail, stroking his wispy beard, and tells how he got started on this environmental crusade: *

> It's a logical thing. All my life I lived outdoors. Twenty years ago my wife and I built a log cabin on the Hudson. I saw what was happening to the Hudson and realized from Thor Heyerdahl that it was the problem of the whole world. What used to be a beautiful stream, is now in places an open sewer. When you lose something you are concerned about the past and the future.

Seeger describes what had happened to the river:

> Lumps of this and that went floating by. You don't see it until you get close—maybe a sewer pipe in the water. It's not very pleasant to describe—condoms, lumps of shit, toilet paper all rippling downstream. Henry Hudson said about this water, "As sweet to taste as any I've ever tasted."

Seeger suddenly begins to sing:

> Sailing up my dirty stream
> Still I love it and I'll dream
> That someday though not this year
> My Hudson River will once more run clear.

Many scientists and teachers have been especially generous in interpreting difficult scientific concepts for laymen. Groups like the Committee for Environmental Information in St. Louis and chapters of the Scientists' Committee for Public Information have supplied speakers to organizations. Scientists, acting as individuals and working with organizations, have fought heroically for local reforms. These men have testified in court, taken water, soil, and air samples to refute official assurances that there was no contamination, found errors in official figures that purported to show no danger existed, and warned of hazards

---

* Personal interview aboard *Clearwater*, summer, 1969.

suppressed by official bodies. The public owes a tremendous debt to these courageous and unselfish people who have often suffered insult and abuse for their efforts on behalf of public welfare.

In Pittsburgh two young college instructors paddled up and down the city's polluted rivers in a canoe gathering evidence against polluters. Through their efforts four companies—a chemical manufacturer and three large steel companies—were accused of violating the 1899 Rivers and Harbors Act, which prohibits the discharge of "any refuse matter of any kind or description" into waterways. The two instructors, David G. Nixon and Dr. John Zavodni, collected evidence by taking samples from stinking and often steaming plant outlets. They used a ten-foot bamboo pole with a plastic bottle attached to gather the more noxious effluents. They were chased by plant guards. A crane operator swung a wrecking ball inches over their heads. Other times workers would quietly point out hidden discharge pipes or whisper, "Come back at five-thirty when they pour out the bad stuff." Under the law the instructors are entitled to half of any fines levied against the defendants. The first conviction, against the chemical company, brought them, along with Michael Watts, a chemist who analyzed the results, the sum of $4,998. The bounty hunters have written a pamphlet on how to investigate polluting industries. They hope to use their rewards to set up a network of concerned people to pool information whenever one of the big companies pours toxic materials into the waterways.

In a similar action United States Attorney Whitney North Seymour, Jr., has asked fishermen in the Hudson River to report on polluters there and collect half the fines under the 1899 law. Fines can range up to thousands of dollars a day.

As people learn what is happening to the world they are demanding more participation in decisions that affect their lives. Many are joining citizen watchdog arms of local government. In New York volunteers are being trained as "smoke watchers" to report offending chimneys. Under a new local law, if the city fails to take action within forty-five days an individual can bring his own complaint to the Environmental Protection Board. The accuser can produce evidence of the violation—photographs, soot samples, or whatever—and the board can impose a penalty of up to $100 for each day's violation; it can also award as much as 50 percent of the penalty imposed to the complaining witness to help defray costs incurred.

An encouraging number of environmental protection groups have sprung up throughout the country, usually made up of women. Among them is Concern, Inc.; it was started by seven prominent Washington women, including Mrs. Richard Helms, wife of the CIA chief, as a result of a "worried dinner-table conversation." Among its services is distributing a purse-size tip sheet that lists products that do the least environmental damage: detergents with the least phosphates, or suggesting switching to soap powder and washing soda. Women are urged to boycott clear plastic containers made of polyvinyl chloride, to seek returnable bottles and simply packaged goods, to avoid DDT, dieldrin, chlordane, and other specific pesticides. The organization also circulates a listing of corporate contacts so consumers can insist directly on more specific labeling, better packaging, and other demands.

"We want people to think more and use less, and we find that the companies don't just send us away with a polite pat on the back," says Mrs. Nancy Ignatius, wife of former Navy Secretary Paul Ignatius.

Many people who have joined citizen groups become disillusioned about the effectiveness of passive protest. The theme of frustration appears repeatedly among those who have tried to stop raids against nature by powerful private interests. They often come away feeling that government is completely unresponsive to public demands. Two different citizen groups in California recently bumped their toes against the powerful industry-government alliance. One, calling itself No Oil, tried in vain to stop the despoliation of a public beach by an oil rig in a fault area near Los Angeles. Even if it were not a fault area they would still protest, said Mrs. Deborah Parducci, a member of No Oil: "There are only about 80 miles of beach accessible to the public along the whole coast of California . . . and this is what we're going to do with it—build an oil rig." Industry's reply was insensitive: people come to the beach to look at the ocean, not to look back at the land and what is built on it.

The second citizen group opposed construction of a freeway that would destroy a coastal scenic route. A second freeway would cut into the Santa Monica Mountains, described as "the only open green space left in the entire Los Angeles Basin." The citizens pointed out that people needed the open space, "particularly poor people, who can't afford to get out of town. And we don't think taxpayers should spend $100 million to build these roads for the convenience of the land developers."

Both struggles left the citizens' groups exhausted and rather bitter, observed *The New York Times*. People who had believed in government no longer believed. "You pay taxes and you hope government will look out for your interest," said Mrs. Shirley Solomon of No Oil. "But now we find we have to give our time, money and effort to fight the things government is supposed to be protecting us from."

"We just can't fight them financially," said Mrs. Parducci. "If we go to court, they can keep our lawyers chasing around until we run out of money. They can deduct all of their costs as a business expense, but our tax comes right out of our own pockets. They can buy their justice in court. . . . It's an incredibly big task. No one can fight an oil company effectively doing it part time. But we can't afford to hire full-time people. What we're really fighting is the American tradition that any money made in any way is okay. We want to appeal to another tradition —the right to have a nice place to live."

Out of such disillusionment come the environmental activists. These activists are, generally, trying to work within the system. Their primary strategy is political and economic leverage. Potentially they are the most powerful force for reform. They strike, not at the effects of pollution and environmental exploitation but at its heart—directly at the political-industrial jugular.

This was demonstrated in the 1970 elections. For the first time the candidate's environmental record became a campaign issue. In at least three cases it is credited with unseating the environmentally insensitive and in others it played some role. In Maryland's primary George Fallon, veteran of thirteen terms in the House of Representatives and chairman of the Public Works Committee, was eliminated almost solely because of his "fanatic devotion to highways and the highway lobby," reports *The New York Times*. In Indiana former Representative J. Edward Roush won after basing his campaign on environmental issues; he had taken the lead in establishing the Indiana Dunes National Lakeshore. Most sensational, perhaps, of the environmentalists' Election Day triumphs was the retirement, after one disastrous term, of Governor Don Samuelson of Idaho:

Eager to turn the magnificent White Cloud Mountains over to the American Smelting and Refining Company and to dredge-mine wild and scenic rivers—even those under federal protection—he framed that immortal creed of the exploiter: "The good Lord never intended us to lock up our resources."

The major role in what the *Times* calls "the timely retirement" of Governor Samuelson and Congressman Fallon (and spearheading other fights of the political-environmental front) was played by the League of Conservation Voters, political arm of the Friends of the Earth, most militant of the conservation groups.

THE MAKING OF AN INFORMED CITIZEN

The political activists have been bruised and toughened by trying to reform a system that does not want to be reformed. Mrs. Hazel Henderson is one. Her education came the hard way, and the lessons she learned are invaluable for the political activist. Mrs. Henderson is English, blond, young, slim, lives in a fashionable townhouse in New York with her husband and young daughter. She has become a deft and skilled political infighter. She believes that reform will come only through aroused citizens challenging corporate might, demanding their rights and being willing to fight for them.

Mrs. Henderson's indoctrination began in 1956 when she came to New York from the Caribbean. She saw with fresh eyes the pollution that New Yorkers had been educated to accept as a natural part of life.

Mrs. Henderson's family had always taken an active role in civic projects. "The idea that people don't have specific projects never occurred to me. That is the way a democracy is supposed to work." She began trying to educate herself about air pollution but found it hard going. She wrote letters and tried to get information. She tried to interest other women, "but they wouldn't really believe someone wasn't taking care of it. They thought I was some kind of nut. But I realized that no one was minding the store."

Hazel Henderson's education included rebuffs from foundations, industry, newspapers, and television stations.* There were discouragement and tears. Gradually she found allies. Citizens

* Personal interview, New York.

for Clean Air, Inc., was formed and she served a term as its president.

Citizens for Clean Air worked hard to get a local air-pollution law passed. They ran into obstacles encountered by citizen groups elsewhere. Federal law called for public hearings to fix air standards.

> But industry pressure groups moved in. They got to the state government before citizens knew what was happening. They would announce a meeting so late that the citizens couldn't get ready, and then change the meeting place at the last minute. There were all kinds of fun and games to thwart citizen participation.
>
> The people in power were helping industry. The company lobbyist can get information on the legislative desk, while the citizen is always three steps behind and unable to catch up. A typical tactic for state and local officials was to make a big hullabaloo to make everybody think everything was taken care of—appointments and committees. But you looked deeper and found that industry people had been appointed.

Mrs. Henderson, out to do good works as an informed citizen, soon learned what the game was all about: "The cards are stacked. It's a subculture—all work together in incestuous little subcommittees for the benefit of those in it, where the regulated industry ends up regulating those who are supposed to regulate it."

She found that the most frustrating part of being an informed citizen was that

> . . . once you're informed you have to become almost para-noiac to implement your concern. The citizen begins to understand that he must do the same thing as the industry lobbyist or pressure groups—he too must organize. But he must do it without funds and after a hard day's work. For the corporation, it's part of the day's business; there are sec-retaries, phones, money, the essential Xerox.

The citizen finds that he must recruit disinterested people, sign petitions, run mimeographs, hold meetings. "For most people it's hard at the end of the day's work. And even if you put together a pressure group it's so easy for industry to counteract it."

She learned that each industry had only its own interests to concentrate on. The informed citizen at any given time must fight a dozen issues. Each issue takes time, energy, money. "How can you do it? Most can't . . ."

The informed citizen at this point is put to the test. Does he or she disappear back into the woodwork to lick the wounds of combat or put on a coat of armor and do battle in the area where the big decisions are made? Mrs. Henderson chose to fight. She became a strategist, a pamphleteer, a lobbyist and lecturer. One of her speeches was put into the *Congressional Record* by Senator Muskie. In it she explained some of the problems of becoming an "informed citizen"—the ideal of the American founding fathers. "Not just citizen, you will note, but *informed* citizen." This is practical advice for all citizen groups:

The first obstacle is the expert. Industry and government say the citizen is not qualified to make important decisions because he does not have the necessary information. And yet they make it almost impossible for him to become an expert. They deny him the essential information, and the citizen lacks all the necessary apparatus available to the experts—researchers, lobbyists, public relations men, and advertising budgets.

The expert himself becomes the biggest threat to the environment. He becomes the victim of his own limited knowledge; Mrs. Henderson observes:

> In our vast, pulsating, computerized America of today, we have thousands of so-called "experts," each with an understanding only of his own narrow discipline, or what is often called "tunnel vision." They are all making momentous decisions on deploying technology in hundreds of new ways, without any real understanding of the big picture. A famous sociologist once studied these people, and called them "technological idiots." . . . They are the highway planners who build roads by destroying neighborhoods and scenic values, the economists who understand fiscal policy but not social values; the computer analysts who only know how to feed their computers with facts and figures that can be quantified, and then wonder why their plans do not work in the real world of people.*

The obstacle of the expert was encountered by Citizens for Clean Air. When the organization tried to get information, "we

* Citizen Action for Clean Air, reprinted in the *Congressional Record*, July 14, 1969. Vol. 115, No. 116, from speech before Citizens Workshop on Air Quality at the University of Massachusetts: Waltham, Mass., June 19, 1969.

were barraged with experts who confronted us with figures and formulas calculated to intimidate and confuse us. We soon caught onto these cheap tricks . . ." Citizens for Clean Air refused to be caught in the trap of being disqualified by their lack of expertise. They not only found their own volunteer experts, such as the Scientists Institute for Public Information, but also insisted upon a broader interpretation of facts:

> When they quoted chemical formulas, we quoted Pericles, Edmund Burke and Thomas Jefferson; when they quoted Adam Smith and the "invisible hand" of the market place we quoted the higher authority of the Almighty. Don't apologize for being laymen. Announce proudly that you are a layman, a generalist and a humanist. For as the country becomes more and more specialized, the generalist who sees the total system becomes the key man.

The citizen pollution fighter is cautioned not to be taken in by diversionary red herring tactics. The expert loves to tell citizen groups to contribute by not burning leaves, to keep their cars well serviced, to minimize pollution and support their local control officials. Citizens for Clean Air soon realized that as long as the local power company generated power in the same old way, and the oil industry continued selling the same high-sulfur fuel oil and leaded gasoline, and the real estate developers kept building apartments with the same old-fashioned incinerators, and Detroit kept producing the same old poisonous cars—"our little individual efforts at controlling our own contributions to the overall pollution, although necessary, were not going to make much difference."

The target has to be the biggest polluter, not the smallest—the automobile industry, the power companies, big industries:

> So a good rule of thumb for any new citizens group is to start by finding out what the biggest sources of pollution are in your area. Then investigate what air pollution laws you already have on the books and how they are enforced. You will often find, as we did, that administration and legal procedures can render the laws useless. Sometimes it is lack of sufficient inspectors; sometimes it is the log jam of court cases; sometimes it is judges who don't take air pollution offenses seriously and only give token wrist-slapping fines. And sometimes, as is currently happening in New York City, a whole industry (in our case the real estate industry)

will openly defy a newly passed law (to control incinerator emissions), and even take the Air Pollution Commissioner to court! *

Citizen groups not only must get laws passed and enforced; they must provide watchdog groups to make sure that large powerful groups do not try to obtain special dispensations from the enforcement agency. Constant vigilance is needed to prevent organized interests from actually rolling back new laws on the books. In New York State, local town officials from upstate districts actually pressured a bill through the state legislature to reestablish open-burning of leaves and garbage because it was too much of a problem to find other disposal methods.

Mrs. Henderson warns against getting too chummy with industry people. "Once you become too friendly with that nice public relations man from the XYZ company or even sometimes with your own control officials, you will become so sympathetic to their problems that you lose sight of the large, public interest."

The key man is the politician. He makes the laws and is now responding to industry pressure because it is the most forceful. If the citizen can exert a greater pressure—the threat of kicking him out of office—he will respond to public pressure. This means mass education of people and requires adroit use of the mass media. So far the mass media, like the government, have been primarily responsive to industry and its enormous advertising budgets.

It is the mass media that, largely, give us our view of reality, that shape our philosophy and attitudes. It is possible for them to present an "objective" news story or feature within a philosophical framework that unconsciously shapes the attitudes of the reader. Business, says Mrs. Henderson, "has succeeded in imposing its largely economic values onto society as a whole to a remarkable degree." She believes it has been so successful, indeed, that it has brought about many forms of backlash: "Might we go so far as to say that business' massive transmitting capability has reached the point of diminishing returns? Has it bred a nation of cynics and a new generation which has already tuned out this one-way communications barrage?"

She and others see much of the unrest among the young as a rebellion against the narrow values being imposed on society by the pressures generated by the mass media. They reject conven-

* Citizens for Clean Air (see page 121).

tional consumer goods in favor of the old and used. The more militant take to force and violence. The weak and overly sensitive withdraw or resort to drugs, in an effort to break the conformist mold into which they are being forced by the weight of social pressure induced by the mass media. "The emotional tyranny in the advertising of personal hygiene products, generating fears of bad breath, body odor and personal rejection may have produced the rebellion of the defiantly human, gloriously unwashed, shaggy-haired hippie."

Control of the mass media is the keystone in shaping society and its values, according to Mrs. Henderson. Whoever controls the mass media will control our destiny. The biggest need now, she believes, is for the citizen to have greater participation in the "politically vital communications media—radio and television."

Mrs. Henderson is now a veteran of many citizen wars on behalf of a more reasonable and healthful climate for human beings. In the struggle she has reached a conclusion: Salvation, in the end, is squarely in the hands of the individual. "The power of the modern corporation has, through mass persuasion, enabled it to impose its narrow economic criteria on society. Only massive public pressure in the form of legislative and economic sanctions will be sufficient to produce the needed business response."

Mrs. Henderson says there is one reason for her activism for corporate responsibility: "because I do believe that capitalism can be modified to serve new needs." And it's starting to work, she says. In the past year "companies who used to dismiss people like me are now asking us to give speeches before their executives."

A WORLD TO SAVE

Environmental destruction, we now know, is not limited to any isolated area. It is global. The Greeks have been warned by a panel composed of their most respected scientists that the capital city of Athens must be abandoned within ten years if air pollution cannot be controlled. "Slow poisoning is overtaking Athenians," according to the panel. It found pollution increasing so rapidly that in a few years the Acropolis, standing on a hill in splendor in the middle of Athens, will be invisible.

Even the remote Andes, soaring higher than any peaks in the Western Hemisphere, are not spared. Smog has replaced the

crystalline air over Santiago to blot out the snowcapped mountains. The politician looks disconsolately into the haze and
muses, "What does politics matter when we can no longer
breathe?"

In Japan the postwar "miracle" has turned into a nightmare.
The pollution problem is so serious that the prime minister has
taken personal command. Traffic snarls the major cities. Tokyo
policemen periodically must get away from the murderous concentrations of carbon monoxide to breathe pure oxygen—an
"oxygen break." For the private citizen there is the vending machine that dispenses oxygen. Astronomical amounts of raw sewage flow into the harbors; industrial wastes and oil flow into the
sea, threatening the vital fishing industry.

The once-picturesque village of Tagonoura on Suruga Bay
with its celebrated view of Mount Fuji, is today described as a

> . . . highly polluted port, clogged by slime from waste dis-
> charged into the water by many papermaking factories; an
> estimated 7,000 tons of sludge, heavily laced with sulphides,
> flow daily into the bay and the Pacific, polluting the waters.
> A dairy farmer must dump his milk into a hole in the
> ground every day because his cows' milk is polluted with
> cadmium from a contaminated pasture. The cadmium, from
> a nearby refinery, has so impregnated the soil that no one
> will buy the rice grown on the farmer's acres or drink the
> milk from his cattle.

In Minamata, on the southern island of Kyushu, fishermen
have lost their eyesight or gone mad after eating mercury-
contaminated fish. North of Tokyo, fifty snowy herons, annual
visitors from southeast Asia, died after eating fish contaminated
by insecticides. Smog is everywhere—not only in major urban
centers but also in some rural areas victimized by the vagaries of
wind and weather.

In Europe the fabled Rhine River, celebrated through the centuries for its exquisite beauty, is now known as "the sewer of
Europe." It begins in the pristine snow runoff from the Swiss
Alps and meanders 820 miles to the Dutch coast. By the time the
Rhine reaches the sea, 20 percent of its content is sewage and
pollutants. "More than 1,000 pollutants are present in the
Rhine's waters at any time, including detergents, pesticides, and
other chemicals," reports *The New York Times*. An international
body studied the river and found up to 2 million bacteria in a

single cubic centimeter of Rhine water in the Netherlands; in the United States, water is considered unsafe if it has more than 50 bacteria per 100 cubic centimeters. The Rhine was so polluted in Holland that its water could not be used to grow tulips.

The French and Italians quarrel over who pollutes the other's beaches. Italy has only 32 water-purification plants, 1 for each 1,000 communities, according to the Council of Europe; most sewage goes directly into the sea untreated, along with millions of tons of industrial wastes. In Milan Province a magistrate personally filed charges against 449 industrial concerns for polluting rivers with industrial wastes; the charges carry possible jail sentences of 6 months to 10 years. The magistrate said that the pollution kills fish within minutes, causes poisonous and infectious illnesses, destroys plants, kills animals, and has laid waste some 2,470 acres on the outskirts of Milan. Only 14 percent of Italy's entire seacoast is effectively free of pollution, according to a Swedish ecologist. A Milan authority estimates that 70.2 percent of Italy's coastal waters are dangerously polluted and 15.2 percent are downright poisonous. All beaches near Rome have been declared unsafe. Beaches around Salerno, Naples, and Genoa, with pollution 100 times the legal limits, have been closed; others would have been closed long ago had not public officials been intimidated when their fellow officers who tried to crack down were ousted by irate townspeople whose income depends on tourists. Health authorities have warned that water extending five to ten miles on both sides of the mouth of the Tiber River and at least three miles offshore is dangerous. "Infections picked up in that beach area have caused the incidence of typhoid fever and hepatitis to climb to near-epidemic proportions," the Italian press reports.

High-rise apartments and industrial concentrations replace Italy's famous vineyards that made fragrant the air of remote villages. Cars clog highways. Cities and suburbs are choked by epic traffic jams that have become a tourist attraction themselves. Only a decade ago, a *New York Times* correspondent recalls nostalgically, a resident of Rome could dash to the seaside by car during the customary three-hour lunch break, have a refreshing swim, dry in the sun, eat a snack, doze off in the shade, and be back at his desk in the city by 4:00 P.M. But now, traffic being what it is, it takes nerve-racking hours to reach the seacoast. And at the end is disappointment:

At the seaside, Romans find an encyclopedia of environmental decay—severely polluted water, oil smears on the gray sand, litter, eyesores in concrete, and dying pine trees that no longer give shade. . . .

The pine groves along the seashore used to look solemn with their umbrella-like crowns that provided protection from the noonday sun. Now the pines look sickly, and many are wasting away. All along Italy's coastlines, pine trees are dying. The breezes from the sea that continually spray the pine groves now carry not only salt water, but also particles of petroleum products and poisonous industrial wastes. The death of the pines is in turn causing erosion of the coastline.

Ritchie-Calder, at Malta, warned that soon "the sea will be out of bounds for bathers" and "the trees will go on dying along the coasts, suffocated or poisoned by the polluted seawinds."

The visitor to Venice today gets a chilling glimpse of Mestre, a "bedroom community, and Marghere, a huge industrial park, just before he sights the lagoon," reports a *Times* correspondent. "With their oil refineries, cracking towers, tank farms, funnels, marshalling yards, motels, service stations and pizza parlors, Mestre-Marghere is a Jersey City-Newark landscape."

The experts say that technology is sapping Venice by undermining its foundations; it rests on mud flats reinforced by wooden pilings, and landfills and a causeway linking Venice to the mainland have upset the complicated tidal system and helped bring about thirty floods in the past ten years. Digging a deep canal across the lagoon to enable oil tankers to reach the industrial area aggravated the water problem, and pumping water from the subsoil for industrial needs has accelerated the sinking into the lagoon. As the city settles and walls crack, industrial fumes violate the air, corroding historic monuments and statues. Millions of gallons of industrial wastes and raw sewage are pumped into the water each year. The Grand Canal, sparkling so beautifully in the sunlight, a romantic vision by moonlight, is an illusion, a deception.

The old Venice disappears into the sea. A new Venice rises, a commercial center with no patience for the old ways: a "wealthy hodgepodge of fertilizer plants, steel mills and belching oil refineries," according to *U. S. News & World Report*. "Refuse floats in narrow canals whose stench fills your nostrils. Lining the canals and the winding streets are rundown, humid, airless tenements which display crumbling plaster and laundry flapping in the wind. Children mill about in dingy courtyards. The lack of greenery, of playfields, is painfully apparent."

What has happened to the Venice of charm and quiet beauty? The lie is given to romantic memory by *U. S. News & World Report:*

> The splendid palaces and mansions along the water's edge [Grand Canal] have a forlorn, abandoned look. Water laps at their foundations, leaving a green, moldlike streak. It floods courtyards and porticos. It makes its way through, or under, rotted wooden doors to invade the lower floors, and from there seeps slowly upward through the masonry. Many shutters are closed, and many of the owners have left. Scattered throughout the city are historic churches, visibly smitten—inside and out—by age and smog. Statues are pitted and covered by grime; like the masonry, they are crumbling. . . .

For a while Londoners reveled in a 70-percent increase in sunshine as a result of clean-air regulations. Trout and pike are being caught again in the London dock area of a cleaner Thames River. "But Britons were shocked a few months ago to learn that the already-heavy discharge of industrial waste and sewage into their ocean and sea waters is increasing."

Hopes that African countries might escape the disaster of pollution by learning from the mistakes of industrial countries seem to have been dimmed by what has occurred in South Africa, where there has been increasing air and water pollution and soil erosion, according to *The New York Times:*

> With soil losses now at 400 million tons a year, waterways polluted and the air of Johannesburg and other industrial areas laden with cyanide sand, coal fumes and diesel exhaust, newspapers and citizens groups have started a campaign to restore the ravaged environment. . . . Cyanide-sand mine dumps that cause a high incidence of nose and throat ailments in Johannesburg are being planted with grass to reduce the amount of wind-blown dust.

In Munich air pollution is so bad (second only to Tokyo's, according to one study) that the city's famed brewery horses had to be banished from the city. In the old days most of the majestic horses that pulled the beer wagons reached the age of twenty before going to the slaughterhouse, reports *Time:*

But of late, said Löwenbräu's Heinz Moelter, "their fur lost its gloss, their eyes their shine, and their pulling power declined." Recently, Munich's local Animal Protection Society confirmed what the brewery had suspected. "They informed us that permitting the animals to continue working in Munich's poisoned atmosphere amounted to sheer cruelty," said Moelter. "Significantly, they didn't mention what the air did to human beings."

In the Near East there are mounds of wind-blown dust and barren fields of stones that lie exposed to the burning sun like so many totems of civilizations that rose in glory, flourished, and fell because of their sins against the natural order; and now they lie buried in the graveyards of their own mistakes, as Lord Ritchie-Calder puts it.

Ours is a global civilization, says Ritchie-Calder:

It is not bounded by the Tigris and the Euphrates nor even the Hellespont and the Indus; it is the whole world. Its planet has shrunk to a neighborhood round which a manmade satellite can patrol sixteen times a day, riding the gravitational fences of Man's family estate. It is a community so interdependent that our mistakes are exaggerated on a world scale.

Pollution is a crime compounded of ignorance and avarice, says Ritchie-Calder. "The great achievement of *Homo sapiens* becomes the disaster-ridden blunders of unthinking manpoisoned rivers and dead lakes, polluted with the effluents of industry which give something called 'prosperity' at the expense of posterity."

Who is in charge of the grand plan? Who is looking out for our welfare? There is no plan. No one knows what we are doing. No one is looking out for our welfare. We walk backwards into the future with our eyes closed.

We are in a mess. More than anything we need a sense of direction. We need a unity, a pulling together of all the diverse parts. We need a clear understanding of where we are and what is at stake. Several limited attempts have been made to deal with pollution and its related problems on a global scale. Almost every action has been too limited and too narrow. None have really come to anything in the climate of distrust and seeking a competitive edge. There is not even agreement on protecting whales from being hunted to extinction or stopping tankers from washing their wastes into the sea. The undeveloped nations are less worried about pollution than starvation; our "necessities" are still their luxuries. They cry for smokestacks while we try to get rid of them. Brazil's Planning Minister was quoted early in 1972 as observing, hopefully, that "Brazil can become the importer of pollution." Endorsing a huge woodpulp project which his country would undertake for Japanese paper producers, he expanded on the theme: "Why not? We have a lot left to pollute. They don't."

The underdeveloped nations do not see pollution. They see inequity. In the United Nations Brazil has made the point that no developing country should be asked to slow down its industrial growth now to make up for what a *New York Times* editorial called "the earlier environmental sins of other nations." The needs of the underdeveloped nations must be met if they are to join in a solution of the environmental problem. A mechanism must be found to deal with the great global problems.

We grasp at straws. One small encouragement came recently when Russian Premier Aleksei N. Kosygin told four visiting Americans that it was crucial for the Soviet Union and the United States to cooperate in a major fight against environmental pollution. Previously the Russians have avoided discussions about pollution with foreigners.

The urgency of finding an international solution without further delay is underscored by Ritchie-Calder. There are no frontiers in present day pollution and destruction of the biosphere, he says. "Mankind shares a common habitat. We have mortgaged the old homestead and nature is liable to foreclose."

CONCLUSION: AT THE CROSSROADS . . .

In 1963 San Diego Bay was a sink. The water, once a clear blue, had a brownish-reddish tinge from dead plankton and eutrophication. It was receiving millions of gallons of raw and inadequately treated sewage and industrial waste every day. In one area sewage solids had formed a sludge mat on the bay floor 900 yards long, 200 yards wide, and 7 feet deep. Coliform counts were dangerously high. Several beaches were quarantined. Water clarity—measured by the distance that an 8-inch disk can be seen below surface—was an average of 6 feet and down to 1 foot in some areas.

Today the warm waters of San Diego Bay are once more blue

and sparkling and clear, reports *The New York Times*. They are also clean, with a purity well above the antipollution standards set by the state of California. The dead phytoplankton have disappeared. Marine life has returned. Dissolved oxygen is again at safe levels. Water clarity ranges from 10 to 30 feet and the sludge beds have been cut in half. Coliform density has so decreased that beaches are once more safe for swimming.

This did not take long to accomplish. Once sewage plants were built, "almost immediately remarkable change occurred," said a spokesman. "We had expected slow improvement, but the bay quickly began to clear as if there had been a few, great tidal flushings. Where the bay had been brown and red, it became blue and sparkling." As early as 1964, only eight months after the new sewage system began operating, "sculpin, sole, sand bass, steelhead trout, silver salmon, bonefish, black sea bass, barracuda, bonito, yellowtail, octopus, shark, seal and porpoise were swarming back into the bay" according to a 1965 report by the United States Public Health Service.

The transformation of San Diego Bay from the 1963 condition is cited by the President's Council on Environmental Quality as one of the two outstanding examples in the nation of water quality recovery by a municipality. The other was Seattle's cleanup of Lake Washington. The remarkable part is that the federal government contributed only about $8.5 million to Seattle's $145 million improvement program and about $2.5 million to San Diego's $60 million cleanup:

> These two cities have demonstrated that water receiving sewage can be cleaned up by a combination of aggressive and imaginative leadership by a few citizens and a willingness by the electorate to pay off the bonds. In Seattle the charge is $2 a month per household; in San Diego it is $1.50.

These are bright spots in the murky pollution picture. In other places there are also encouragements. They are told in headlines: "A Study in Pollution Controls . . ." "Tough Rules Saving a Dying Oregon River . . ." "Several Communities Haul Firms into Court in Bid to Clean Up Air . . ." "To Save 54 Trees, Paris Discards Parking Lot Plans . . ." "Antipollution Laws Are Forcing Steelmakers to Close Old, Dirty Open Hearth Furnaces . . ." "Legislatures Across U.S. Seek Tough Laws Against Pollution . . ." "New Jetports Held Up by Protest Movements . . ."

"Ecology: Pesticide Makers' Bête Noire . . ." "L.I. Town Moves to Control Waste . . ." "City in Michigan Restricts Autos . . ."

All to the good. While the Nixon Administration has been overzealous in protecting business, it has also made significant gains in protecting the environment in some respects. The President himself seems to be taking a more firm stance. In his 1971 message to Congress on the environment, he asked for federal legislation to reduce noise, to modify environmental effects of both strip and underground mining operations, to ultimately ban unregulated ocean dumping, a new approach to the location of sites for power plants, and legislation to keep developers away from vulnerable areas such as wetlands. Question: is this more Presidential loop-hole rhetoric or really reform? In some important areas, such as pesticide control, the Administration has faltered. A new pesticides control bill is being progressively watered down. Funding for many projects was improved but still was inadequate compared with the size of the job to be done.

In 1970 the President asked Congress to allocate $4 billion for water treatment facilities to be spent over four years; in 1971 he asked for $2 billion for each of the next three years, a more realistic figure if still short of the amount needed. The 1971 Presidential budget called for appropriations totaling $2.5 billion for programs gathered into the new Environmental Protection Agency (water, air, solid waste, pesticides, and radiation), almost double the 1971 congressional appropriation of $1.3 billion for the various programs. The budget also provided for enlarging EPA's staff from 6,223 to 8,863—an increase of 2,640—but still short of the army necessary to do the big enforcement job ahead.

The Nixon Administration has, to its credit, established a modus operandi to deal with environmental matters: implementation of the National Environmental Policy Act and the establishment of the Council on Environmental Quality and the Environmental Protection Agency. As against his support of the SST and giving too willing an ear to the voice of another Nixon creation—the National Industrial Pollution Control Council (made up of big business, many of them the biggest polluters)—there have, admittedly, been pluses. Work was halted on the Everglades jetport and the cross-Florida barge canal (after $50 million in federal funds had already been spent on the latter project). The action is being challenged in the courts by the Florida Canal Authority. The Alaska Oil Pipeline is also still indefinitely

postponed, although there is no doubt about the ultimate outcome. Whether the President was racking up environmental
points for the 1972 Presidential election or was expressing a
genuine concern was not as important as the result.

There was tough talk from William D. Ruckelshaus, head of
the Environmental Protection Committee, a Nixon appointee, in
a speech to the National Industrial Pollution Control Council. He
told industry not to wait for "ultimate" solutions to pollution but
to invest in the technology at hand. Ruckelshaus warned foot-
draggers that he would be fair, but "I will make accusations when
in my opinion they are warranted and I will seek court action
when there is no other reasonable recourse." Moments later the
dismayed executives were reassured by the President that nobody was going to "beat industry over the head" or make it a
"scapegoat" for pollution, and the council members breathed
easily again.

Ruckelshaus has promulgated some stiff air and water standards for industry to meet. But when industry said it could not
meet the deadline to report fifty-one types of industrial waste
discharge, among sixty-five pollution measurements required
under the law, it was given an indefinite delay. The delayed reports included radioactivity, sulfur compounds, chlorides, cyanide, arsenic, lead and other heavy metals, mercury, oil and
grease, phenols, DDT and other chlorinated hydrocarbons, plus
other pesticides and bacteria. Until the deadline is met these
compounds will continue to flow into the waterways in unknown
amounts.

In his first major test over pesticides, Ruckelshaus disappointed some of his early supporters. He had a chance to ban
DDT by declaring it an imminent hazard that required immediate suspension. Instead he ordered cancellation of DDT's registration, along with four other pesticides. This merely set in motion the long appeals process that enables DDT to continue its
charmed life.

Ruckelshaus disappointed again in September, 1971, when he
replaced three career men temporarily in charge of three regions
with political appointees, an action apparently dictated by the
White House. Regional administrators are crucial to the success
of federal environmental programs. They issue permits under
the 1899 Refuse Act, the primary tool for reducing water pollution; they also deal directly with the district's U.S. attorney in
prosecuting violators of air and water laws. Two of the appoint-

ments were made in EPA's Atlanta office. *Environmental Action* stated: "Strom Thurmond [Senator from South Carolina], the man responsible for bringing Spiro Agnew to the White House, is still collecting debts from President Nixon." The men replaced were identified as John Thoman, a pioneer in federal water pollution control programs, and Paul Traina, a sanitary engineer who worked with water pollution problems for sixteen years. The two replacements were said to have no environmental experience; one had been an administrative assistant to Thurmond. A *New York Times* editorial described one of the appointments as "downright flagrant." The Atlanta *Constitution,* the city's leading paper, showed less restraint. It blasted the appointments:

> President Richard M. Nixon bears the responsibility for as irresponsible and crass a political decision as any in living memory.
> The Nixon Administration's action this week pegs the EPA as perhaps the most political (in the old school, regular garden variety, county-unit type political hack sense) of all federal agencies. The next step, probably, will be to put the agency under the direct supervision of the Republican National Committee.

Elsewhere appear scattered bright spots. In an action believed to be the first of its kind by any local government in the country, Suffolk County, Long Island, approved a ban on the sale of virtually all detergents in the county, and other communities have followed its lead. In New York City, Jamaica Bay still has not been turned into a sanitary landfill or a jet runway, as often threatened. The city passed a tough air-pollution bill, and the Court of Appeals upheld the city's law controlling the emission of pollutants from apartment house incinerators and fuel burners. In New England seven major universities agreed to pool their training and research talents in the fight against air pollution and other environmental problems. In Gary, Indiana, some 1,500 students, suburban liberals and conservatives are creating a working-class pressure group to gain political, social and economic power at the grass-roots level; their aim is to bring pressure on the giant steel mills to end pollution, and on political organizations that have controlled Gary for years. Several national consumer organizations are forming to fight the government-industrial complex. If people could get together and force government to behave, this could be a tremendous force for reform.

More far-reaching, perhaps, is the fact that several states are considering tough pollution laws with harsh fines. One of the first to be hit was the Anaconda Wire and Cable Company, a subsidiary of the Anaconda Company. It was fined $200,000 in late 1971 by a federal judge for polluting the Hudson River; the fine is said to be the largest ever imposed under the 1899 Refuse Act.

Industry can also see ominous warnings in a stern new approach to pollution control by federal prosecutors who are trying to jail and fine the owners and executives of companies that discharge industrial waste into rivers and harbors. What is believed to be the first criminal conviction in the country was obtained in November, 1971, against the president of a woolen company in a federal district court in Boston. He faces up to five years in jail and $12,500 in fines. Shortly thereafter a federal grand jury in Baltimore returned a 100-count criminal indictment against the president of a company charged with pumping solvents and petroleum residues into a tributary of the Chesapeake Bay for nearly two years. Charges are pending against the president of a ceramics concern in Massachusetts and the manager of a steel plant in Chicago.

These actions constitute a significant departure in government pollution control efforts, which until now have relied on civil actions. *The New York Times* observed:

> It is understood that the new strategy has not come from the Justice Department in Washington, but has been developed separately in various states by the local federal prosecutors, who exercise considerable autonomy. It has not been discouraged by Washington, but in some districts criminal action has been considered and rejected for various reasons.

A few enlightened industries are making an effort to control pollution. Even the most progressive, however, are still so cost-oriented that they are spending less than one percent of gross profits to clean up their mess. They are finding that the solution to pollution is not dilution. Rather, it is not to pollute in the first place. In various ways, they are trying to "close the loop"—to reuse materials and not let them escape into the environment. This is sometimes called "source control." The pollutant is attacked as close to its source as possible. A process may be eliminated altogether, or changed so that the pollutant is reduced or less toxic; or the waste may be treated in concentrated form with an effort to extract the pollutant and, if possible, to re-use it. With imagination and sometimes without excessive expenditures, companies have found that they can not only eliminate or reduce pollution but can even make control devices pay for themselves or sometimes even turn a profit. Many are finding that conservation can be good business—an appeal that seems to strike more fire in the corporate heart than the most impassioned plea or sermon.

From abroad comes encouragement. Environmental awareness bolstered by action is reversing pollution in some places. In Great Britain the amount of smoke poured into the country's atmosphere has been drastically cut from 2,360,000 tons in 1951 to 930,000 tons in 1968. In small ways the English seem to have concluded that some things are more important than money in deciding what makes life worth living. The government, for example, rejected a plea to increase the maximum length of trucks from 32 to 44 feet, which would have added to their efficiency. But the public made clear its feelings that the cost in noise and fumes and intrusion on the English countryside was too high, and the Conservative government respected their wishes.

Public opinion also made itself felt when it was disclosed that a Dutch chemical company planned to dump 600 tons of chemical wastes into the North Atlantic. The compounds, used in making plastics, could build up in the ocean's food chain. Six governments protested vigorously. The Irish sent out minesweepers to intercept the vessel. The company relented and the waste material will now be stored in a sealed tank for two years while the company builds an incinerator to dispose of the poisons.

Here is a lesson for Americans. It is possible for resolute human beings to fight back. Given the necessary help, or just left to herself, nature will try to heal her wounds. The rest is up to us. Will it be done? Many of the answers must come from Congress. But there are few hopeful signs that politicians have ceased to play the old power game of protecting private interests at the expense of public welfare. Government continues to spend absurdly small amounts on research and pollution control. Our priorities remain distorted. Air pollution cost United States citizens $13.5 billion in 1969, according to the National Wildlife Federation. Projected expenditures for air-pollution control by government, industry, and 100 metropolitan areas through 1975 are only about $7 billion. In other words, air pollution will cost U.S. citizens some $67.5 billion over the next five years. But only about one ninth of that amount will be spent to remedy the situation.

When it comes to making a personal commitment to do what they can to help the environment, few people indeed are willing to make minor sacrifices in their daily lives. A *Times* survey revealed that the

> . . . American consumer is concerned about pollution—but apparently not when it comes into sharp conflict with price, convenience and quality. It is recognized that dyes in colored tissue are an added pollutant to the nation's waterways, but the Boston lady told her grocer that she had spent $1,000 to make her bathroom blue "and I'm not going to put white tissue in there."

A Los Angeles service station owner reported that since unleaded gas costs more and has a lower octane rating and thus lower performance, there is no demand for it. The Coca-Cola Company, one of the largest users of nonreturnable containers, concedes that many people say they would prefer returnable bottles. But in actual practice the customer is not translating that preference into buying habits, according to a company official. A supermarket manager noted that returnable bottles are a headache. "People do not want to be bothered . . ."

On the public level there is little cause for optimism. People, generally, have not rallied to the environmental cause. Many appear disinterested. Others are still uninformed about the severity of the threat, or they choose to ignore it. Many are happy to dismiss it as another false crisis exploited by conservation groups for their own benefit—a happy rationale for those who want an excuse to disbelieve. Others warn of an "environmental backlash"—industry, threatened with controls, moving to less restrictive areas, cutting jobs, reducing wages, going out of business. "Environmental backlash is a myth propagated by lobbyists for the polluters," according to Representative Reuss of Wisconsin.

Unfortunately, good politics is usually bad ecology. One authority claims that the environmental problem is 90 percent political and only 10 percent ecological. This may or may not be true. The essential fact is the result, and this reflects a greater problem. The degradation of the environment, it seems to me, is basically a moral problem. Everything else stems from that central issue.

It begins with our view of ourselves as individuals, as a society, as a civilization. In the past we had an excuse for what we were doing because of our ignorance. Now we have knowledge never before available to people. We have a growing realization that the world is not made up of many isolated parts but is one single unit, an intricate, interlocked, interrelated global ecosystem. Previous generations had no way of grasping this enormous truth.

We talk of progress but our attitudes and ideas still belong to the past, a primitive world of artificial barriers among people and an economic system that is out of harmony with any effort to preserve the environment, or to develop it wisely. Inherent in free enterprise is the right of the individual to commit acts against the public good. We have taken only a few feeble steps away from the laissez-faire abuses of the nineteenth century. Public relations has taken greater strides than public protection.

Private rights, as we continue to interpret them, enable an individual to fill in the marsh, build up his shore property, sell woodlands to developers at personal whim, build high-rise apartments in overcrowded areas, put up a factory that may be a public nuisance, kill whales and other sea life, and wipe out land creatures or gouge the earth and leave open wounds. The few restraints placed upon individual rights have not been sufficient to meet the problem. We have been so concerned with individual rights or private property that we often overlook the fact that private rights often lead to public wrongs.

The almost unrestricted rights of the individual were possible in a frontier nation where power to affect the environment adversely was limited. Then there was space to flee from danger and margin for error. This is no longer true in a complex, industrialized society, where continents are only hours apart.

People today have awesome power at their fingertips and there is no real control over them. One deranged individual can blow up an airliner in flight, or a building housing thousands. He can poison an entire city's water supply. A well-intentioned gardner in Kansas City, spraying his roses, can poison the Eskimos in Alaska. A single individual, in the exercise of his rights and using the power available to him, can inconvenience or damage an entire community and diminish the world for all of its people.

We have an option to improve the quality of life and restore our surroundings, but we refuse to do so because of our values. Pollution could be greatly reduced almost overnight by applying the technology available. We could "buy time" to conduct the vital research that must be done. But we continue to place greater importance on money than on our own welfare.

When a society determines to preserve its habitat only on economic grounds it is an immoral society. But that is the measure by which we are now determining our fate. The businessman will recycle only if he can be shown that it is profitable; he is not concerned with dwindling resources, except as they influence his profits. The lawmaker says the only way to get industry to clean up is by imposing taxes on effluents or offering monetary incentives for controls. Industry is able to challenge most pollution laws in court on economic grounds. Most pollution restrictions carry the provision that they may be overruled if it is not "economically feasible" to enforce them.

Maine has a law protecting its coastal wetlands. Yet the state's highest court ruled in a recent case that land, unfilled, was of no commercial value to the owners, and that this consideration outweighed the "compensation" the owners would have by sharing in the benefits of the state's regulation of wetlands. In other words, individual profit and not public welfare was the measure of whether or not to preserve the marsh.

Those who shape and administer the laws are beholden to that monetary goal. Society's basic protection, in pursuit of profit, is twofold: First, the corporate conscience. This is almost wholly a public relations myth; if it were not so, pollution would long ago have been largely corrected. The headline tells the story: "Round-Up of Bottles Improves Only the Manufacturers' Image." The bottle manufacturers, fearing restrictive laws and bad publicity, mounted a nationwide promotional campaign to get people to turn in their old bottles rather than throwing them away. But 92 glass collection stations in 25 states received no more than 2 to 3 percent of the 36 billion glass containers produced in the country last year. Even at their best, a spokesman told *The New York Times*, "the collection centers will never get above 5 percent." At best, the collection drive would barely stay even with the increase in production. (In one case, brought to my attention, a company established a bottle recycling station and then had the collected bottles hauled to a local dump.)

Second, we must look to public agencies for any protection we have. Repeatedly we have seen that such protection is meager indeed.

We have not been serious in the effort to correct pollution. The public remains almost wholly at the mercy of the polluters. We have seen that such laws as exist are weak to begin with, not enforced, or interpreted in such a way that they can be ignored or violated at will. Most penalties are wrist slaps. Most protection is illusory. Present laws are more protective of industry than the public—they prevent "imminent hazard"—the embarrassment of corpses in the streets with the incriminating fingerprints of the polluting industries on them.

We still await the kind of environmental research that put a man on the moon. If the research, so widely proclaimed by industry, is being done, where are the results? Where is the smog-free automobile? Where are the devices that trap sulfur, nitrogen oxides, and the various scarce resources going out of smokestacks in gases? Why no real improvement in sewage disposal methods for forty years? By now the answer is obvious: it is cheaper to pollute than to correct, it is more profitable to waste than conserve. Pollution has now been woven into the economy. It is one of our economic pillars, like crime, poverty, sickness and war. Antipollution technology is fast becoming more of a parasite feasting on the problem rather than a real effort to eliminate it. Control begins with regulating growth, not looking desperately for another means of handling its wastes.

We are at the crossroads, says Dr. Szent-Gyorgyi. Man can do anything he wants. "You have only to wish it and you can have a world without hunger, disease, cancer and toil—anything you can wish, wish anything and it can be done. Or else we can exterminate ourselves." So now the question is, which course will man take? "Toward a bright future or toward exterminating himself? At present we are on the road to extermination."

We have refused to see ourselves realistically. We remain blind to the evidence around us. A large part of the world sees us as monsters devouring the earth's limited resources and brutally slaughtering defenseless civilians to promote our own self-interests under the holy banner of "Democracy and Human Decency."

Democracy can work only as long as there is an informed public. At this point democracy in this country has ceased to function as it was intended. In place of a democracy of the people we have an oligarchy of industry, as Mrs. Henderson points out. In place of an informed public we are a manipulated mass. People cannot make honest and intelligent choices as long as they are deceived and deprived of essential truths. If the people prefer to go to their doom in a 400 cubic engine car spewing pollutants, and have knowingly made that choice, that is fair enough. If they have been lied to or not been informed about the risks they are being forced to take and the price that they and their children must pay, that is wrong. We cannot escape many of the pollutants that plague us, but we should have the right to know what those identifiable are and what they might be doing to us and our descendants. We should have the enlightened option of saying, "I'll take this risk as the price of that benefit." We deserve to make the choice and not have it rammed down our throat or slipped to us unknowingly.

There are several things that can and should be done. Many have been pointed out. To say we must develop a new ethic is nonsense. People do not develop a new ethic because they are told they must. They do what they have to do, no more and no less. We must have hard rules about what is necessary and they must be enforced. If people are to have an ethic or morality it comes through example and imitation, not being told one thing while the opposite is practiced.

No individual should have the right to diminish the world of another individual by a selfish or destructive act. Any threatened person should be able to halt another from such an act by going to a special environmental agency that could issue a restraining order to prevent the act, pending adjudication. Such agencies should exist at every level of government—local, county, state, and federal. They should be staffed by people with knowledge of ecology. They should have power to prevent irreversible acts from being performed until they have been properly considered. These agencies would be under a process of review that would progress from local level to county, state, and finally federal. The federal would, in effect, be an environmental Supreme Court.

Gradually a new body of environmental law would evolve. Its force would be preventive rather than punitive. It would, hopefully, lead to a new ethic. The personal ethics of the people would tend to follow the values of their society. It would require public leadership that we are not getting at this time. Ultimately, it is hoped, life would come to have value because of its intrinsic nature—simply because it represented life, and not because of a dollar figure set in the marketplace.

In time, if we could change our direction and goals, we might come to a new perspective about the land. We might come to think of ourselves, as the Indians did, as its stewards rather than exploiters. We might even come to think of our own bodies differently, not as sewers or experimental vats, but as stewards of a genetic heritage in the chain of life. This is another of the big ideas we have yet to absorb. People of childbearing age do not degrade their bodies for their own immediate gratification at their own risk. They are also asking unborn generations to take the risk with them and help pay the price. The person who smokes and absorbs cumulative poisons and needlessly exposes himself to chemicals that may cause birth defects and mutations is imperiling the children he may give birth to. This is a large responsibility. It is one we have not even started to face up to, but it is essential that we do.

There should immediately be formed an environmental fact-finding commission similar to the Kefauver Commission that exposed organized crime. This commission would hold hearings all over the nation and call expert witnesses. It would determine exactly the extent of the threat and where the greatest menace lies. It would establish what has to be done and what urgency exists. The results could be dramatic and powerful. It would enable people to hear at first hand the concerns of knowledgeable biologists who are most sensitive to the present hazard. If we took such a lead, other nations might follow.

It is too late to point the finger of blame, to wring our hands in hopelessness or cry out in despair. It is useless to waste time speculating how imminent the end may be, whether we have a decade, twenty years, thirty or whatever. It is pointless to predict whether the end, if it comes, will be triggered by the abrupt failure of air, water, or soil to function, singly or in harmony, or the flooding or gradual gasping of a dying planet. Or, as Rene Dubos suggests, life sinking to a meaningless level. There would be alive only the prudent and durable herdsman, barely afloat in a sea of poisons. In such a world only the meanest and most primitive of creatures could survive—the cockroach, the sparrow, the pigeon, unfeeling, uncaring man. The greatest tragedy would be that it was not necessary. We could have done otherwise. But we did not care enough. Most civilizations have passed because they failed to think. Our disaster would be that we had forgotten how to feel.

Common
Environmental
Poisons

MERCURY

Mercury is now recognized as an international pollutant and possibly the most dangerous of all. It is a cumulative poison. In small amounts it can cause kidney and nerve damage, blindness, loss of speech, diarrhea, deafness, and emotional problems. It can break chromosomes and deform. Ten percent of the amount ingested goes to the brain, where it kills brain cells. It also concentrates in the fetus. Like DDT, mercury poisoning can masquerade as other diseases; its symptoms are often misdiagnosed as encephalitis, senility, or mysterious brain damage.

Mercury exists in tiny amounts, naturally, in seawater. But levels now being found are unquestionably due to man-made contamination during the past century. Mercury enters the environment primarily through industrial wastes. The waters of the world now appear to be contaminated or on their way to being contaminated. This could have disastrous and possibly lethal long-term implications. Like DDT, mercury has been found to reduce photosynthesis. It probably enters the food chain through the usual route. Once more the complexity of nature was underestimated. Instead of the liquid metal sinking harmlessly to the bottom of the sea, as was once thought, it is believed that it is transformed by bacteria from metallic mercury to the highly poisonous methyl mercury, absorbed by plankton and then consumed by fish; the mercury concentration in some fish can be 3,000 times as great as in the water they inhabit. The fish, in turn, are eaten by birds, larger marine life and, finally, by man, who gets the full concentrated dose. Mercury has been found in birds and fish, often in incredibly high amounts, above even allegedly "safe levels" set by government and often in places far removed from the point of discharge. Eagles in Wisconsin contained up to 130 parts per million of mercury in their kidneys. Seals that live in the Pacific Ocean hundreds of miles from shore had accumulated "astonishingly" large quantities. Concentrations up to 116 times the amount considered "safe" for humans were found in Alaskan fur seals. High amounts of mercury turned up in coho salmon from the Great Lakes, where there is only man-made mercury pollution.

More than a million cans of tuna were recalled from store shelves because of excessive contamination. Some had almost twice the permitted level of a half part per million (0.5 ppm). People who ate large amounts of tuna for dietary reasons were found to have as much as five times the mercury in their bodies as others. Soon after the tuna recalls, FDA warned the public to stop eating swordfish. Nearly every brand of frozen swordfish was recalled from the market after 89 percent of all samples tested were found to be highly contaminated (Americans were eating some 26 million pounds of swordfish a year); the samples averaged almost double the permitted level; some ran to almost five times the 0.5 tolerance. The recalls and warnings were accompanied by the usual pap about there being no health hazard. FDA did not answer the obvious question: why bother to ban the sale of fish if no risk is involved? The truth is that FDA has no idea what risks are involved. Neither does it know where disaster will strike next or how bad it will be. Other types of fish and food products have not yet been tested.

The tests performed are hardly reassuring. Products made from seal livers (iron supplement pills sold in health food shops) were contaminated with sixty times the permitted level of mercury. Natives of the Pribilof Islands of Alaska, who eat the seal meat, had up to ten times as much mercury in their bodies as legal tolerance levels permit. Several physicians who practice in Alaska said there was growing suspicion that mercury in fur seals, sea lions, and fish eaten by Alaskans "might be contributing to some diseases in the natives."

Only the most limited tests have been made to learn the effects of mercury on humans at less than lethal levels over various periods of exposure. It is known that mercury can kill at 100 ppm in food, and chronic mercury poisoning can cause the physical and emotional disabilities previously mentioned. The present permitted level of 0.5 ppm, set by the FDA for mercury residues on foods, is ten times higher than the 0.05 limit considered safe by the World Health Organization. The present FDA standard is "probably very inadequate," according to Dr. Barry Commoner. He suggests that the FDA has more optimism than data, and "appears to be minimizing the hazard" of mercury pollution, as reported in *The New York Times*. He called for a national "total survey" of all foods in the American diet, to test for the presence of mercury and whether it is present in dangerous amounts: "In this connection, he said that out of about 50 million pounds of mercury consumed and processed by industry between 1945 and 1958, about 24 million pounds was still 'missing' and thus unaccounted for."

Dr. Commoner said that mercury pollution might "emerge as the most serious acute problem among the dismal array of environmental problems which constitute the environmental crisis."

For many years American industries dumped more than a half-million pounds of mercury annually into the Great Lakes. Canadian industries discharged half as much. More was dumped into coastal waters in the United States. Other industrial nations have been doing the same thing. Now we are collecting the bill. In Japan more than 100 people died in agony from eating fish contaminated by industrial mercury wastes. The most serious outbreak in the United States befell a New Mexico family that ate pork from pigs fed mercury-treated seeds in 1969. Three of the children went into coma. One, after eighteen months of hospitalization, is almost recovered. A second remains blind and partly paralyzed, confined to a wheelchair and he speaks with difficulty. The third, after more than a year in coma, emerged blind and can do little except roll over and move her arms a few inches. The baby that the mother was carrying when she ate the tainted meat seemed normal at birth but later turned out to be blind and retarded and at fourteen months had the motor development of a month-old baby. Curiously, the parents and four of their other children that ate the tainted meat remained unaffected as far as anyone can tell. Once again this underlines the different responses of individuals to poisons—how treacherous it is to set the same tolerances for all.

In many places in the United States warnings have now been issued against eating game birds, and in some places fishermen are warned not to eat their catch. Lake George, high in New

York's Adirondacks, was found to have fish (lake trout, rainbow trout, and smallmouth bass) with mercury concentrations well above the FDA tolerance. This is the kind of world we are creating.

As long ago as 1966 Sweden banned the use of phenyl mercury in the manufacture of paper. This is a less toxic form than the methyl mercury now being found in fish and birds in the United States. The American government has taken only mild steps to halt the dumping of mercury wastes. Fifty plants were found guilty of the practice, but injunctions were filed against only ten. No suits were filed against the others because they reportedly had reduced their discharges from a total of 187 pounds a day to 40. Now the companies are required to get permits from the Army Corps of Engineers to continue dumping. Incredible as it may seem, said Representative Henry S. Reuss, FWQA has not even referred all fifty cases to the Corps of Engineers so that the corps could require the discharger to apply for a permit under its new so-called full-enforcement policy. The registration of several companies using mercury compounds has been suspended by the USDA, but industry has challenged this ban in the courts; meanwhile the sale, use, and discharge of mercury continue.

How the mercury gets into the environment has been something of a mystery. For a long time it was believed to come primarily through water discharges. Now it is also thought to be released to the environment through air pollution. It is in coal and certain industrial products, such as paper. When they are burned the mercury is released into the atmosphere. The burning of coal around the world may be adding almost as much mercury to the environment as the waste from all industrial processes combined, according to Oiva I. Joensuu, a geochemist at the University of Miami's Institute of Marine and Atmospheric Sciences. Concentrations of mercury have been found in the fumes from large smokestacks. Dr. Joensuu estimates that at least 3,000 tons of mercury are being released into the atmosphere each year from the smokestacks of coal-burning furnaces used in electric power generators, ore refineries, heating and other enterprises. Once circulating in the atmosphere, the mercury vapor might be able to rain back to earth, reaching streams and lakes that might otherwise have escaped mercury contamination.

Millions of pounds of the metal already in the water and atmosphere will remain indefinitely. Mercury does not disappear.

It is now part of our environmental legacy to the young, and no one knows what the long-term effects of ingesting small amounts will be. After the poison was found in seals that live 50 to 100 miles off the California coast, an industry official said, "You can just figure from this that there isn't any place in the whole earth that isn't contaminated."

The threat of long-term contamination was underscored by Dr. John M. Wood, associate professor of biochemistry at the University of Illinois. He told a Senate subcommittee that 200,-000 pounds of inorganic mercury was deposited in the St. Clair River system (between Michigan and Ontario, Canada) in the past twenty years. But the problems of mercury in fish that have already been discovered might only be the tip of the iceberg, he said. In the St. Clair Basin, only 1 percent of the mercury has been picked up by the food chain. While hazardous levels of mercury are already being found in fish in the area, the vast majority of the metal has yet to pass through the food chain.

"There is now from 50 to 100 times as much mercury in fish from the St. Clair River as there was in 1935," he said. He estimates that it might take 5,000 years for natural processes to dispose of all the mercury presently in the St. Clair system.

The most remarkable part of the entire mercury scandal is that it was not uncovered by the official sources responsible for our protection. Rather it was discovered, by chance, when a Canadian graduate student performed a routine biopsy on a wall-eyed pike in March, 1970, and the preliminary studies were made by the Canadian government. Where were the officials who were supposed to be protecting the U.S. food supply? They were busy reassuring the public that all is well and there is no danger in permitting "small amounts" of poison to remain on every bite we eat. They were shaken by the mercury disclosures —the extent and intensity of the contamination. But the mercury scandal is only the beginning. How many other unpleasant revelations await their turn to "astonish" the advocates of poisons in our foods?

LEAD

One of the most feared and widespread of all contaminants is lead. Some authorities are convinced that lead poisoning was in part responsible for the collapse of the early Roman Empire, and it might be undermining us in the same way. Lead is highly toxic, but the effects of low-level contamination are subtle and not easily detected. But subtle effects are not necessarily small, warns Dr. Paul P. Craig of the Brookhaven National Laboratory. "Furthermore, the sum of low-level damage to many individuals may lead to total costs to society of substantial magnitude." He and other investigators believe there are two basic causes of ill health: genetic weakness and medical neglect in childhood. A potential toxic risk such as lead in the common environment may act either to cause such a genetic abnormality or to serve as a contributing factor to produce ill health. Thus, inborn weakness and the variable susceptibility of people to disease explain why all individuals in a similarly exposed population do not react identically. Individual differences are lost in averages. They make tolerances a cruel deception, virtually meaningless in protecting the individual from damage.

The medical effects of lead range from anemia to insomnia, blindness, and insanity. "Crazy as a painter" was a common phrase when many artists were really suffering from lead poisoning. Lead can cause sterility in men and women (one reason given for the collapse of the Roman nobility who used lead pots for cooking, lead pitchers for wines, and lead plates and cutlery; they also inbred intensively). It can cause miscarriage, stillbirth, premature labor (children born to such women reportedly are likely to die shortly after birth). It is an enzyme inhibitor, impairing cell metabolism and thereby disrupting the natural function of certain organs. When fed to test animals, lead in the amounts now being taken in by most human beings reduced lifespans by 20 percent, increased infant mortality rates, sterility, and birth abnormalities.

Americans ingest about 20,000 tons of lead a year. The average American is suffering "severe chronic lead insult, probably with adverse effects," according to Dr. Clair C. Patterson, a geochemist at the California Institute of Technology and an authority on lead. We do not level off our storage of lead, says Patterson. Instead we continue to accumulate more. In adults most of this storage is in the bones "where it acts as a sluggish reservoir in contact with the blood, accumulating or draining." In children it is distributed throughout the body. The government estimates that 400,000 children a year living in dilapidated housing contract lead poisoning from eating paint chips. Only 16,000 are treated. Some 200 die and 4,000 suffer brain damage. Many of the young victims who escape death are mistakenly thought to be suffering from mental afflictions while the true cause goes unrecognized. Some children have been found to have mental problems from lead poisoning when the lead in their blood was in the so-called safe, normal range.

In 1970 Congress authorized $30 million over two years to combat lead poisoning, the government supposedly picking up 75 percent of the cost of all screening of high-risk children. The President signed the bill into law. "Fifty-two cities, including New York, applied for a total of $45 million under the provisions of the bill," reports *The New York Times*. "They waited, but the Administration did not ask for funds and none were appropriated."

Many groups and politicians began questioning the sincerity of the Administration's new health reform proposal when no appropriation was requested for lead poisoning prevention. The Administration responded by asking for $2 million in 1971. Congress raised the appropriation to $7.5 million, barely enough to touch the problem.

Most of the lead in the air comes from automobiles, approximately 310,000 tons a year. It also appears in pesticides and other products. We take it into our bodies through our lungs, from food and water. In the atmosphere it joins all the other pollutants. Concentrations in the atmosphere of remote locations in the Pacific and Arctic are "hundreds of times above natural levels," researchers report. In urban areas the excess of industrial lead in the atmosphere is said to be about 1,000 times above natural levels. At the beginning of the Industrial Revolution two centuries ago the lead concentration in Greenland snow was 25 times above "natural levels." Today it is 500 times above natural levels.

Air samples taken seven years apart in three big cities showed sharp increases in lead concentrations, according to a 1971 report released by the Environmental Protection Agency. Samples were taken in Los Angeles, Philadelphia, and Cincinnati in 1961–62 and again in 1968–69. The increases ranged from 33 to 65 percent in Los Angeles, 2 to 36 percent in Philadelphia, and 13 to 33 percent in Cincinnati. Most of the lead was said to come from motor vehicle exhausts. The Los Angeles study was withheld for more than a year. The delay was attributed by the government to the need to "study and analyze" the findings. But a conservationist group that obtained a stolen copy of the study before publication, and demanded the release, said it was suppressed because publication was against interests of the oil and lead industries and those industries had "undue influence over the federal people."

Lead air contamination was given a new dimension soon after the study was released when it was learned that lead from paint, and possibly from the atmosphere, apparently caused serious illness among some animals at the Staten Island Zoo in New York. The researchers said that animals living in outside areas had higher lead levels than those living inside, suggesting that lead fallout from the atmosphere might explain the difference. One leopard became paralyzed and died; on autopsy there were found abnormally high levels of lead and zinc in its vital organs, despite normal amounts in the zoo food and water. Other animals became seriously ill. Several species had high lead levels. The researchers also found high lead levels inside the zoo. Outside they found similarly high levels of lead in soil and leaves. Analysis of samples of paint offered for sale in New York City for interior use revealed that about 10 percent contained more lead than is allowed by law. Amounts ranged from 2.6 percent to 10.8 percent. The legal limit under the city code is 1 percent.

Particulate lead in the atmosphere follows the universal route of all air contaminants. An estimated 430,000 tons of industrial lead annually finds it way to the sea, most of it from automobiles. It is washed off streets by rains and in sweepings and washed into the oceans by snow and rain. In the sea it enters the food chain. It enters and re-enters the atmosphere as wind-stirred dust and through the burning of lead-contaminated materials. It combines and interacts with other airborne and water-borne chemicals; it returns to the ocean and its vulnerable estuaries and coastal regions to be dispersed again and again. No one knows the ultimate effect. The "young layers" of the ocean now contain five times more lead than is natural. Lead pollution of the sea is highest near coastal regions where industrial activity is most intense. But it appears to be polluting waters 200 miles or more offshore and to depths of 30,000 feet, according to Dr. T. J. Chow of the Scripps Institution of Oceanography. This is 18 times greater than in the middle of the Mediterranean and 50 times greater than in the Atlantic 15 miles upwind from Bermuda, Dr. Chow says. High percentages of lead appear in rainfall, especially in California where nearly 10 million gallons of gasoline are consumed annually; this is 70 percent greater than in New York State. In drizzles the lead content is much higher because there are more small water particles for the lead to adhere to. In downpours little lead is washed down.

Recently lead has been found in the livers of fish. It is not yet known if lead is concentrating in the fleshy part of the fish. So lead joins mercury as a contaminant of marine life. Dr. Chow said that fish caught near the California coast "near smog-plagued Los Angeles showed an average content of 22 parts of lead for each million parts of liver tissue—two to three times the normal amount."

Why has the hazard of lead in the environment been permitted to exist when it has been recognized among experts for many years? Because the risk has been hidden from the public, according to Dr. Patterson. He accuses public health officials, notably the U.S. Public Health Service, of working with industry-hired investigators to hide the risk involved. Their studies were designed to prove that no hazard existed when their own figures contradicted this, he charges:

> Public health agencies show a dismal lack of leadership, vision, and resolve in defending the public's health, as might be expected, since they are staffed with compromisers, the most successful being those who have displeased the fewest in power. The U.S. Public Health Service did not rush to consider the hazards of lead alkyls, as one might infer from the propaganda commonly available. Lead tetraethyl was manufactured and marketed without a twitch from the agency until poisoning scandals associated with its manufacture brought legislative pressure to bear on the service, and it was forced to act. . . .
>
> Public health agencies are patently unable to conduct evaluative investigations that turn out to be truly defensive of health.*

Dr. Patterson points out that in 1939 the USPHS was involved in an extensive investigation of public health hazards originating from lead arsenate insecticide on apples. Ultimately it concluded that there was no harm to the public at a level of 7 ppm lead in apples (the level still permitted today). At the same time, he says, a private investigator was obtaining data on lead ingestion "showing that 1.5 ppm on food should give rise to classical lead poisoning in a matter of months. The conclusions of

* *Lead Poisoning*, from manuscript in preparation, by C. C. Patterson, Senior Research Fellow in Geochemistry, Division of Geological Sciences, California Institute of Technology, Pasadena, California.

the PHS were dictated by matters of economy then, and it is likely they always will be." He observes that the official view persists that federally approved lead levels are within the accepted range and do not pose a health threat to humans, a view that has prevailed for decades. This is based upon a threshold for damage concept applied to industrial workers, he says. It involves the axiom that a worker must be either perfectly healthy or classically intoxicated but there is no middle ground. "This is a seriously unfortunate situation . . ." He and other investigators insist that there is no sharp dividing line like this, and to try to make such a line where symptoms appear and where they allegedly do not exist obscures the real question: to determine what damage, however slight, occurs at lower levels and may be difficult or impossible to detect or to ascribe to the real cause.

Patterson believes that the basic danger of lead poisoning is underestimated. Like "other kinds of technological filth" it may do more than bring "agony into our existence and shorten our lives." It may also affect our nervous system and thought processes in the concentrations that now appear in the environment:

> The course of human events is determined by the activities of the mind. Intellectual irritability and disfunction are associated with classical lead poisoning, and it is possible, and in my opinion probable, that similar impairments on a lesser but still significant scale might occur in persons subjected to severe chronic lead insult. It has recently been maintained on the basis of experimental evidence from animals, that pathologic and histologic changes of the brain and spinal cord together with functional shifts in the higher nervous activity are induced by exposures to atmospheric lead concentrations corresponding to those exposures now experienced by dwellers in most large American cities.*

This is a bold and far-reaching thesis. Recent investigations have disclosed that brain damage can lead to violence; this has been confirmed with both human and animal experiments. In the past it was believed that crime and violence were psychological or social in origin. But recent evidence suggests that such behavior may originate from actual physical damage in the depth of the human brain. Researchers have triggered violent behavior merely by stimulating certain cells of the brain, and

* C. C. Patterson, "Contaminated and Natural Lead Environments of Man." *Archives of Environmental Health*, Vol. 11, Sept., 1965.

curbed it by deadening other areas. In studies of prisoners who committed crimes of violence chromosomal deficiencies were found. In some patients treatment with hormone-type drugs had an effect on aggressive behavior. This suggests the possible consequences that DDT, lead, and other poisons and drugs commonly pumped into the environment may be having not only on health but on human behavior.

The brain is so sensitive, so mysterious, that no one begins to understand it. Recently a drug was synthesized that makes rats fear the darkness, upsetting an instinctive response. Still we continue to expose this vital organ in humans to the most violent compounds without even a basic understanding of the possible consequences. The subject is so complex that it discourages investigation. Damage might not be direct but indirect: a substance could harm a distant organ of the body that deprives the brain of some minute nourishment required for normal function or emotional balance (just as DDT destroys the ability of birds to reproduce by damaging the mechanism that regulates the thickness of the eggshell). It could come from malnutrition or from a supposedly adequate diet lacking in unsuspected vital minerals due to soil deficiencies. In any controlled laboratory experiment the increased violence and irrational behavior in our society would be considered phenomenal. Investigators would be searching frantically to find the cause of so widespread and dramatic an effect. Instead we look for new drugs to control the symptoms of a society already in a process of collapse.

ARSENIC

Arsenic (from the Greek, meaning "male," because of its strength) is violently poisonous and cumulative. Most of that which is absorbed stays in the body, building up in volume. It is taken through the whole skin, the lungs, from food or drink. The long-term effects may not be detectable for two to six years. Short of death it can cause abdominal pain, thirst, nausea, vomiting, diarrhea, pallor, cold moist skin, weak pulse, shallow breath, and collapse. It can also cause cancer of the skin and liver.

Arsenic is found in some forty cosmetic products. Organic arsenic (less toxic than inorganic arsenic) is added to poultry feed to make the birds grow faster. FDA tolerances permit 1 ppm of arsenic in the liver and other organs of poultry after slaughter and 0.5 ppm in muscle tissue. The government found illegal residues of organic arsenic in one fourth to one sixth of the poultry samples taken in 1968 and 1969, according to the Associated Press. ". . . officials say they doubt the residues pose any serious human health hazards." "Doubt" is a disturbing word when used in connection with so toxic a substance in a food eaten regularly by millions.

The primary source of arsenic in the environment is from detergents. It exists in many brands, the amount differing widely with the product; this ranges from negligible to astronomical. Levels of 10 to 17 ppm of arsenic were found in several common American presoaks and detergents, according to a study by the State Geological Survey in Kansas, reported in *Environment* (Jan.-Feb. 1970). From the mountains of detergents used and poured down drains and sewers, arsenic has seeped into water supplies. Present treatment plants are not designed to remove such poisons. Arsenic has also been dumped into the ocean in industrial wastes. Surprisingly high levels of arsenic in the Baltic Sea led recently to the discovery that 7,000 tons of arsenic had been dumped almost forty years ago in concrete containers, reportedly enough to kill the population of the world three times over, according to *Marine Pollution—Potential for Catastrophe*, a United Nations publication.

Since 1958, when detergents became popular, the accumulation of arsenic in human hair has been increasing. No one knows the ultimate effect of arsenic in the body, but even tiny amounts are considered dangerous. The greatest concern is for water supplies that are reused repeatedly. Among these is the Kansas River basin. Kansas researchers reported what they considered alarming levels of arsenic in the Kansas River and the Lawrence, Kansas, sewage system, *Environment* reported. Dr. Samuel Epstein, a professor at Case-Western Reserve University, and one of the nation's foremost toxicologists, pointed out that the amount found in the Kansas River is only slightly below the recommended government danger level of 10 parts per billion. The government's "level of safety" represents another act of faith rather than precise scientific knowledge.

One of the hazards associated with arsenic in laundry soaps is the possibility of contamination through residues on clothing or from contact while using the soap product, cautions *Environment*. "Arsenic may be absorbed into the body through unbroken skin. The metal is cumulative, meaning that most of the absorbed arsenic stays in the body. Long-term arsenosis may not be detectable for from two to six years or more."

NTA

Another contaminant found in detergents is NTA (sodium-nitrilotriacetate acid), which was used as a replacement for phosphates. Then it was found to be even more hazardous, especially for pregnant women. The chemical picks up mercury, cadmium, lead, and other metals from water pipes; it could possibly circulate them through the human body if they got into drinking water, according to U.S. Surgeon General Jesse L. Steinfeld. Animal studies disclosed that NTA altered the toxicity of the metals and contributed to a tenfold increase in fetal abnormalities and fatalities in the animals. NTA contains arsenic. It also is suspected that bacterial action can turn the acid into cancer-causing chemicals, according to experiments conducted by Dr. Samuel Epstein. Up to 200 million pounds of NTA are being used annually, and it was predicted that it would rise to a billion or more pounds in the next few years. But we will be spared that increase, it seems. Under federal prodding, the nation's major detergent producers agreed in December, 1970, to remove NTA from their products, *The New York Times* reports. Howard Morgens, president of Procter & Gamble, the nation's largest detergent manufacturer, had this to say: "We believe there has been more research supporting the safety of NTA than there has been supporting the safety of most, if not all, of the materials going into the nation's food products." That is a sobering statement. Despite extensive testing the danger apparently was long overlooked or misinterpreted. Either way it does not say much for the testing procedures or for the safety of those hundreds of millions of pounds of food additives in common use. Nor does it get rid of those hundreds of millions of pounds of NTA already released into the environment.

There has been much confusion over detergents and which brands are the best to use. The basic ingredient that makes them effective in cleaning clothes and that consitutes an environmental problem is the same: the phosphates. Approximately 2.5 billion pounds of phosphates are used annually in detergents. Phosphates remove dirt by holding solid materials in suspension and emulsifying oils during the washing process. They are more effective in getting clothes cleaner than soap, especially in hard water. Some well-meaning authorities urge women to use water softeners or presoaks and switch to soap, but many water softeners and presoaks are primarily composed of phosphates and little is accomplished. Most manufacturers do not list what goes into their softeners and presoaks. Consumers are urged to write and demand to know what materials are used. The best water softeners are those permanently installed that remove minerals (that make water hard) through ionization rather than by chemical action. On a community level phosphates can be removed from waste water with advanced methods of sewage treatment, but this is expensive and few communities will spend the money necessary. Housewives are often urged to buy brands of detergents with the lowest phosphate content. This too can confuse, as different lists vary. Such variations can be due to experimental error in analysis: single brands vary 10 to 20 percent on different lists, according to one investigator. Variations may also be due to different testing methods or measurements of phosphate content. Weight can change according to the amount of water absorbed or other materials present; and percentage of phosphate can mislead because of the amount of detergent used per load. The most reliable measure is the amount in grams of phosphate used per wash load.

Several states have passed laws banning or restricting the sale of detergents. In the spring of 1971, Indiana became the first. Connecticut followed, and New York announced that it was planning similar action.

AGRICULTURAL POLLUTANTS

Almost every edible meat animal receives a variety of sprays, dips, vaccinations, food supplements, tranquilizers, antibiotics, and artificial hormones. Cows are dehorned, castrated, artificially bred, confined in a tiny space, mechanically fed, and slaughtered at a time dictated by a computer. All practices have one common purpose: to put on weight as cheaply as possible. One of the more infamous drugs given to meat animals is diethylstilbestrol, better known simply as stilbestrol, a chemical with an effect in the body similar to that of a female sex hormone. For a long time stilbestrol pellets were implanted into the necks of chickens; but they migrated to other parts of the animals' bodies and were eaten, causing profound effects. The practice was finally discontinued in chickens. But increasing amounts of the drug are now being fed to cattle and other animals. An admitted 80 percent—and undoubtedly more—of cattle get stilbestrol, an acknowledged carcinogen. Varying amounts of it may remain in meats that go to market. By the time its presence is discovered by testing, the meat has long since been eaten.

Many cancer experts have warned against the use of stilbestrol, including Dr. Wilhelm Hueper. The drug causes cancers among many species of animals and in many sites—breast, uterus, kidneys, bladder, testes, and blood-forming organs. There is strong suspicion that estrogens (hormones) can cause cancer in humans and accelerate some established cancers in women, especially breast cancer in younger women. Three Harvard physicians reported in May, 1971, a rare form of vaginal cancer in several young women whose mothers were treated with stilbestrol during pregnancy. The doctors, according to *The New York Times*, implicated the drug after studying the cases of eight young women, aged fifteen to twenty-two. Before 1966, the doctors said, they had not diagnosed this cancer in any young women. "The time of birth of these [cancer] patients—1946 to 1951—coincides with the beginning of the widespread use of estrogens in support of high-risk pregnancy." Stilbestrol illustrates our shizophrenic approach to poisons. It can be bought in a drugstore only with a prescription, but a farmer can buy it freely in any feedstore.

ANTIBIOTICS

Antibiotics are fed routinely to various animals on the factory farm—cattle, chickens, turkeys, and hogs. Allegedly these drugs prevent disease and thus promote cheap weight gains. Both propositions are challenged. Some 2.7 million pounds of the drugs are used annually in animal feed. In addition, antibiotics are sprayed on a variety of crops, on golf courses and ornamental shrubs to prevent bacterial and fungal diseases. Many researchers object to them because they tend to build up resistance in bacteria so the drugs are not effective in emergencies. Some of the resistant strains include highly pathogenic organisms that can cause a variety of diseases such as typhoid fever, gastroenteritis, dysentery, plague, and undulant fever.

The biological community these antilife drugs are directed against is bewildering in its complexity and virtually unknown.

Most of the bacteria are not even identified. Margaret E. Duffy writes about the mysterious realm—the gut—in *Environment*, how easily it can be damaged or disrupted to bring on disease and death. In the small world of the animal intestine, where nutrients not completely digested in the stomach are further broken down and absorbed into the bloodstream, bacteria, fungi, and protozoa live together in delicate balance.

> Their world reflects an order parallel to the order we observe in the world around us. Any alteration of this environment affects the organisms which live in it. The relationships among the various organisms which live in the intestine are not completely understood. It is a complex place where the same species of bacteria act differently in different parts of the intestine, act differently according to the number present and respond differently to chemicals introduced from the outside, depending upon their condition. Some bacteria aid in the process of digestion. Other bacteria which can cause disease inhabit the intestine but do not cause overt symptoms of disease unless something happens to upset the ecological balance. In addition to bacteria, there are fungi which live in the intestine and are kept in check if bacteria are allowed to grow freely.

The FDA, in February, 1972, called for strict limitations on the use of antibiotics in livestock feed, describing them as "a very real potential health hazard." This after almost a quarter-century of usage. The agency estimated that 80 percent of the meat products on the American dinner table come from animals fed medicated feeds during at least part of their lives; $85 million a year is spent on antibiotics for animal food in this country. Data on antibiotics in animal feeds are "very frankly grossly inadequate and therefore incomplete," said FDA Commissioner Dr. Charles C. Edwards. Another FDA spokesman stated that more than 1,000 combination drug products already faced removal from the medicated feed market because of inadequate safety and efficacy data.

ASBESTOS

A known cause of lung cancer in humans, as well as causing other respiratory ailments, asbestos is also suspected of causing cancer in other sites. Among asbestos workers the cancer rate is about seven times higher than among the general population. In one study one out of five asbestos workers died of cancer. The risk is believed to be greater among those who also smoke cigarets. Studies suggest that cigaret smoke and asbestos are synergestic—they increase the cancer-causing power of one another beyond their individual values.

Because of the increasing exposure of the public to asbestos contamination, especially through air pollution, this has sinister implications. Eventually asbestos will rival cigarets as a cause of lung cancer, some investigators believe. Asbestos fibers were found in half of 2,000 lungs autopsied in a recent study; as many as an estimated half-million fibers had been inhaled by some people. The fibers are especially dangerous as they are almost indestructible. But proof is not easy to come by, as the cancers do not develop until after twenty years. Present studies are based on exposures going back two decades when asbestos use was relatively light. In recent years its use has increased eightfold. The United States is the world's largest user of asbestos, importing one billion tons annually from Canada. Asbestos has been used since ancient times. Only recently was it found to be a cause of cancer.

Asbestos is in the air, especially in cities, often in high concentrations. A 1969 report by Litton Systems, Inc., estimates that 100,000 asbestos workers, 3.5 million workers in related industries, and approximately 50 million other Americans have been exposed to asbestos and could have its fibers in their lungs. When inhaled, asbestos dust forms deposits on the lungs that are never absorbed, blocking the ability of tissues to absorb oxygen. This can result in breathing difficulty—a condition called asbestosis; it is incurable and can be fatal.

Asbestos comes from the destruction of automobile brake linings and clutch facings, from street sweepings, from industrial use, and particularly from buildings being demolished or constructed. Industrial publications estimate that half the asbestos used in spraying to insulate new buildings is lost into the air. In industry it has an estimated 3,000 uses, appearing in everything from plastic frying-pan handles to manhole covers. It is in floor tiles, cement, roofing materials, paints, children's cosmetics, and jewelry kits; kindergartens use raw asbestos mixed with water as an inexpensive form of play dough. Surgeons have sprinkled it on wounds to promote healing. People are exposed to small amounts from torn ironing-board covers, potholders, do-it-yourself projects such as sawing up wallboard, mixing crack-filler compounds, insulating pipes, and repairing furnaces with asbestos cement. It is used to line the inside of air-conditioning ducts, and the dust is picked up by the moving air and borne throughout the building. Asbestos is even used in baby talcum powder. Jerome Kretchmer, Environmental Protection Administrator in New York City, recently disclosed that initial tests of two brands of talcum powder showed 5 to 25 percent pure asbestos content; other popular brands were also being tested for asbestos content.

Several cities have now banned the spraying of asbestos on new structures. Boston and Chicago were the first, in May of 1971. New York, Los Angeles, Detroit, and Philadelphia have banned it from city projects or restricted its use. Other areas are considering bans or regulations on spraying.

CADMIUM

A metal-derived poison, cadmium is common in Japan, but found less in the United States. However, recently (like mercury) it has been found to pose an environmental hazard. Scientists are concerned because the metal can accumulate in the body over a long period of time and displace materials vital to the life process.

Cadmium poisoning causes severe pain, particularly in the back, shoulder or knee joints. It can mimic rheumatism, neuralgia, or neuritis. As the disease progresses it weakens the bones, which become soft from lost minerals such as calcium; the slightest stress can produce painful bone fractures, according to a *New York Times* report. "A victim's height may be shortened. Often, he must waddle, or stay in bed. Kidney damage can result."

Cadmium disease causes the bones to ache severely. The Japanese call the ailment "Itai, Itai"—their equivalent of "Ouch! Ouch!" Itai, Itai killed about half of its more than 200 victims among 24,000 Japanese who lived along a stream of the Jintsu River in Toyama Prefecture. For unknown reasons the disease has a predilection for 50-to-60-year-old mothers who have borne many children. The poison came from a cadmium mining plant upstream that discharged the metal wastes into a river that deposited cadmium particles in rice paddies. Japanese people absorbed the cadmium from the contaminated food and water.

In the United States cadmium poisoning is a rarely diagnosed industrial hazard. Its victims are primarily welders, smelters, and other metallurgic workers exposed to cadmium in plating processes. When these workers inhale its fumes, they can suffer severe lung inflammation, producing chest pain, coughing, and difficult breathing. Inhalation cadmium poisoning kills one in five victims. These toxic fumes reportedly have been considered for chemical warfare use.

Americans have suffered acute episodes of nausea and vomiting immediately after drinking cadmium-contaminated citrus liquids such as lemonade. When placed in cadmium-plated containers these acid foods can dissolve the metal. The body usually defends itself by producing immediate vomiting. "Yet tiny amounts can be silently absorbed and accumulated over a long period of time."

Several states banned sales of cadmium-plated containers following outbreaks of cadmium poisoning. This type of food poisoning is now rare, but it may result from using war surplus cadmium-plated containers.

Investigators have studied environmental cadmium in the United States. An average concentration of cadmium in the air of twenty-eight American cities showed a marked correlation with death rates from high blood pressure and arteriosclerotic heart disease, according to Dr. Robert E. Carroll, in the *Journal of the American Medical Association.*

"Other doctors have reported that patients with high blood pressure had increased cadmium levels in their kidneys. Still others believe cadmium pollution is a major factor in producing high blood pressure which affects 23 million Americans."

Fear of universal cadmium poisoning developed after 1968 when the U.S. Public Health Service reported levels of the metal's concentration in shellfish. The USPHS recommended guidelines that would limit the total amount of four heavy metals (cadmium, lead, chromium, and mercury) to 2 parts for each million, according to *The New York Times:* "But at the same time the Public Health Service reported that the average level of cadmium alone in oysters was 3.1 parts for each million."

No guidelines were set in 1968. Instead more studies were recommended. The studies, completed in 1971, revealed toxic cadmium metal in concentrations. Concentrations "some federal health officials feel to be well above safe levels have been found in oysters taken from the Atlantic Coast from Maine to North

Carolina." An Environmental Protection Agency official said the studies also showed that cadmium contamination is apparently worsening. Cadmium levels in oysters, taken from 100 locations, ranged from 0.1 to 7.8 parts per million. So now we have proposed guidelines setting the permissible amount of cadmium in oysters, and already the average contamination exceeds the guidelines. And no one has any idea of how much—or how little—cadmium in a person is dangerous.

The situation became more ominous in June, 1971, when three separate species of fish caught near the outwash of a federally built battery factory on the Hudson River were found to contain up to 1,000 times as much cadmium as might normally be expected. The case was believed to be the first in which the heavy metal has rendered fish unfit for consumption in the United States, the *Times* reported.

PCBs

PCBs (polychlorinated biphenyl compounds) are related to the polychlorinated phenolic chemicals and are said to be the potential source of the extremely powerful birth-deforming agents called dioxins. PCBs have many of DDT's characteristics. They are incredibly persistent in the environment and exert similar physiological effects; one is to stimulate enzymes in the liver that break down sex hormones. PCBs cause a variety of health effects. They cause lesions in the liver, jaundice, nausea, vomiting, weight loss, edema, and abdominal pain. Following severe liver damage "the patient may pass into a coma and die." In Japan, PCB poisoning (from a pesticide) affected more than ten thousand people. There were miscarriages and stillbirths, widespread skin disorders, jaundice, birth defects, and other liver damage. The greater danger is the long-term low-level exposure to PCB and other poisons, singly or in combination. The authors of one study (Risebrough-Brodine) say "there is also the more indirect, but no less real danger of destroying other forms of life—part of the vast interconnected web of species of which man is but one part and on which he depends." *

PCB has been produced for forty years. Until 1966 it was not suspected that it was being released into the environment. By that time it had furtively pervaded the world. Now it is found in

* Dr. Robert Risebrough with Virginia Brodine, "More Letters in the Wind," *Environment*, Jan.–Feb., 1970.

the bodies of fish, birds, and men; it is in conifer needles, cow's milk, and mother's milk. Levels in wildlife near industrial centers may be very high, sometimes startlingly higher than those of DDT. PCBs have been found in the body fat of North American peregrine falcons in amounts as great as 1,980 ppm and in the fat of eagles in Sweden in the amount of 17,000 ppm. It has been found, along with DDT, in unhatched eggs of peregrine falcons. Recently it contaminated vast quantities of fish meal used as poultry feed when the chemical leaked from the heating machinery in a North Carolina plant and 77,000 broiler chickens had to be destroyed. The chickens had up to 40 ppm PCB in their fat. FDA sets a tolerance of 5 ppm for PCB in food. A short time before the North Carolina incident, 146,000 more chickens were destroyed in New York State when they were contaminated by PCB, presumably from paint that flaked off a storage bin or when absorbed by chicken feed. Millions of other chickens contaminated by high levels of PCB were probably eaten. After the chicken incident a Senate commerce subcommittee disclosed that fish in Alabama were contaminated with PCB in amounts as high as 360 ppm—72 times the FDA guideline.

Despite the optimistic FDA tolerance level, almost nothing is known about PCB. It is not known whether it enters the body through water, food, or air. As little is known about its breakdown products. Even the way in which it is released into the environment is a mystery. It is used in a wide variety of products, from closed systems, such as insulating fluids in transformers, to use in some forty pesticides and plastic containers. PCB is fire-resistant but at sufficiently high temperatures it will burn, releasing into the air highly toxic fumes. It is sold in 50-pound cans, 600-pound steel drums, and by the tank carload. All of this, it appears, comes back to us through the usual channels, remaining indefinitely and relentlessly increasing in total.

PCNB

A fungicide, PCNB (pentachloronitrobenzene), has produced tumors of the liver, lung, and lymphoid organs in animals. It has also caused birth defects such as missing kidneys, cleft palates, and single or missing eyes; liver enlargement was observed in the mothers. More than 5 million pounds of PCNB is produced annually. Some 3 million acres of cotton and a half-million acres of peanuts are treated with the compound in Texas and Arkansas alone. It is also used on potatoes, lettuce, crucifers, wheat, beans, tomatoes, peppers, and ornamentals.

CAPTAN, FOLPET, AND CARBARYL

These are all pesticides shown to be teratogenic in animal studies. They have caused such bizarre effects as skeletal malalignment, bones that do not fuse or form, cleft palate and gross facial malformations. Each of these pesticides is used commonly on foods and large residues are permitted, based on their toxicity alone. Their teratogenic effects are not considered.

BERYLLIUM

A common contaminant of air, water, and human tissue, beryllium can be mined as an ore or produced as an atomic energy by-product. It is a carcinogen; it can cause tissue changes if breathed into the lungs, inflammation, chronic ulcers (if it gets in open cuts), and dermatitis. Treachery is added to toxicity as it can have delayed effects following exposure. A small percentage of the population (about 2 percent) is extremely sensitive to it. Beryllium is used in a variety of industrial processes. Beginning in the mid-1920s it was widely used in the manufacture of fluorescent lamps, despite known dangers if the tubes broke and people were exposed to the poison. The practice was discontinued in the mid-1950s, but it's anybody's guess as to what happened to all of the beryllium phosphors released into the environment when old tubes were disposed of. As early as 1943 beryllium was recognized by Europeans as a cause of illness among workers, but at that time it was "described as harmless by a U.S. Public Health Service report," according to Frank Graham, Jr.

DIOXINS

Unknown until recently, dioxins had been virtually unstudied. They are highly toxic and the most potent teratogenic substances ever discovered. Researchers still have not found the lower limit where they do not cause birth defects. The substances can cause a variety of ailments: an acnelike skin eruption, fatigue, lassitude, depression, mental problems, disturbed sleep, irritability, reduced libido, impotence, weight loss; they can also affect the

internal organs and nervous system. Dioxins are suspected of accumulating in the body, in the manner of DDT, but are far more
toxic. They are found in water samples and on foods, including
nearly all vegetable oils.

Dioxins are found in a variety of products derived from the so-
called polychlorinated phenolic compounds. These range from
pesticides to deodorants, including the widely used antiseptic
hexachlorophene that is used in hundreds of products. In 1968
alone 27 million pounds of polychlorinated phenolic compounds
were used just to preserve wood. Dioxins contaminate almost
every chlorinated hydrocarbon pesticide, germicide, and fungicide, adding their own threat to the existing poison. Dioxins develop when almost any chlorinated hydrocarbons are heated.
Even the heat of the sun can form them in some compounds.
Little or nothing is known about how they break down or what
happens to them in the environment.

Dioxins appear in the notorious herbicide 2,4,5-T, which has
been widely used since developed in World War II as a defoliant.
For more than twenty years the government insisted it was safe.
It has been used on millions of acres of fields, forests, and food
crops. But it turned out that this assurance of safety was based
only on a lack of testing. In 1966, 2,4,5-T testing found it to be
teratogenic to mice, rats, and chickens. Some of the effects are
horrendous. Only 50 parts per trillion in fertilized eggs left
chickens walking on their knees and with cleft palates and beak
deformities. It took three years to pry out of the government that
2,4,5-T is teratogenic. It is also suspected of causing cancer and
genetic damage. Similar compounds, including the defoliant 2,4-
D, are also considered potentially dangerous. The government
recently cut down the approved use of 2,4,5-T on crops and
around lakes, ponds, and irrigation ditches, but this accounts for
only about 10 percent of its total use. Hundreds of millions of
pounds are still doused on rangeland where cattle and sheep
graze and millions of pounds of meat are produced. The government has been completely indifferent to the potential effects of
the herbicide or its dioxins getting into the meat or into water
supplies from runoff. This attitude is commented on by Thomas
Whiteside in *The New Yorker:*

> In Texas alone, more than a million acres of rangeland
> and pastureland are being sprayed with 2,4,5-T this year;
> probably at least a quarter of a million head of cattle will
> graze on that sprayed land; and the cattle will produce

something like a hundred and fifty million pounds of meat that will be sold to Americans as edible—all in the absence of a solitary meaningful restriction imposed by the federal government on either the spraying or the grazing and also in the absence of a solitary scientific study, either by industry or by any government agency concerning the stability, the persistence, and the cumulative capacity of the dioxin contaminant in the bodies of living creatures.

In the past nine years the U.S. Defense Department has dumped some 40 million pounds of 2,4,5-T on Vietnam and its people. There were many reports (often suppressed) of Vietnamese women giving birth to deformed babies. Only an aroused public opinion in Vietnam and in the United States finally forced the government to halt or cut back on its use. This is expected to make more 2,4,5-T available for use in the United States. Meanwhile, other defoliants are being used in Vietnam. They too are considered potentially dangerous (and remain untested). What happens to all of these poisons dumped on Vietnam? Do they become more or less toxic? What happens to the dioxins? They appear to be almost indestructible; they may remain long after the herbicide is gone. There are indications that when products containing them are burned they produce new dioxins. In one form or another the 2,4,5-T and its metabolites will enter the world environment. Will the woman in Iowa or Maine or France or India who gives birth to a deformed baby think of what the United States is doing in Vietnam, or will she feel guilty and wonder how this terrible thing happened to her? We have not conducted chemical warfare just against the Vietnamese but against ourselves, all of humanity, the yet unborn, and all of nature.

MILITARY NERVE GASES

In August, 1970, the latest consignment was dropped into the deep: 66 tons of the most toxic gases inspired by Hitler's chemists, encased in steel containers embedded in blocks of concrete; they were sunk in 16,000 feet of water off the coast of South Carolina. The whole disposal was a black comedy of tragic errors. The Army began by assuring us that the dumping was completely safe. Investigation revealed that it had not the slightest idea of what the gases might do or the risk the public might be exposed to. The Army even lost track of which containers held the most lethal gases. No one could predict the ultimate effect on marine or human life when the gases escape. The only certainty is that eventually they will escape. Then someone in Washington will probably order a study to find out why more species of birds and fish are disappearing, and why is the ocean behaving so peculiarly?

The fracas revealed other outrages that have been quietly perpetrated by the military over the years. Thousands of tons of poison gases, rockets, and other armaments have been quietly dumped at sea. In 1967 and 1968 more than 50,000 obsolete M-55 rockets were buried off the New Jersey coast in 7,200 feet of water. All over the world this same ocean-dumping has been going on for decades. After World War I the Allies sank 20,000 tons of German mustard gas into the Baltic Sea off Sweden. Fishermen recently have been burned by the gas while handling contaminated fish. Three Polish fishermen had to be hospitalized with burns after they accidentally netted one of the leaking containers.

The Army was recently set to dump 3,071 tons of obsolete mustard gas, stored since World War II, into the ocean. It was stopped by public outcry. Now the gas is being destroyed by being burned at high temperatures. The Army claims the waste gases contain none of the poison mustard gas but quantities of sulfur dioxide and hydrogen chloride. Still on hand are huge stockpiles of nerve gas that were also scheduled for ocean dumping but will now be disposed of in other ways.

Getting rid of unwanted germ-warfare residues marks another military misadventure. In 1968 President Nixon renounced this form of homicide (which we were pledged not to engage in at the same time we were busy manufacturing the agents—the mockery of making war humane). The unloved and unwanted germs could cause deadly diseases such as anthrax, encephalomyelitis, botulism and other food poisoning, as well as rabbit fever and other nonlethal diseases that can incapacitate without killing ("It's more humane," said an army officer). At first the plan was to dump the germs in the ground, until some level head decided that was not such a good idea; it was recalled that a bird had picked up anthrax after a cup of germs was tossed on the desert many years before. Plans were then made to sterilize and dump the germs into the Arkansas River, a tributary of the Mississippi, a source of drinking water. Federal water officials were uneasy over this procedure. "Presumably any germs will be gone from the stuff that comes out at the end," said one. "But how can you be categorical about it when you don't know what they started with?"

An alternative plan was to bury the material in the earth. This too was received with misgivings. Rain conceivably could carry chemical constituents through the soil outside the military reservation.

The year after the Nixon ban, the United States authorized as much money for biological research as was appropriated in the previous year: $21.9 million. Additionally, a *New York Times* dispatch explained, the Geneva Protocol of 1925, which banned the first use of chemical and biological agents, still had not been approved.

Among the biological agents were 45,000 poison bullets and darts. The Army reportedly wanted no witness to their destruction, "because it is apparently unwilling to have anyone know they existed." Then this interesting paragraph from *The New York Times:*

> The bullets and darts reportedly contain botulinum, a toxin that produces a swift and fatal disease of the nervous system. Knowledgeable sources indicate that the weapons could serve only one purpose: assassination. To kill an enemy leader, it would be necessary to do no more than nick him.

Destruction of the germs, which were produced between 1962 and 1969 at a cost of hundreds of millions, finally began at the Pine Bluff Arsenal in July, 1971. The cost of the procedure was put at more than $10 million over a 48-week period. The biological agents of death were to be sterilized at high heat in special furnaces in what was described as a brick, windowless ten-story building beyond a barbed-wire fence and several guards. After initial sterilization, the germs would be tested to confirm that they were indeed destroyed. The material then would be inoculated with sewage organisms and biodegraded in a manner similar to the treatment of domestic sewage. Another sterilization treatment at high heat would be conducted, with more verification tests. The material then would be placed in drums, sterilized once more, and put through a small commercial sewage treatment plant where it would be further biodegraded. It would be discharged into a sealed, covered evaporation pond and allowed to dry. Finally it would be collected, spread on arsenal property, disked into the soil to a depth of four inches, and over it would be planted a cover crop of grass.

Bibliography

The complexity and scope of this book's subject matter made enormous research necessary. It included personal interviews and correspondence; encyclopedias; special-interest publications; speeches; published and unpublished research papers; federal, state and municipal documents; pamphlets, leaflets and booklets; court proceedings; and individual reports and communications representing a variety of points of view. Every source either added factual information that went directly into the text or supplied background information that gave perspective and shaped my thinking.

Some of this material goes back several years, but most of it appeared within the last two or three years. Many experts in the environmental field were generous beyond belief in giving of their time and knowledge. I am indebted to them and to others who have written on this subject. I especially acknowledge the importance of *The New York Times* as a source of running information. It has done an admirable job of reporting and interpreting news developments in this difficult and often sensitive area, as well as taking a strong and often courageous editorial viewpoint. I have also depended heavily on various environmental publications, among them *Environment, Environmental Action,* and the Sierra Club's publications.

To list each article and document used would require a volume in itself. Because of the thousands of newspaper articles in my files, I have simply listed the various papers from which the items came. Because many of the magazines, newsletters, and other publications are devoted wholly to the environmental field, to list each separate item would be impossible. To give an indication of the range of source material, and as a guide to the person who would like to pursue in greater depth aspects of the subjects covered here, the major sources of printed information follow:

Books

Albrecht, William A. *Soil Fertility and Animal Health*. Webster City, Iowa: Fred Hahne Printing Co.

Ardrey, Robert. *The Social Contract*. New York: Atheneum, 1970.

Arendt, Hannah. *The Human Condition*. Chicago: University of Chicago Press, 1958.

Balfour, E. B. *The Living Soil*. New York: Devin-Adair, 1948.

Bates, Marston. *The Forest and the Sea*. New York: Vintage Books, 1960.

Beiser, Arthur. *The Earth*. New York: Time-Life Books, Time Inc., 1963.

Bergamini, David B. *The Universe*. New York: Time-Life Books, Time Inc., 1966.

Bronowski, J. *The Identity of Man*. New York: Harper & Row, 1935.

Carson, Rachel. *Silent Spring*. Boston: Houghton Mifflin Co., 1962.

Clark, John C. *Fish and Man*. Highlands, N.J.: American Littoral Society, 1967.

Collis, John Stewart. *The Triumph of the Tree*. New York: William Sloan Associates, 1954.

Columbia Encyclopedia. New York: Columbia University Press, 1950.

Commoner, Barry. *Science and Survival*. New York: The Viking Press, 1963.

Cromie, William J. *Secrets of the Sea*. Pleasantville, New York: Reader's Digest Association, 1971.

De Bell, Garrett (ed.). *The Environmental Handbook*. New York: Ballantine Books, Inc., 1970.

Dubach, Howard, and Taber, Robert. *Questions about the Ocean*. Washington, D.C.: Naval Oceanographic Office, 1968.

Dubos, René. *So Human an Animal*. New York: Charles Scribner's Sons, 1968.

Dubos, René. *The Torch of Life*. New York: Simon and Schuster, 1962.

Ehrlich, Paul. *The Population Bomb*. New York: Ballantine Books, 1968.

Encyclopaedia Britannica. Encyclopaedia Britannica, Inc., 1967.

Encyclopedia Americana. Americana Corp., 1969.

Encyclopedia of Popular Science, The. New York: Little Ives, 1963.

Esposito, John. *The Vanishing Air*. New York: Grossman Publishers, 1970.

Farb, Peter. *Ecology*. New York: Time-Life Books, Time Inc., 1963.

Galbraith, John K. *The New Industrial State*. Boston: Houghton Mifflin Co., 1967.

Graham, Frank, Jr. *Since Silent Spring*. Boston: Houghton Mifflin Co., 1970.

Hofstadter, Richard. *American Political Traditions*. New York: Vintage Books, 1957.

Howard, Sir Albert. *An Agricultural Testament*. Oxford University Press, 1939.

Howard, Sir Albert. *The Soil and Health.* New York: Devin-Adair, 1947.

Howard, Louise E. *The Earth's Green Carpet.* Emmaus, Pa.: Rodale Press, 1947.

Hyams, Edward. *Soil and Civilization.* London-New York: Thames and Hudson, 1952.

Illustrated Library of the Natural Sciences. American Museum of Natural History, Simon and Schuster, 1958.

Leopold, Aldo. *A Sand County Almanac.* Oxford University Press, 1939.

Leopold, A. Starker. *The Desert.* New York: Time-Life Books, Time Inc., 1961.

Leopold, Luna B., and Davis, Kenneth S. *Water.* New York: Time-Life Books, Time Inc., 1965.

Longgood, William. *The Poisons in Your Food.* New York: Simon and Schuster, 1960.

Marine, Gene. *America the Raped.* New York: Simon and Schuster, 1969.

Marx, Wesley. *The Frail Ocean.* New York: Ballantine Books, 1967.

Matthiessen, Peter. *Wildlife in America.* New York: Viking, 1959.

McLuhan, H. M. *Understanding Media: The Extension of Man.* New York: McGraw-Hill, 1964.

Milne, Lorus and Margery. *Water and Life.* New York: Atheneum, 1964.

Mumford, Lewis. *The Highway and the City.* New York: Harcourt, Brace and World, 1963.

Novick, Sheldon. *The Careless Atom.* New York: Dell Publishing Co., 1969.

Olsen, Jack. *Slaughter the Animals, Poison the Earth.* New York: Simon and Schuster, 1970.

Osborne, Fairfield. *Our Plundered Planet.* Boston: Little, Brown and Co., 1948.

Petry, Loren C., and Norman, Marcia. *A Beachcomber's Botany.* Chatham, Mass.: Chatham Conservation Foundation, 1963.

Picton, Lionel, M.D. *Nutrition and the Soil, Thoughts on Feeding.* New York: Devin-Adair Co., 1949.

Price, Weston A. *Nutrition and Physical Degeneration.* New York: Paul B. Hoeber, Inc., 1939.

Reich, Charles. *The Greening of America.* New York: Random House, 1970.

Resources and Man. A Study and Recommendations by the Committee on Resources and Man of the Division of Earth Sciences, National Academy of Sciences, National Research Council. San Francisco: W. H. Freeman and Co., 1969.

Rienow, Robert, and Train, Leona. *Moment in the Sun.* New York: Ballantine Books, 1967.

Sears, Paul B. *Deserts on the March.* University of Oklahoma Press, 1959.

Simeons, A. T. W. *Man's Presumptuous Brain.* New York: E. T. Dutton and Co., 1960.

Soil, the 1957 Yearbook of Agriculture. U.S. Department of Agriculture.

Sterling, Dorothy. *Captain of the Planter.* New York: Doubleday & Co., 1958.

Sterling, Dorothy. *The Outer Lands.* New York: Natural History Press, 1967.

Storer, John H. *Men in the Web of Life.* New York: New American Library, 1968.

Szent-Gyorgyi, Albert. *The Crazy Ape.* New York: Philosophical Library, Inc., 1970.

Thompson, Philip D., and O'Brien, Robert. *Weather.* New York: Time-Life Books, Time Inc., 1965.

Tidal Marshes of Old Lyme, Connecticut. Old Lyme Conservation Commission, 1968.

Turner, James S. *The Chemical Feast.* New York: Grossman Publishers, 1970.

Udall, Stewart L. *The Quiet Crisis.* New York: Holt, Rinehart & Winston, Inc., 1963.

Went, Frits W. *The Plants.* New York: Time-Life Books, Time Inc., 1963.

Wiley, H. W. *History of a Crime Against the Food Law.* H. W. Wiley, 1929.

Zurhorst, Charles. *The Conservation Fraud.* New York: Cowles Book Co., 1970.

Specific Sources

Included here are reports, speeches, official government documents, court transcripts, magazine articles, newspapers, newsletters, etc. Many of the periodicals and publications are devoted wholly to environmental matters; others of a more general nature have regular departments or special issues devoted to the subject, so the bulk and diversity of material is too great for detailed listing. The major sources that follow are presented in broad categories:

Newspapers

Atlanta Constitution; Arizona Republic; Asbury Park (N.J.) *Evening Press; Australian; Barrons; Barrytown* (N.Y.) *Explorer; Beacon* (N.J.); *Bridgeport* (Conn.) *Post; Bridgeport Telegram; Capital Times* (Madison, Wisc.); *Chemical and Engineering News; Chicago Tribune; Christian Science Monitor; Cleveland Plain Dealer; Cleveland Press; Courier Post* (Camden, N.J.); *Daily Express* (Detroit); *Daily Town Crier* (Las Vegas, Nev.); *Dental Digest; Denver Post; Detroit Free Press; Earth Times; Evening News* (Newark, N.J.); *Family Weekly; Financial Post; Grand Rapids Press; Herald News* (N.J.); *Kalamazoo Gazette; Kansas City Star; Law Journal; Long Island Advance; Long Island Press; Los Angeles Times; Miami Herald; Medical Tribune; Milwaukee Journal; Milwaukee Sentinel; Morning Herald* (Maryland); *National Fisherman; National Observer; New Brunswick* (N.J.) *Sunday Home News; New York Daily News; New York*

Herald Tribune; New York Post; New York Times; New York World-Journal-Tribune; Newsday; Observer (London, England); *Outdoor Life; Passaic Herald; Register-Herald* (Pine Plains, N.Y.); *Rocky Mountain News; San Francisco Examiner; Saint Louis Globe Democrat; Saint Louis Post Dispatch; State Journal* (Michigan); *Suffolk Sun-Times; The* (N.J.) *Record; Trenton* (N.J.) *Sunday Times Advertiser; Trenton Times; Wall Street Journal; Washington Post; Washington Star.*

AIR POLLUTION

Abelson, Philip. "Progress in Abating Air Pollution." *Science,* March 1970.

"Aerial Lead: Minds Blown for Profit." *Environmental Action,* August 29, 1970.

Air and Water News. McGraw-Hill's Weekly Report on Environmental Pollution. March 30, 1970.

Air Conservation, A Progress Report. Standard Oil Co. (N.J.), undated.

Air Pollution Implementation Manual for a High Air Pollution Alert and Warning System. Dept. of Air Resources, Environmental Protection Administration, N.Y.C., 1968.

Air Pollution in the Queens-Midtown and Brooklyn Battery Tunnels. Scientists' Committee for Public Information, Inc., New York, N.Y. (Statement) 1968.

"Air Pollution." A Scientists' Institute for Public Information Workbook, 1970.

"Air Pollution: Peril from the Skies." *Senior Scholastic,* October 27, 1969.

Air Pollution Publications—A Selected Bibliography, 1963–1966.

"Air Pollution—the Facts." National Tuberculosis Assn. (Leaflet) undated.

Air Quality Act of 1967. USDHE&W.

"Air Quality Criteria for Particulate Matter." USDHE&W, January 1969.

Air Quality Criteria for Sulfur Oxides." USDHE&W, January 1969.

Alexander, Tom. "Burning Questions About Combustion." *Fortune,* February 1970.

Annual Data Report Aerometric Network. Dept. of Air Resources, Environmental Protection Administration, N.Y.C., 1969.

"Another Problem Solved." New York State Action for Clean Air Committee. (Pamphlet on car care) undated.

"Approaching the Clean Air, 1967–1968 Progress Report." Dept. of Air Resources, N.Y.C. Environmental Protection Administration.

Blade, Ellis, and Ferrand, Edward F. "Sulfur Dioxide and Air Pollution in New York City. Statistical Analysis of 12 Years." *Air Pollution Control Assn. Journal,* November 1969.

Boesen, Victor. "The Poisoned Air Around Us." *West Magazine, Los Angeles Times,* April 16, 1967.

"Bridge to Clean Air, The." Prepared for the Technical and Executive

Committees of the Southwestern Ohio–Northern Kentucky Air Pollution Survey. National Center for Air Pollution Control. USDH-E&W. (Booklet) undated.

Brodine, Virginia. "Running in Place." *Environment,* January–February 1972.

Broecker, Wallace. "Enough Air?" *Environment,* September 1970.

Falconer, Raymond E. "Air Pollution and the Temperature Inversion." *The Conservationist,* October–November 1970.

Gould, Donald. "Battle for Smokeless Air." *New Statesman,* January 23, 1970.

Gordon, C. C. "Air Pollution—Montana Style." *Montana Business Quarterly,* Summer 1968.

Graham, Frank, Jr. "The Breath of Death." *Audubon,* July–August 1968.

Heller, Austin N. New York City Congressional Hearing on Automotive Air Pollution. December 8, 1969.

Hersey, Irwin. "We Will Have to Run Very Hard Just to Stay Even." *Engineering Opportunities,* May 1969.

Macinko, John. "The Tailpipe Problem." *Environment,* June 1970.

Middleton, John J. "The Government and the Polluters." *Actual Specifying Engineer,* November 1967.

Middleton, John J. "We Can Have Clean Air." *Country Beautiful,* 1969.

Miller, Jim. "Clean Air Dirty Politics." *Environmental Action,* October 16, 1971.

"Missouri Air Quality." Missouri Air Conservation Commission, January–June 1969.

Muskie, Sen. Edmund S. Statement Opening Senate Consideration of the National Air Quality Standards Act. September 21, 1970.

Nadler, Allen A., et al. "Air Pollution." Scientists' Institute for Public Information Workbook. Scientists for Public Information, 1970.

"Needed: Clean Air. The Facts About Air Pollution." Channing L. Bete Co., Inc. (Booklet) 1969.

New Air Pollution Control Rule Part 190. "Open Fires," State of New York Department of Health Air Pollution Control Board.

New Jersey Air Pollution Control Commission. Motor Vehicle Committee. September 2, 1964.

"One Problem Solved." New York State Action for Clean Air Committee. (Auto care pamphlet) undated.

"Our Dirty Air." *Connecticut Conservation Reporter,* August–September 1965.

"Perspectives on Air Pollution by Sulfur Oxides. Action Now for Cleaner Air." Report, Standard Oil Co. (N.J.), undated.

Peterson, Eugene K. "The Atmosphere: A Clouded Horizon." *Environment,* April 1970.

"Pollution Threatens Desert Air." *Environmental Action,* October 3, 1970.

"Progress in Abating Air Pollution." *Science,* March 1970.

"Progress Toward Cleaner Air." USDHE&W, NAPCA, April 15, 1964.

Rediske, J. H. "The Human Environment–Air Quality." *Journal of Forestry,* January 1969.

Report For Consultation on the Buffalo Air Quality Control Region. USDHE&W, February 1969.

Report to New Jersey Air Pollution Control Assn. by Motor Vehicle Committee, September 21, 1964.

Roueché, Berton. "A Woman with a Headache." *The New Yorker,* January 31, 1970.

Schaefer, Vincent J. "The Threat of the Unseen." *Saturday Review,* February 6, 1971.

Seymour, Whitney North, Jr. "Cleaning Up Our City Air." *Urban Affairs Quarterly,* September 1967: reprinted *Congressional Record,* December 13, 1967.

"Smokeless Jet Engines Really Aren't." *Environmental Action,* October 31, 1970.

Statement, Mayor John Lindsay of New York on Air Pollution. October 30, 1969.

"Sullivan Gives Overall View of Air Control." *New Jersey Air, Water and Waste Management Times,* February 1969.

"Take Three Giant Steps to Clean Air." USDHE&W. Public Health Service Publication No. 1551, 1967.

Teller, Aaron J. "Air Pollution." Science and Technology Yearbook. New York, 1966.

West, Wallace. "Clearing the Air." Committee on Public Affairs, American Petroleum Institute. (Booklet) undated.

"Where Does Jersey Go from Here in Air Pollution Control?" N.J. Air Pollution Control Commission. July 18, 1966.

Williams, Thomas F. "Polluted Air. A Social Challenge." *National Tuberculosis Assn. Bulletin,* January 1965.

Williams, Thomas F. "Polluted Air. What Can be Done?" *National Tuberculosis Assn. Bulletin.* January 1965.

"Your Children's Chances of Dying from Polluted Air Are Better than Yours." Citizens for Clean Air, Inc., New York, N.Y.

AUTOMOBILE DESPOILATION

Aaronson, Terri. "Tempest Over a Teapot." *Environment,* October 1969.

Armstrong, Garner Ted, and Kroll, Paul. "The Ever-Present Automobile." *The Plain Truth,* November 1970.

"The Automobile—Does It Have a Future?" *Science World,* October 20, 1969.

"Case of the Invisible Killer, The." *Science World,* September 15, 1969.

Craig, W. S. "Not a Question of Size." *Environment,* June 1970.

Demaree, Allan T. "Cars and Cities on a Collision Course." *Fortune,* February 1970.

"Freeways—Tearing Up the Cities." *Environmental Action,* October 17, 1970.

Gooding, Judson. "How Baltimore Tamed the Highway Monster." *Fortune,* February 1970.

Hayes, Dennis. "Can We Bust the Highway Trust?" *Saturday Review,* June 5, 1971.

Heller, Austin. N.Y. Congressional Hearings on Automotive Air Pollution. (Talk) New York City, December 8, 1969.

Hollyman, Tom. "The Race That Went to the Cleanest." *The Lamp*, 1970.

Lear, John. "The Skeleton in the Garage." *Saturday Review*. June 5, 1971.

Macinko, John. "The Tailpipe Problem." *Environment*, June 1970.

"Nation's Highway Building Program." *The Conservation Foundation Letter*, June 1970.

"No More Highways—Cry for Mass Transit." Source unidentified, April 16, 1970.

Rjazanov, V. A. "The Automobile—A Cause of Unbalance." *World Health*, December 1964.

"What is the Role of the Highway in Society and the Environment?" *The Conservation Foundation Letter*, June 1970.

CITIES

Armstrong, Garner Ted, and Kroll, Paul. "Can Our Cities Be Saved?" *The Plain Truth*, December 1970.

Armstrong, Gardner Ted, and Kuhn, Robert. "Civitis . . . What Cities Do to Us, And What We Do to Cities." *The Plain Truth*, July 1971.

"Citizens Need Services." *World Health*, December 1964.

de Vries, E., and Thijsse, Jac P. "City Congestion and City Spread." *World Health*, December 1964.

de Vries, E., and Thijsse, Jac P. "Mental Health in the City." *World Health*, December 1964.

"Dynamic City, The." *World Health*, February–March 1966.

"Engineering and the Urban Crisis." *Design News*, March 16, 1970.

Graz, L. "Children in Cities." *World Health*, February–March 1966.

Graz, L. "Magnetic City." *World Health*, February–March 1966.

Graz, Liesl. "The Artificial Life." *World Health*, February–March 1966.

Griffin, C. W., Jr. "Frontier Freedoms and Space Age Cities." *Saturday Review*, February 7, 1970.

Jove, Jose A. "The Lure of the City." *World Health*, December 1964.

Lear, John. "Cities on the Sea." *Saturday Review*, December 4, 1970.

Leroux, Alfred Gustav. "City Health." *World Health*, February–March 1966.

"Man and His Cities." *World Health*, February–March 1966.

McQuade, Walter. "Downtown Is Looking Up." *Fortune*, February 1970.

Querido, A. "Alone in the Crowd." *World Health*, February–March 1966.

Ragon, Michael. "Impact." *World Health*, February–March 1966.

Ragon, Michael. "Towards a Better City." *World Health*, February–March 1966.

Senn, Charles L. "Cities Are Thirsty." *World Health*, December 1964.

"Sick, Sick Cities, The." *Newsweek*, March 17, 1969.

von Eckard, Wolf. "The Perils of Concentration." *Saturday Review*, May 2, 1970.

Watson, Emmett. "This, Our City." *The Northwest Technocrat*, November–December 1969.

Wolfe, Tom. "Oh Rotten Gotham—Sliding into the Behavorial Sink." *World-Journal-Tribune Magazine*, 1966.

CONSERVATION

"About the Conservation Foundation." The Conservation Foundation (Pamphlet) undated.

Association of New Jersey Conservation Commission's Newsletter, October 1969.

"Birds Today, People Tomorrow. What You Can Do to End New York's Environmental Crisis." The Mayor's Council on the Environment. *The New York Times* (advertising supplement), April 18, 1971.

Brief Outline of the Inadequate Economic Justification of the Cross-Florida Barge Canal Project, A. Citizens for the Conservation of Florida's Vital and Economic Resources, Inc., March 1965.

Brooks, Paul. "Notes on the Conservation Revolution." *Sierra Club Bulletin*, January 1970.

Brooks, Paul. "The Plot to Drown Alaska." *Atlantic Monthly*, 1965.

Citizens Advisory Committee on Environmental Quality. Report to the President and to the President's Council on Environmental Quality. August 1969.

Clement, Roland C. *A Look at the New Conservation*. Wilmington, Del.: Fourth Annual Meeting, Delaware Nature Educational Center, November 7, 1970.

Collins, Robert. "Florida's River of Grass." *The Lamp*, Spring 1970.

"Compromise on Grand Canyon." Sierra Club Letter. May 2, 1967.

"Conservation. A New Boy in Court." *Time*, October 24, 1969.

"Conservation Bill of Rights." *Argonaut*, October–November 1969.

"Conservation Landmark: The Private Trust." *Connecticut Conservation Association*. April 1968.

Dasmann, Raymond E. "Conservation—Bland or Biting." *Pacific Discovery*, July–August 1962.

Ehrlich, Henry. "The Ladies Save the Lakes." *Look*, April 21, 1970.

Environmental Defense Fund. "One Anti-Pollution Device That Works." (Pamphlet)

Evans, Brock. "The Battle for the Lost Wilderness." *Sierra Club Bulletin*, May 1969.

"Everglades Coalition Combats Corps Flood Control Program." *Environmental Action*, April 16, 1970.

"Gifts of Land to the Nature Conservancy." *The Nature Conservancy*. (Pamphlet) undated.

Hano, Arnold. "The Battle of Mineral King." *New York Times Magazine*, August 17, 1969.

"Hooker Dam and the Gila's Gentle Wilderness." *Sierra Club Bulletin Annuals*, 1966–68.

Jennings, Gary. "Let's Save a Place to Walk." *Reader's Digest* (reprinted from *Empire*), May 1968.

"Jetport or Everglades Park. The Leopold Report." *Audubon*, November 1969.

Klenck, Richard. "Conservation Is Unselfishness." *Scholastic Teacher*, April 29, 1966.

Magon, Allen H. "Land Use Yardstick." Connecticut Conservation Assn.

Main, Jeremy. "Conservationists at the Barricades." *Fortune*, February 1970.

Miller, James Nathan. "Conservation Is Everybody's Battle." *Reader's Digest* (reprinted from *Nation Civic Review*, July 1964), August 1964.

Mnara, H. Hassovna. "Green Corsets." *World Health*, December 1964.

"Natural Areas." *The Nature Conservancy*. (Pamphlet) undated.

Ottinger, Rep. Richard L. "Conservation Bill of Rights." House of Representatives, June 13, 1968.

"Regional Approach to Conservation Education, A." *The Conservationist*, June–July, 1970.

Roberts, Steven V. "The Better Earth." *New York Times Magazine*, March 29, 1970.

Statement, Ad Hoc Committee to Save Hilton Head.

"Stewardship." Open Space Action Committee, 1965.

"Summary of New York State's New Environmental Conservation Law, A." *The Conservationist*, June–July, 1970.

Tupling, Lloyd W. "Washington Report (Alaska pipeline)." *Sierra Club Bulletin*, June 1969.

"Victory on San Francisco Bay." *Fortune*, February 1970.

"When a Project Becomes a Preserve." *The Nature Conservancy News*, Spring 1969.

"Wilderness: The Last Refuge." *Senior Scholastic*, October 27, 1969.

Yannacone, Victor. "Master Planning: The Aviation Environment." Speech, 1969.

ECOLOGY

"Blackout." *Environmental Action*, June 26, 1970.

Bookchin, Murray. "Toward an Ecological Solution." *Ramparts*, May 1970.

Boyle, Robert H. "The Nukes Are in Hot Water." *Sports Illustrated*, January 20, 1969.

Bulloff, Jack. "A World Safe for Rhinos Is Not Best for Man." *University Review*. State University of New York, Summer 1970.

Clark, John R. "Thermal Pollution and Aquatic Life." *Scientific American*, March 1969.

Cole, LaMont C. "Can the World Be Saved." *New York Times Magazine*, March 31, 1968.

Cole, LaMont C. "Man's Ecosytem." *Bioscience*, April 1966.

Cort, Abraham H. "The Energy Cycle of the Earth." *Scientific American*, September 1970.

"Dawn for the Age of Ecology." *Newsweek,* January 26, 1970.

"Dimension USA—Industry's Road Show." *Environmental Action,* November 14, 1970.

"Diversity and Stability in Ecological Systems." *Brookhaven Symposia in Biology,* No. 22. Brookhaven National Laboratory. May 26–28, 1964.

"Earth's Web of Life." *Citizen,* November 9, 1967.

"Ecology of Man, The." *Northwest Technocrat.* Technocracy, Inc., Rushland, Pa., October 1, 1969.

"Electric Power and the Environment." *Conservation Foundation Newsletter,* March 1970.

"Environmental Effects of Weapons Technology." Scientists' Institute for Public Information Workbook, 1970.

"Fermi Reactor Plant Flops Back to Life." *Environmental Action,* April 9, 1970.

Gentry, Jerry. "Dam Construction, or Why the Beaver Has a Better Idea." *The Plain Truth,* October–November 1970.

"Heat Can Hurt—Better Water for America." FWPCA. September 1969.

Hofner, Everett M., et al. "Environmental Education, 1970." Scientists' Institute for Public Information Workbook.

Jensen, "Thermal Pollution in the Marine Environment." *The Conservationist,* October–November 1970.

Johnson, Charles C., Jr., Commissioner Protection and Environmental Service. Human Ecology Symposium, Warrenton, Va., November 24–27, 1968.

Lear, John. "Clean Power from Inside the Earth." *Saturday Review,* December 5, 1970.

Martell, E. A. "Fire Damage." *Environment,* May 1970.

McKinley, Daniel. "Who Needs a Rhinoceros? We Do." *University Review,* Summer 1970.

Odum, Eugene P. "Dawn for the Age of Ecology." *Newsweek,* January 26, 1970.

"Power for a Continent." *The Northwest Technocrat,* October–November–December 1969.

"Public Feels Energy Pinch in Wallet." *Environmental Action,* October 3, 1970.

Stock, Robert W. "Saving the World the Ecologists' Way." *New York Times Magazine,* October 5, 1969.

"Thermal Pollution." *Science World,* December 8, 1969.

"Thermal Pollution of Water." *Sports Fishing Institute Bulletin,* January–February 1968.

"We Have Met the Enemy and He Is Us." *Life,* July 10, 1970.

"What is Ecology?" Connecticut Conservation Association. October 1969.

Wheeler, Harvey. "The Politics of Ecology." *Saturday Review,* March 7, 1970.

"Who Gives Anyone the Right to Dump Radioactive Wastes into Our Air and Water?" Anti-Pollution League, Allendale, N.J. (Leaflet) October 1969.

Woodwell, George M. "The Energy Cycle of the Biosphere." *Scientific American*, September 1970.

"You and the Ecology Movement." *Natural History Magazine*, June 1970.

ESTUARIES

Act Concerning the Preservation of Wetlands and Tidal Marsh and Estuaries and Systems. (Substitute for Senate Bill 419) Public Act No. 695 (reprint) undated.

Act Increasing Certain Penalties for Filling and Dredging Salt Marshes, S737, State of Rhode Island, Providence and Providence Plantations, April 15, 1968.

"Artificial Fishing Reefs Study." Sandy Hook Marine Laboratory, U.S. Bureau of Sport Fisheries and Wildlife, Highlands, N.J., undated.

"Artificial Reefs as Tools of Sport Fishery Management in Coastal Marine Waters," *SFI Bulletin*, January 1966.

Barlow, Elizabeth. "Keeping Jamaica Bay for the Birds." *New York Magazine*, December 1969.

"Brookhaven Town Board Report on Acquisition of Properties in Mt. Sinai Harbor and West Meadow Brook," November 2, 1962.

Clark, John R. "Ecology of Anadromous Fish of the Hudson River, Hearings before the Subcommittee on Fisheries and Wildlife Conservation of the Committee on Merchant Marine and Fisheries of the U.S. House of Representatives," June 24, 1969.

Clark, John R. "Salt Water Fish Prefer Estuaries." Bureau of Sport Fisheries and Wildlife. *In-Sight*, March 1968.

Clark, John R. "Why Salt Water Fishes Need the Shallow Coastal Estuaries." U.S. Bureau of Sport Fisheries and Wildlife (Report) undated.

Clement, Roland C. "Marshes, Developers and Taxes." *Audubon*, November 1969.

"Connecticut Coastal Wetlands Crisis." *Connecticut Conservation Reporter*, December 1968.

Connecticut Wetlands Conference. Summary. Esso Exploration and Production, Australia, Inc., March 19, 1968.

"Estuaries . . . Cradles or Graves. Better Water for America." FWPCA (Pamphlet) undated.

Estuaries and Their Natural Resources. Hearings before the Committee on Commerce. U.S. Senate, June 4, 1968.

Feldman, Maurice. "Sanitary Landfills: Swamps + Engineering + Muscle = Parks." *Sweep*, Winter 1959.

Galveston Bay: Test Case of an Estuary in Crisis." *Science*, February 20, 1970.

"Gem to Save, A." *Nature Conservancy* (Pamphlet) 1969.

Gilliam, Howard. "Battle of the Bay." *The Lamp*, Fall 1969.

"Jamaica Bay Wildlife Preserve." New York City Parks, Recreation and Cultural Affairs Administration (Booklet) undated.

Jensen, Eugene T. "The Edge of the Ocean." Speech presented at 22nd annual meeting, National Association of Soil and Water Districts, Dallas, Texas, February 6, 1968.

Jensen, Eugene T. "The National Estuarine Study." Speech presented at Interstate Conference on Water Problems, Hartford, Conn., December 14, 1967.

"Length and Breadth and the Sweep of the Marshes of Glynn, The." *Life,* 1969.

"Marshlands Are Food Lands." Rhode Island Dept. of Natural Resources. *AgReCon,* Summer 1969.

Marx, Wesley. "Nurseries of the Sea." *UNESCO Courier,* March 1969.

Matthiessen, George C. "Tidemarshes: A Vanishing Resource." Connecticut Conservation Assn. (Pamphlet) undated.

Murray, Linda A., and Johnson, Peter L. "Why Wetlands?" Open Space Institute Report, August 1969.

National Estuarine Pollution Study. July 1, 1967 (revised, March 1968.)

The National Estuarine Study. Volumes 1–2–3. U.S. Dept. of the Interior, FWPCA, November 3, 1969.

"National Goal: Using Estuaries Wisely. A Clean Water Fact Sheet." FWCPA, May 1968.

"Nation's Coastline—Another Dwindling Natural Resource, The." *Conservation Foundation Newsletter,* May 1970.

"Ocean's Edge," *Sierra Club Bulletin,* May 1970.

"Oceans of Life—The Web That Keeps It Going." *Science World* (Tearsheet) undated.

Proposal to the Town of Brookhaven for the Development of Mt. Sinai Harbor. Advisory Committee for Mt. Sinai Harbor, September 23, 1966.

"Red Tide." Sandy Hook Marine Laboratory, U.S. Bureau of Sport Fisheries and Wildlife, May 1, 1969.

Ryther, John H. "Photosynthesis and Fish Production." *Science,* October 3, 1969.

Schuster, Carl N., Jr. "Tidal Marshlands Are Precious." *Estuarine Bulletin,* Autumn 1958.

Shepherd, Jack. "The Disappearing Beauty of the Salt Marsh." *Look Magazine,* April 21, 1970.

Singer, S. Fred. "The Federal Interest in Estuaries." Speech before Oyster Institute of America. Washington, D.C., July 25, 1968.

Stone, Richard B. "Artificial Reef Study." Sandy Hook Marine Laboratory, Highlands, N.J. (Research paper) undated.

"Study of Estuarine Pollution and Water Quality Distribution in the New York City–New Jersey Metropolitan Area." Prepared for Interstate Sanitation Committee, September 1967.

Teal, John and Mildred. "Ribbon of Green, The Epic of a Salt Marsh." *Audubon,* November 1969.

"Today's Challenge: Inland Waterways." Massachusetts Audubon Society (Pamphlet) undated.

Walford, Lionel. "Fish Ecosystem Research." Atlantic Marine Game Fish Research Program, Highlands, N.J. and Narrangansett, R.I. (Report) undated.

Walford, Lionel. Problems of Conservation in the Atlantic Estuarine Zone. Transactions, 32nd North American Wildlife and Natural Resources Conference, March 1967.

"What Good is a Tidal Marsh?" Save the Wetlands Committee, Inc., New Haven, Conn. (Pamphlet) undated.

"The Zone." American Littoral Society Newsletter, April 1970.

FORESTS

Eber, Ron. "The Mighty Redwoods—Debunking Madison Avenue." *Environmental Action*, November 13, 1971.

Evans, Brock, "The Battle for the Last Wilderness." *Sierra Club Bulletin*, May 1969.

Forest Service Research. "Solving Problems on Forest and Related Lands." Forest Service, USDA, February 1970.

"Grazing Fees on National Forest Range, Past History and Present Policy." Forest Service, USDA, June 1969.

"Growing Trees to Meet the Nation's Demands." American Forest Products Industries, Inc., Washington, D.C. (Booklet) 1966.

High Yield Forest, The. Weyerhaeuser Co. (Statement) April 13, 1967.

Linford, Lloyd. "The Great Tree Theft." *Earth Times*, April 1970.

Margolin, Malcolm. "The Habit of Waste." *The Nation*, March 2, 1970.

Mellem, Roger. "Industry Plans National Forest Lumber Raid." *Environmental Action*, April 3, 1971.

"Multiple Use Management—Forest Conservation for a Growing Nation." USDA Forest Service (Pamphlet) September 1966.

"National Forest Wildernesses and Primitive Areas." The National Forests. Forest Service, USDA (Leaflet) July 1964.

"Raiding the Forests." *The New Republic*, December 13, 1969.

Robinson, Gordon. "Whatever Happened to Tree Farms?" *Sierra Club Bulletin*, May 1969.

Stoddard, Charles H. "O & C Formula." *Sierra Club Bulletin*, June 1970.

"Telling It like It Isn't." Excerpts from a speech by William J. Moshofsky, assistant to the President of Georgia–Pacific Corp., given at Pendleton, Oregon, January 19, 1970. *Sierra Club Bulletin*, March 1970.

"Timber Supply Act—Anatomy of a Battle." *Sierra Club Bulletin*, March 1970.

Weyburn, Peggy and Edgar. "What the Redwood Industry Isn't Telling." *Sierra Club Bulletin*, March 1967.

Witter, Robert N. "Relationships of Forests to Oxygen Supply." Weyerhaeuser Co. (Interoffice communication) January 13, 1970.

GENERAL

Aaronson, Terri. "Mystery." *Environment*, May 1970.

A Bill to Establish an Office of Technical Assessment for the Congress as an Aid to the Identification and Consideration of Existing Impacts of Technology Application. H.R. 17046, House of Representatives, April 16, 1970.

ABM–ABC. Union of Concerned Scientists, Cambridge, Mass., April 15, 1969.

"Act Now on Pollution—Don't Just Talk." *The Plain Truth*, February 1970.

"Affluence and Effluence." *Saturday Review*, May 2, 1970.

"After Six Months the Rhetoric Fades." *Environmental Action*, July 16, 1970.

"Age of Effluence, The." *Time*, May 10, 1968.

"America's Everyday Dreariness." (A portfolio) *Fortune*, February 1970.

"Annual Report, Atlantic Marine Game Fish Research Program." U.S. Dept. of Interior Fish and Wildlife Service. Bureau of Sport Fisheries and Wildlife, 1968.

"Army Engineers Rape Land with Projects." *Environmental Action*, April 22, 1970.

Atlantic Marine Game Fish Research. U.S. Bureau of Sport Fisheries and Wildlife (Pamphlet) undated.

Baity, H. G. "An Engineer Looks at Man's Changing Environment." *World Health*, February–March 1966.

Ball, Richard H. "The Grand Canyon Dams, A Brief Analysis." Sierra Club, December 1966.

Barkley, Katherine, and Weissman, Steve. "The Eco-Establishment." *Ramparts*, May 1970.

Barlow, Elizabeth. "The New York Magazine Environmental Teach-In." *New York Magazine*, March 1970.

Berg, George G. *Breakdown in the Mechanism*. Scientists' Institute for Public Information Workbook, 1970.

Berry, Philip S. "Corporate Responsibility and the Law." *Sierra Club Bulletin*, May 1970.

"The Biologist Looks Ahead." *The Nation*, March 2, 1970.

Blackaby, Frank. "History's Greatest Dead End" (The World War Industry). *Saturday Review*, March 14, 1970.

"Boycott Reynolds!" The Emergency Committee to Save Our Environment, Congers, N.Y. (Leaflet) undated.

Branscome, James. "Appalachia—Like the Flayed Back of a Man." *New York Times Magazine*, Dec. 12, 1971.

Brooks, Paul. "The Plot to Drown Alaska." *Atlantic Monthly*, May 1965.

Brown, Harrison. "Human Materials Production as a Process in the Biosphere." *Scientific American*, September 1970.

Brown, Lester H. "Human Food Production as a Process in the Biosphere." *Scientific American*, September 1970.

Calder, Ritchie. "Mortgaging the Old Homestead." *Foreign Affairs Quarterly*, January 1970.

Can Man Survive? American Museum of Natural History, New York, N.Y. (Display and booklet) 1969.

"Can Venice Survive Pollution?" *U.S. News and World Report*, January 26, 1970.

Cancro, Robert, and Slotnick, Daniel L. "Computer Graphics and Resistance to Technology." Conference on Energy Concepts in Computer Graphics. University of Illinois, November 19, 1967. Also personal interview, 1969.

Carlin, Alan. "A Question of Value." *Sierra Club Bulletin,* August 1969.

"Changing the American Environment." *Look,* April 21, 1970.

"Chemical and Engineering News Lists Top 50 Firms in Chemical Sales." *Chemical and Engineering News,* June 16, 1969.

City of New York, Chapter 41 Administrative Code as Amended by Local Law 14/66 and Local Law 14/68. May 1, 1968.

"Cleaning Humanity's Nest." *Saturday Review,* March 7, 1970.

"Cleaning Our Environment. The Chemical Basis for Action." A report. American Chemical Society, 1969.

Clement, Roland C. "A Look at the New Conservation." Talk given at the Fourth Annual Meeting, Delaware Nature Education Center, Wilmington, Del., November 7, 1968.

Clement, Roland C. "The Environmental Crisis." Talk, Pitt–Carnegie–Mellon Teach-In. Pittsburgh, Pa., April 2, 1970.

Cloud, Preston, and Gibor, Aharon. "The Oxygen Cycle." *Scientific American,* September 1970.

Cole, LaMont C. "Can the World be Saved?" *Bioscience,* July 1968.

Commoner, Barry, et al. "The Causes of Pollution." *Environment,* April 1971.

"Conservation: A New Say in Court." *Time,* October 24, 1969.

"Conservation Law I: Seeking a Breakthrough in the Courts." *Science,* December 19, 1969.

"Conservation Law II: Seeking a Breakthrough in the Courts." *Science,* December 26, 1969.

"The Corps of Engineers Has a Strengthened Mandate to Protect Environmental Value." Conservation Foundation Newsletter, August 1970.

Cousins, Norman. "New York's Fight Against Pollution." *Saturday Review,* March 7, 1970.

Crowe, Beryl L. "The Tragedy of the Commons Revisited." *Science,* November 28, 1969.

Cunningham, John T. "He Fights the Polluters." *New Jersey Business,* October 1969.

Cusack, Michael. "SST—Boom or Bust." *Science World,* November 3, 1969.

Daddario, Emilio. "Technology Assessment." Statement, Subcommittee on Science, Research and Development. Committee on Science and Astronautics. U.S. House of Representatives, August 1968.

Dale, Edwain L., Jr. "The Economics of Pollution." *New York Times Magazine,* April 19, 1970.

Dasmann, Raymond F. "An Environment Fit for People." Public Affairs Pamphlet No. 421, July 1968.

Davenport, John. "Industry Starts the Big Cleanup." *Fortune,* February 1970.

Davis, Wayne. "The Stripmining of America." *Sierra Club Bulletin,* February 1971.

Deevey, Edward S., Jr. "Mineral Cycles." *Scientific American,* September 1970.

Diamond, Robert S. "What Business Thinks." *Fortune,* February 1970.

Dingell, John. Statement, "Creation of National Environmental Information Bank Proposed by Bipartisan Group of Representatives." April 8, 1970.

Dingell, John. Statement, "Importance of Funding Environmental Enforcement Programs." November 15, 1970.

Dingell, John. Speech before Fontana Conservation Roundup, May 16, 1969.

"Dollars and Cents," *Moneysworth,* December 28, 1970.

"Ecological Doomsday? A People in Trouble with Its Environment." *The International Teamster,* March 1970.

"Education vs. Pollution." *Scholastic News Explorer,* February 16, 1970.

Ehrlich, Paul R. "Eco-Catastrophe." *Ramparts,* September 1969.

Ehrlich, Paul R., and Holdren, John P. "Dodging the Crisis." *Saturday Review,* November 7, 1970.

Ehrlich, Paul R., and Holdren, John P. "Starvation as a Policy." *Saturday Review,* December 4, 1971.

Eipper, Alfred W. "Pollution Problems, Resource Policy, and the Scientist." *Science,* July 3, 1970.

"Emerging Science of Survival, The." *Time,* February 2, 1970.

Environmental Action Coalition Letter. March 26, 1970.

"Environmental Crimes." *Sierra Club Bulletin,* June 1970.

"Environmental Defense Fund: Yannacone Out as Ringmaster." *Science,* December 26, 1969.

"Environmental Folly." *Connecticut Conservation Reporter,* August–September 1968.

"Environmental Pollution: Researchers Go to Court." *Science,* December 22, 1967.

"Environmental Protection in Sweden. Fact Sheet on Sweden." Swedish Institute for Cultural Relations with Foreign Countries.

"Europe Confronts Environmental Crisis." *The Plain Truth,* April–May 1970.

Evans, Brock. "Battle for the Lost Wilderness." *Sierra Club Bulletin,* May 1969.

Evans, David M., and Bradford, Albert. "Under the Rug." *Environment,* October 1969.

"Eye on the Ecosystem." American Museum of Natural History, 1970.

Farvar, Taghi, et al. "The Pollution of Asia." *Environment,* October 1971.

Fife, Daniel. "Killing the Goose." *Environment,* April 1971.

"Fight for Survival, A." *Northwestern University Magazine,* 1970.

"Fighting to Save the Earth from Man." *Time,* February 2, 1970.

"Five Who Care." *Look,* April 4, 1970.

Fisher, R. Frederic. "Environmental Law." *Sierra Club Bulletin,* January 1971.

Flateboe, Connie. "Environmental Teach-In." *Sierra Club Bulletin,* March 1970.

Forest, Herman S. "Reveille in Rochester: The Education of a Community." Prepared in 1970.

Forest, Herman S. "Survival Is Not Enough." *American Scientist,* March 1958.

Forest, Herman S., and Greenstein, Harold. "Biologists as Philosophers." *Bioscience,* November 1966.

Forest, Herman S., and Morrill, Thomas. "Biological Expansion—A Perspective on Evolution." *The Monist,* April 1964.

Gardner, Richard N. "U.N. as Policeman." *Saturday Review,* August 7, 1971.

Gellen, Martin. "The Making of a Pollution–Industrial Complex." *Ramparts,* May 1970.

Gofman, John W., and Tamplin, Arthur R. "Radiation: The Invisible Casualties." *Environment,* April 1970.

"Global Environment." *The Plain Truth,* June 1971.

Gordon, C. C. "Environmental Destruction, Montana Style," Speech, 1970.

Gordon, C. C. "Prof. Gordon on Ecology Law." *Montana Law Forum,* May 1970.

"Government Hits Environmental Lawsuits." *Environmental Action,* October 31, 1970.

"Grand Canyon National Monument Is Hereby Abolished." Sierra Club advertisement, 1969.

Griffin, C. W., Jr. "Frontier Freedom and Space Age Cities." *Saturday Review,* February 7, 1970.

Haider, Michael L. "The Pollution Problem." *The Lamp,* Spring 1967.

Hannan, Bruce, and Cannon, Julie. "The Corps Out-Engineered." *Sierra Club Bulletin,* August 1969.

Harrar, J. George. "Human Ecology and a New Ethic of Responsibility." Statement, personal correspondence, June 25, 1970.

Hayes, Dennis. Statement, Subcommittee on Conservation and Natural Resources. House Government Operating Committee, March 13, 1970.

Henderson, Hazel. "Citizen Action for Clean Air." *Congressional Record,* July 14, 1969. (Also personal interview and correspondence.)

Henderson, Hazel. "Computers, Hardware of Democracy." *Forum,* February 1970.

Henderson, Hazel. "Does Business Listen as Well as Transmit?" Talk, Workshop on Planning and Managing Corporate Communications in the '70s, New York, November 5–6, 1969.

Henderson, Hazel. "Politics by Other Names." *The Nation,* December 14, 1970.

Hofner, Everett. "Ecography: A New Scientific Discipline." Statement, Symposium on Undergraduate Studies in Environmental Science. AAAS Annual Meeting, Boston, 1969 (also personal correspondence).

Hofner, Everett M., Fowler, John M., and Williams, Curtis A. *Environmental Education 1970.* Scientists' Institute for Public Information Workbook, 1970.

Hohenemser, Jurt. "Onward and Upward." *Environment,* May 4, 1970.

"Housing and the Quality of Our Environment." Research Report,

Middlesex–Somerset–Mercer (N.J.) Regional Study Council, January 1970.

"How Hurricanes Are Born." *Saturday Review,* October 4, 1969.

"How to Control Pollution." *U.S. News and World Report,* January 19, 1970.

Howe, Sydney. "Making the Most of the Environmental Crisis." Remarks delivered at Annual Meeting, Connecticut River Watershed Council. Woodstock, Vt., May 8, 1970.

Hozberg, Gene H. "Europe Confronts Environmental Crisis." *The Plain Truth,* April–May, 1970.

Hutchinson, G. Evelyn. "The Biosphere." *Scientific American,* September 1970.

"Industrialist's Concern for Our Environment, An." Speech by Norton Clapp, Weyerhaeuser Co., Western States Farm Forum, Yakima, Wash., January 20, 1970.

Jacobs, Foster. Personal interview, Princeton University, Spring 1970.

Jennings, Dana. "The National Grasslands." Forest Service, USDA. *The Dakota Farmer,* May 1970.

"Law Students Training in Environmental Action." *Environmental Action,* April 9, 1970.

Lear, John. "The Enemy Is Us." *Saturday Review,* March 7, 1970.

Lear, John. "Global Pollution: The Chinese Influence." *Saturday Review,* August 7, 1971.

Lear, John. "Science: The Endless Search." *Saturday Review,* March 28, 1970.

"Let's Fill the Bay." *Sierra Club Bulletin,* July 1969.

Levy, Gerald F., and Folstad, John W. "Swimmer's Itch." *Environment,* December 1969.

Lindsay, Sally. "Cleanup-Man Maurice Strong." *Saturday Review,* August 7, 1971.

Lobel, Marty. "Oligopolies: Repealing the Law of Supply and Demand." *Environmental Action,* November 13, 1971.

"Local Wars Against Pollution." *U.S. News and World Report,* February 9, 1970.

Love, Sam. "After Six Months the Rhetoric Fades." *Environmental Action,* July 16, 1970.

Love, Sam. "The Failures of an Act That Once Sparked Hope." *Environmental Action,* January 9, 1971.

"Major Kite's Fawley." *The Lamp,* Spring 1968.

Mander, Jerry. Speech, San Jose State College. "Survival Fair Teach-In." February 18, 1970.

Mander, Jerry. Testimony, Hearings Before House Committee on Education and Labor, May 1, 1970.

"Man Examines Environmental Perception, A." Report from Rutgers. April 1970.

"Man's Environment—Can We Save It?" *Senior Scholastic,* October 27, 1969.

Manzel, Charles. "We Can Save Our Towns." *Look,* April 4, 1969.

Martell, E. A., et al. "Fire Damage." *Environment,* May 1970.

McGowan, Alan. "Getting Their Feet Wet." *Environment,* November 1968.

McInnes, Noel. "Gestalt Ecology: How Do We Create Our Space?" Talk, Aspen Center for Environmental Studies, August 26, 1968.

McMahon, Peg. "Karl Menninger Looks at Our Changing Life." *The Kansas City Star Magazine,* March 8, 1970.

Michaelson, Mike. "The Bounty Hunter Wore Sneakers." *Today's Health,* April 1971.

Michaelson, Mike. "Sweden's Youth Says 'Nej' to Pollution," *Today's Health,* January 1971.

Mills, Stephanie. "Kenneth Watt—the Opinions of an Ecologist." *Earth Times,* May 7, 1970.

"Mineral King at the Crossroads." *Sierra Club Outdoor Newsletter,* May 19, 1969.

Morgan, Allen H. "Land Use Yardstick." Connecticut Conservation Assn., undated.

Moss, Laurence L. "Taxing U.S. Polluters." *Saturday Review,* August 7, 1971.

"Muir and Friends." Friends of the Earth and John Muir Institute, January 1970.

Muskie, Edmund S. "Legislation Dealing with the Environment." *Congressional Record,* January 23, 1970.

Muskie, Edmund S. "Our Polluted America—What Women Can Do." *Ladies' Home Journal,* February 1970.

Nader, Ralph. "No Escape from Corporate Embrace" (excerpts), *Environmental Action,* March 16, 1970.

"National Environmental Policy Act." *Conservation Foundation Newsletter,* April 1970.

"Needed: A Rebirth of Community." *Newsweek,* January 26, 1970.

Nelson, Gaylord. "Detergent Pollution Control Act of 1970." *Congressional Record,* February 26, 1970.

Nelson, Gaylord. "Earth Day." *Congressional Record,* April 30, 1970.

Nelson, Gaylord. "The Environment." *Congressional Record,* April 1, 1970.

Nelson, Gaylord. "Introduction of Environmental Quality Act." *Congressional Record,* November 18, 1969.

Nelson, Gaylord. "Introduction of a Joint Resolution Relating to an Environmental Agenda for the 1970s." *Congressional Record,* January 19, 1970.

Nelson, Gaylord. "Introduction of the Marine Environment and Pollution Control Act of 1970." *Congressional Record,* February 19, 1970.

Nelson, Gaylord. "The National Pollution Scandal." *Congressional Record,* November 18, 1969.

Nelson, Gaylord. "The New Citizenship for Survival." *Progressive Magazine,* April 1970 (reprinted *Congressional Record,* April 1, 1970.)

Nelson, Gaylord. "Water Pollution Control." *Congressional Record,* November 18, 1969.

Nevin, David. "These Murdered Old Mountains." *Life,* January 12, 1968.

"Nixon and the Environment." *Conservation Foundation Newsletter,* February 1970.

Bibliography 523

Nixon, Richard M. "A Call for Cooperation," September 1970.
Nixon, Richard M. "What's Right with America," June 25, 1970.
"Nixon Shift, Criticism Increases." *Environmental Action,* August 16, 1970.
"Notes and Comment" (Shortcomings in press coverage) *The New Yorker,* February 28, 1970.
Novick, Sheldon. "Do-It-Yourself." *Environment,* December 1969.
Olds, Jerome. "The Nicest People Protest Pollution." *Organic Gardening and Farming,* February 1970.
"Omnibus Bill Bypasses Crucial Step" (National Environmental Policy Act of 1969) *Environmental Action,* October 31, 1970.
Ottinger, Richard L. Address, National Wildlife Federation, 33rd Annual Meeting, March 1, 1969.
Ottinger, Richard L. "Battle for the Earth." Speech, June 1970.
Ottinger, Richard L. "The Congress and Environmental Deterioration." *Bioscience,* June 1969.
Ottinger, Richard L. "The Conservation Bill of Rights." Remarks in the House of Representatives, June 13, 1968.
Ottinger, Richard L. "Legislation and the Environment: Individual Rights and Government Accountability." (Author's copy) Speech (audience not identified and no date given).
Ottinger, Richard L. "Our Present Responsibility to Our Future Environment." Speech before American Institute of Industrial Engineers, Tappan Zee Chapter, October 18, 1967.
Ottinger, Richard L. Remarks Before Atlanta Chapter, Sierra Club, February 2, 1970.
Ottinger, Richard L. Testimony before the Water and Power Subcommittee of the Senate Interior Committee, March 13, 1970.
Ottinger, Richard L. "To Improve the Quality of Life: A Program for Environmental Action." House of Representatives, June 13, 1968, and January 3, 1969.
"Our Polluted Planet." Ambassador College (Booklet) 1968.
"Our Threatened Environment." *The Electrical Workers Journal,* November 1970.
Perlman, David. "America the Beautiful." *Look,* November 4, 1969.
Platt, John. "What We Must Do." *Science,* November 28, 1969.
"Pleasantness Made to Order." (A portfolio) *Fortune,* February 1970.
Pollution-Control Laws of New York State. Air Pollution Control Board, New York State, undated.
"Pollution: Growing Menace—What U.S. Is Doing About It." *U.S. News and World Report,* June 9, 1969.
"Prepare to Meet Thy Doom." *Argonaut,* January 1970.
"Progress in Sport Fishery Research 1967." U.S. Department of the Interior Fish and Wildlife Service, Bureau of Sport Fisheries and Wildlife.
Ramsey, James. "So What (SST)." *Sierra Club Bulletin,* January 1970.
Rappaport, Roger. "Catch 24,400 (or Plutonium Is My Favorite Element)." *Ramparts,* May 1970.
"Ravaged Environment, The." *Newsweek,* January 26, 1970.
"Reorganizing to Fight Urban Pollution." *Outlook,* July 1968.

"Report Card." Environment Staff Report. *Environment,* September 1970.

"Reshuffling the Bureaucracy." *Conservation Foundation Newsletter,* September 1970.

"Restoring the Quality of Our Environment." President's Science Advisory Committee. Government Printing Office, 1965.

Rickover, Hyman S. "Liberty, Science and Law." *Massachusetts Audubon,* 1966.

Ridgeway, James. "Para-Real Estate. The Handing Out of Resources." *Ramparts,* May 1970.

Roberts, Steven V. "The Better Earth." *New York Times Magazine,* March 29, 1970.

Rogin, Gilbert. "All He Wants to Save Is the World." (Source not identified.)

Rose, Sanford. "The Economics of Environmental Quality." *Fortune,* February 1970.

Royko, Mike. " 'The Fox' Picks U.S. Steel." *Environmental Action* (reprinted from *Chicago Sun Times*), December 12, 1970.

Royko, Mike. " 'The Fox' Strikes Again." *Environmental Action* (reprinted from *Chicago Sun Times*), October 3, 1970.

Shayon, Robert Lewis. "Dangerous Environments." *Saturday Review,* April 11, 1970.

Shepherd, Jack. "The Fight to Save America Starts Now." *Look,* April 21, 1970.

"Should We Also Flood the Sistine Chapel So Tourists Can Get Nearer the Ceiling?" *Sierra Club Bulletin.*

Sierra Club Bulletin. "A Handbook." December, 1967.

"Sierra Club Goes to Court for Mineral King." *Sierra Club Bulletin,* June 1969.

"Smith, Adam." "The City Politic. What Does the Fizzle in the Ecology Boom Tell Us?" *New York Magazine,* March 1970.

Snow, Joel A. Statement Before Subcommittee on Science, Research and Development, Committee on Science and Astronautics, House of Representatives, 1970.

Soucie, Gary A. "The Everglades Jetport—One Hell of an Uproar." *Sierra Club Bulletin,* July 1969.

Special Report of New York Board of Trade Environmental Study Tour of Sweden. June 6–13, 1970.

Stahr, Elvis J. "The Death of Life." *The Sunday Record Call,* November 30, 1969.

"The States, Playing a Crucial Role in Environmental Management, Try a Variety of Innovations in 1970." *The Conservation Foundation Newsletter,* November 1970.

Stebbins, G. L. "Prospects for Spaceship Man." *Saturday Review,* March 7, 1970.

Steck, Henry. "Why Does Industry Always Get What It Wants?" *Environmental Action,* July 24, 1971.

Stonier, Tom. "Man and His Environment." *Current Topics in Plant Science,* 1969.

"Storm King Pumped-Storage Project." Sierra Club, Atlantic Chapter, undated.

"Students and the Environment." *Trail Hiker,* January–February 1970.

"Study of Technology Assessment, The." Report of the Committee on Public Engineering Policy, National Academy of Engineering, Committee on Science and Astronautics. U.S. House of Representatives, July 1969.

Sullivan, Richard A. "A View from the Firing Line." Speech, New Jersey Clean Air Council, New Brunswick, N.J., March 5, 1970.

Szent-Gyorgyi, Albert. "The Third Environment." *Saturday Review,* May 2, 1970 (Also personal correspondence, June 19, 1970).

Taormina, Anthony J. "Journey Down the Nissequogue." *The Conservationist,* February–March 1966.

Technology Assessment Seminar. Proceedings before the Subcommittee on Science, Research, and Development of the Committee on Science and Astronautics. U.S. House of Representatives, August 1968.

"Technology: Processes of Assessment and Choice." Report of the National Academy of Sciences, Committee on Science and Astronautics. U.S. House of Representatives, July 1969.

"Transportation Department's Hoax." *Environmental Action,* November 28, 1970.

"TV Towers May Be Radiating Harm." *Medical World News,* December 2, 1966.

Udall, Morris K. "Our Spaceship Earth—Standing Room Only." *Congressman's Report,* July 30, 1969.

Udall, Stewart L. "Environmental Action—What You Can Do." *Newsday* (Reprint) 1970.

"Unconfident, The." *Time,* November 15, 1971.

Wald, George. Speech reprinted in "Talk of the Town, Notes and Comment." *The New Yorker,* March 22, 1969.

Walford, Lionel. "Implications of the Report by the Environmental Pollution Panel, President's Science Advisory Committee." Transactions of the 31st North American Wildlife and Natural Resources Conference, March 14–16, 1966.

Wallach, Henry C. "Paying for the Cleanup." *Newsweek,* January 26, 1970.

Walter, Eugene M., and Hogbert, Gene M. "Arid Lands: Can They Be Reclaimed in Time?" *The Plain Truth,* October 1969.

Ways, Max. "How to Think About the Environment." *Fortune,* February 1970.

Wehrheim, John. "Paradise Lost." *Sierra Club Bulletin,* February 1970.

Wheeler, Harvey. "The Politics of Ecology." *Saturday Review,* March 7, 1970.

Whyte, Lynn, Jr. "The Historical Roots of Our Ecological Crisis." *Science,* March 10, 1967.

Williams, Steve. "Utilities Push 'Ecology' with $300 Million in Advertising." *Environmental Action,* April 17, 1971.

Willis, Thayer. "Antarctic Ice on the Move." *Science World,* November 3, 1969.

Wilson, David J. Vanderbilt University, Nashville, Tenn. Personal correspondence, June 23, 1970.

Wolff, Anthony. "Miss Stephanie Mills vs. Motherhood." *Look,* April 21, 1970.

Woodward, Joanne, as told to Terry Schaertel. "What Our Family Is Doing About Environmental Pollution." *Family Weekly,* September 6, 1970.

Woodwell, George M. "Science and the Gross National Pollution." *Ramparts,* May 1970. (Also personal interview.)

Yannacone, Victor. "Environmental Litigation." Remarks prepared for a program presented by the Committee on Problems Arising from Environmental Litigation and Legislation of the Section of Insurance, Negligence and Compensation Law, American Bar Association. 1971 Annual Meeting, New York City, July 7, 1971.

Yannacone, Victor, Jr. "Environmental Legislation and Political Reality." Speech, 1970.

Yannacone, Victor A. "Defending the Environment." Speech, 1969.

NOISE

Bailey, Anthony. "The Sound of Madness." *New York Times Magazine,* November 23, 1969.

Baron, Robert Alex. "Noise: The Audible Pollutant." *Nation's Cities,* September 1969.

"Deafening Crescendo of Noise, The." *The Plain Truth,* August–September 1969.

Ljarshij, P. P. "The Rising Tide of Noise." *World Health,* December 1969.

"Noise Dr." *Life,* May 5, 1970.

"Noise Pollution. A Review of Its Techno–Sociological and Health Aspects." Biotechnical Program, Carnegie–Mellon University, February 1, 1968.

"Quiet!" Citizens for a Quieter City. New York, N.Y., Spring 1969.

Shurcliff, William A. "SST and Sonic Boom Handbook," February 15, 1969.

"Ssh!" *Newsweek,* September 8, 1969.

Stewart-Gordon, James. "We're Poisoning Ourselves with Noise." *Reader's Digest,* February 1970.

"Toward a Quieter City." Report of Mayor's Task Force on Noise Control. New York, N.Y., 1970.

NUCLEAR POLLUTION

"A New River." Staff report. *Environment,* January–February 1970.

"Autopsy at Cannikin." *Time,* November 22, 1972.

Clark, Wilson. "AEC May Put 'Hot' Liquid Into Rocks." *Environmental Action,* October 3, 1970.

Clark, Wilson. "AEC Suppresses Critical Safety Report." *Environmental Action,* May 14, 1970.

Gofman, John W., and Tamplin, Arthur. "Radiation: The Invisible Casualties." *Environment.* April 1970.

Lapp, Ralph. "The Four Big Fears About Nuclear Power." *New York Times Magazine*, February 7, 1971.

Leiternberg, Milton. "So Far, So Good." *Environment*, July–August 1970.

McClintock, Michael, et al. "Environmental Effects of Weapons Technology." Scientists' Institute for Public Information Workbook, 1970.

Metzger, Peter. "Project Gasbuggy and Catch 85." *New York Times Magazine*, February 22, 1970.

Novick, Sheldon. "Earthquake at Giza." *Environment*, January–February 1970.

"Nuclear Hazards." *Environmental Action*, June 11, 1970.

"Nuclear Power: A Social Cost." *The Conservation Reporter*, October 1968.

"Nuclear Tests Threaten Severe Ecological Damage." *Environmental Action*, April 16, 1970.

Pesonen, David E. "Science and Public Policy: Bodega Head." *Medical Opinion and Review*, December 1967.

Progress Report on Atomic Energy Research. Hearings before the Subcommittee on Research and Development of the Joint Committee on Atomic Energy. Congress of the U.S., 1956.

Radford, Edward P., et al. "Statement of Concern." *Environment*, September 1969.

Seaborg, Glenn T. "Energy and Environment—A Rational Outlook." Remarks, Ninth Annual Meeting, Southern Interstate Nuclear Board. Lake Eufaule, Okla., April 20, 1970.

Seaborg, Glenn T. "Energy and the Future." Talk before the Chemical Institute of Canada and the American Chemical Society," May 25, 1970.

Seaborg, Glenn T. "The Human Side of Energy." Talk before National Electric Manufacturing Assn., Chicago, November 11, 1969.

Seaborg, Glenn T. "The New Alchemy." Remarks, Ninth Annual Meeting of the Manufacturing Chemists Assn., White Sulphur Springs, W. Va., June 4, 1970.

Shea, Kevin P. "Old Weapons Are Best." *Environment*, June 1971.

"Sixteen Reasons Why Nuclear Power Plants Endangering Millions Should Be Stopped." Citizens Committee for the Protection of the Environment. Ossining, N.Y. (Leaflet) undated.

Taylor, Avery. "Cannikin, Biggest U.S. Blast Ever, Set for October." *Environmental Action*, July 10, 1970.

Taylor, Avery. "Cannikin: The Bomb No One Wanted." *Environmental Action*, November 13, 1971.

"Underground Nuclear Testing." *Environment*, August 1969.

"The Western White House Is Located About 4,400 Yards from This Nuclear Power Plant." *Environmental Action*, June 11, 1970.

OCEAN

"Another Garbage Pile Rejects Technology." *Environmental Action*, May 14, 1970.

Bloomfield, E. Jervis. "The World's Greatest Dump." *The National Fisherman*, August 1969.

Cahill, William T. Statement, "Sludge Barging to Sea." New Jersey, February 14, 1970.

Cornwell, John. "Is the Mediterranean Dying?" *New York Times Magazine*, February 21, 1971.

"Dying Oceans, Poisoned Seas." *Time*, November 8, 1971.

Ehrlich, Paul and Anne. "The Food from the Sea Myth." *Saturday Review*, April 4, 1970.

Emery, K. O. "The Geology of the Atlantic Continental Shelf and Slope." *Underwater Naturalist*, Spring 1968.

Fonselus, Stig H. "Stagnant Sea." *Environment*, July–August 1970.

"Food from the Sea." Alpine Geophysical Association, Inc., Annual Report, 1969.

Gorsky, Nicolai (member USSR Geographical Society). "The Pollution of the Ocean." (Tearsheet from U.N., original publisher not listed) dated July–August 1959.

Hedgepath, Joseph W. "The Oceans: World Sump." *Environment*, April 1970.

Hirdman, Sven. "Weapons in the Deep Sea." *Environment*, April 1971.

Hood, D., ed. *Impingement of Man on the Oceans*. Chapter on lead by C. C. Patterson.

Livingston, Betty. "Froth on Foam: The Ocean Will Never Be the Same." *Environmental Action*, June 26, 1971.

Marx, Wesley. "How Not to Kill the Ocean." *Audubon*, July 1969.

"Ocean Pollution." *The New Yorker*, January 31, 1970.

"Offshore Dumping." *Environmental Action*, July 1970.

Olla, Bori L. "Biological Rhythms in Fishes and Other Aquatic Life." *Underwater Naturalist*, Vol. 3, No. 4.

Pearce, Jack. "Saucers in the Sea." *Underwater Naturalist*, Spring 1968.

Pell, Clairborne. "The Oceans: Man's Last Resource." *Saturday Review*, October 11, 1969.

"Progress in Sport Fishery Research, 1967." *Sport Fisheries and Wildlife*, May 1968.

Unger, Iris. "Artificial Reefs." American Littoral Society. Special Publication, No. 4, 1966.

"Water-Shock at Sea." *Time*, August 15, 1969.

Winchester, James H. "The Sea Around Them." *The Lamp*, Winter 1968.

OIL

"Alaska—Land Ownership and the Pipeline." *Environmental Action*, November 28, 1970.

Boyd, Paul, and Fritsch, Albert. "Hail Shale! The Gang's All Here." *Environmental Action*, October 1, 1971.

Cowan, Edward. "Mankind's Fouled Nest—Oil on the Waters." *The Nation*, March 10, 1969.

"Dirty Dilemma of Oil Spills, The." *Life*, March 6, 1970.

"Disaster at Sea." Enjay Chemical Co., Cranford, N.J.

Linford, Lloyd. "The Story of Oil: Let Them Eat Ethyl." *Earth Times,* May 1970.

Marx, Wesley. "Oil Spill." *Sierra Club Bulletin,* April 1970.

Maxness, Ron. "The Long Pipe." *Environment,* September 1970.

McCaull, Julian. "The Black Tide." *Environment,* November 1969.

"Moratorium on Alaska Oil Exploration, A." *Sierra Club Bulletin,* December 1969.

"Oil Is Beautiful . . . Or Is It?" *Environmental Action,* October 3, 1970.

"Oil Pollution." A Report to the President from the Secretary of the Interior and Secretary of Transportation, February 1968.

Oil Pollution Act. FWPCA, 1966.

"Oil vs. Water: Machiasport Dilemma." *Environmental Action,* October 3, 1970.

Pollak, Richard. "Oil on Ice. *Sierra Club Bulletin,* January 1971.

"Questions and Answers. Alyeska Pipeline." Alyeska Pipeline Service Co., Bellevue, Washington.

"Refinery in Kite's Country, A." *Esso Magazine,* Winter 1967–68.

Rutter, Richard M. "Salt and Pepper, Oil and Birds." *The Lamp,* Winter 1968.

"Supertankers Increase Oil Spill Threat." *The Plain Truth,* 1970.

Weisberg, Barry. "The Ecology of Oil. Raping Alaska." *Ramparts,* January 1970.

POISONS

Aaronson, Terri. "Gamble." *Environment,* September 1971.

Aaronson, Terri. "Mercury in the Environment." *Environment,* May 1971.

"Allergist Warns on Risks of Food Additives, American Medical Section on Allergy." *Medical Tribune,* August 18, 1969.

Biskind, Morton S. "DDT Poisoning and the Elusive Virus X—a New Cause for Gastro-Enteritis. *American Journal of Digestive Diseases,* March 1948.

Biskind, Morton S. "DDT Poisoning and X Disease in Cattle." *Journal of American Veterinary Medical Assn.,* January 1949.

Biskind, Morton S. "On the Alleged Harmlessness of DDT for Man." *Journal of Applied Nutrition,* Spring 1957.

Biskind, Morton S. "On Certain Aspects of the Origin and Treatment of Communicable Disease with Particular Reference to Viruses." (Published by author) 1955.

Biskind, Morton S. "Public Health Aspects of the New Insecticides." *American Journal of Digestive Diseases,* November 1953.

Biskind, Morton S. "Some Medical Aspects of Aerial and Ground Spraying with Carbaryl ('Sevin.')." Published by Connecticut Conservation Assn., March 19, 1971.

Biskind, Morton S. "The Technic of Nutritional Therapy." *American Journal of Digestive Diseases,* March 1953.

Biskind, Morton S., and Bieber, Irving. "DDT Poisoning—a New Syndrome with Neuropsychiatric Manifestations." *American Journal of Psychotherapy,* April 1949.

Carper, Jean. "Danger of Cancer in Food." *Saturday Review,* September 5, 1971.

Carter, Luther. "DDT: The Critics Attempt to Ban Its Use in Wisconsin." *Science,* February 7, 1969.

"Cause of Cancer, The." *Environment,* June 1970.

Chemicals in Food Products. Hearings before the House Select Committee to Investigate the Use of Chemicals in Food Products. House of Representatives, Parts 1, 2 and 3, 1950 and 1951.

Chemical Additives in Food. Hearings before a Subcommittee of the Committee on Interstate and Foreign Commerce, House of Representatives, 1956.

Clement, Roland C. "The Safe Use of Pesticides." National Audubon Society, May 9, 1968.

Collins, Stephen. "Biocide Blunder." *Massachusetts Audubon,* Autumn 1965.

Color Additives. Hearings before the Committee on Interstate and Foreign Commerce, House of Representatives, 1960.

Conway, Gordon, et al. "DDT on Balance." *Environment,* September 1969.

Craig, Paul P. "Correlation Between Exposure to Atmospheric Lead and Blood Lead Levels." Brookhaven National Laboratory, research paper, undated.

Craig, Paul P., et al. "DDT in the Biosphere: Where Does It Go?" Environmental Sciences Committee, Brookhaven National Laboratory. Work performed under auspices of AEC, research paper, undated.

Craig, Paul P., and Berlin, Edward. "The Air of Poverty." *Environment,* June 1971.

Dankenbring, William F. "Food Additives, Are They Really Safe?" *The Plain Truth,* February 1970.

"DDT. Declining Deadly Toxin," *Connecticut Conservation Reporter,* May–June 1969.

"DDT: The Poison That Lost Its Punch." *Science World,* January 26, 1970.

"DDT—The Wrong Insecticide." National Audubon Society, position paper, undated.

"DDT: Universal Contamination of the Earth." Symposium, *Roche Medical Image and Commentary,* March 1970.

"Deficiencies in Administration of Federal Insecticide, Fungicide and Rodenticide Act." Committee on Government Operations. Government Printing Office, May 7 and June 24, 1969; also November 13, 1969.

"Defoliants, Deformities: What Risk?" *Medical World News,* February 27, 1970.

Dietrick, Everett J. "Integrated Pest Control Theory and Practice." Paper, National Meeting, Entomological Society of America, Portland, Ore., November 1966.

"Does Environmental Pollution Cause Cancer?" *Environmental Action,* November 3, 1970.

Duffy, Margaret E. "Penicillin by the Pound." *Environment,* October 1969.

"EDF Takes DDT Appeals to Court." Environmental Defense Fund Newsletter, January 8, 1971.

"Effects of Pesticides on Fish and Wildlife, The." 1964 Findings of the Fish and Wildlife Service, 1965.

Effects of 2,4,5-T and Related Herbicides on Man and the Environment. Hearings, Subcommittee on Energy, Natural Resources, and the Environment. Committee on Commerce, U.S. Senate. June 17–18, 1970.

Ehrlich, Paul R., and Holdren, John P. "The Co-evolutionary Race." *Saturday Review,* December 6, 1970.

"Environmental Pollution: Scientists Go to Court." *Science,* December 22, 1967.

Epstein, Samuel S. "A Family Likeness." *Environment,* July–August 1970.

Epstein, Samuel S. "NTA." *Environment,* September 1970.

Epstein, Samuel S., et al. "Eye On Our Defenses." *Environment,* April 1971.

Evans, David M., and Bradford, Albert. "Under the Rug." *Environment,* October 1969.

"Fact Sheet on Sevin (Carbaryl)." Citizen–Scientist Committee on Pesticide Use. White Foundation, Morris, Conn. (Leaflet) undated.

"FDA Geneticists Raise New Doubt on Cyclamates." *Medical World News,* November 15, 1968.

Fitzsimmons, K. R. Statement to the Environmental Health Subcommittee, New York State Assembly, Buffalo, N.Y., July 8, 1969.

Food Additives. Hearings before a Subcommittee of the Committee on Interstate and Foreign Commerce, House of Representatives, 1958.

"Food Additives—3 Pounds per Person per Year." *Chemical and Engineering News,* October 10, 1966.

Frost, Justin. "Earth, Air, Water." *Environment,* August 1969.

Gentry, Jerry. "Mercury Pollution—Threat to Global Environment." *The Plain Truth,* June 1971.

"German Pesticide Law May Raise Problems for U.S. Farm Trade." *National Agricultural Chemicals News and Pesticide Review,* June 1967.

Gessner, Teresa. "The Hazards of Using DDT." Statement, N.Y. State Assembly Environmental Health Subcommittee, July 3, 1969.

Gilfillan, S. C. "Lead Poisoning and the Fall of Rome." *Journal of Occupational Medicine,* February 1965.

Gillette, Robert. "The Economics of Lead Poisoning." *Sierra Club Bulletin,* September 1970.

Good, F. D. T. "Pesticides and the Smarden Affair." Tenterden, Kent (G.B.), reprint, February 14, 1964.

"Government Ignores Mrak Commission." *Environmental Action,* February 18, 1970.

"Governor Announces Strict Controls on Harmful Pesticides." *The Conservationist,* December–January 1970–71.

Grant, Neville. "Mercury in Man." *Environment,* May 1971.

"Growing Crisis of Pesticides in Agriculture, The." *The Plain Truth,* May 1970.

"H&W, USDA Hold Firm; 2,4,5-T Ruling Postponed." *Chemical and Engineering News,* February 16, 1970.

Haynes, N. Bruce. Statement, N.Y. State Assembly Environmental Health Subcommittee (on use of DDT and related compounds), July 8, 1969.

Henkin, Harmon. "DDT and the Constitution." *The Nation,* March 10, 1969.

Higdon, Hal. "Well, If Not DDT, Then What?" *New York Times Magazine,* January 11, 1970.

"Higher Levels for Heavier Cattle" (stilbestrol). Doan's, October 1967.

Hueper, W. C. "Carcinogens in the Human Environment." *American Archives of Pathology,* March 1961.

Hyman, M. H. "Timetable for Lead." *Environment,* June 1971.

"Is Mother's Milk Fit for Human Consumption?" Environmental Defense Fund (Advertisement), *New York Times,* March 29, 1970.

Jerard, Elise. Testimony for the N.Y. State Assembly, Environmental Health Subcommitee re Pesticides, July 9, 1969.

Klendshoj, Niels C. Statement to Environmental Health Subcommittee, N.Y. State Assembly, Buffalo, N.Y., July 8, 1967.

Laycock, George. "The Beginning of the End for DDT." *Audubon,* July 1967.

"Lead Removal Is Not the Only Solution." *Environmental Action,* June 11, 1970.

"Leading Environmentalists Petition USDA to Ban DDT." Environmental Defense Fund (Press release), October 31, 1969.

Lofroth, Goran. "Pesticides and Catastrophe." *New Scientist,* December 5, 1968.

Longgood, William F. "Pesticides Poison Us." *The American Mercury,* 1960.

Magidson, Daniel T. "Half Step Forward." *Environment,* June 1971.

"Man vs. Gypsy Moth." *Connecticut Conservation Reporter,* February–March 1971.

Marier, J. R. "Fluoride Research." *Science,* March 29, 1968.

McCaull, Julian. "Building a Shorter Life." *Environment,* September 1971.

McCaull, Julian. "Know Your Enemy." *Environment,* June 1971.

"Meat You Eat, The." *The Plain Truth,* October–November 1971.

"Mercury—a Maddening Menace." *Science World,* September 29, 1961.

"Mercury in the Air." *Environment,* May 1971.

"Mercury—Major New Environmental Problem." *The Conservationist,* August–September 1970.

"Mercury Mess, The." *Time,* September 28, 1970.

Milby, Thomas F. "Prevention and Management of Organophosphate Poisoning." *Journal of the American Medical Assn.,* June 28, 1971.

Millar, Nathan. "The Alarming Case Against DDT." *Reader's Digest,* October 1969.

Miller, Morton W., and Berg, George G., eds. *Chemical Fallout—Cur-*

rent Research on Persistent Pesticides. Springfield, Illinois: Charles C Thomas, pp. 497–500, copied, no date.

"Monitoring and Residues." *Toxicology and Pharmacology,* 1950.

Montague, Peter and Katherine. "Mercury: How Much Are We Eating?" *Saturday Review,* February 6, 1971.

Moorman, James. "DDT—the Ban That Isn't." *Sierra Club Bulletin,* 1970.

Mrak, Emil. "Report of the Secretary's Commission on Pesticides and Their Relationship to Environmental Health." USDHE&W, December 1969.

National Agricultural Chemical News and Pesticide Review, February 1968 and February 1970.

Niering, W. A. "The Effects of Pesticides." *Bioscience,* Vol. 18, No. 9, 1968.

Odum, W. E., Woodwell, G., and Wurster, C. F. "DDT Residues Absorbed from Organic Detritus by Fiddler Crabs." *Science,* May 2, 1968.

"Of Mice and Men. Cancer and Pesticides." *Medical World News,* March 14, 1969.

Patterson, Clair C. "Contaminated and Natural Lead Environments of Man." *Archives of Environmental Health,* September 1965.

Patterson, C. C. "Lead," chapter in *Impingement of Man on the Oceans,* D. Hood, ed. (Excerpt, undated).

Peakall, David B. "Effect of Chlorinated Hydrocarbon Pesticides on Metabolic Pathways." Statement, N.Y. Assembly Subcommittee on Environmental Health," July 9, 1969.

Peakall, David B., and East, Ben. "Let's Not Kill Ourselves." *Outdoor Life,* September 1966.

Peet, Creighton, "The Effluent of the Affluent." *American Forests,* May 1969.

"Pesticides, and Catastrophe." *New Scientist,* December 5, 1968.

Pesticides and Wildlife. The Wildlife Society (Pamphlet), 1968.

Pesticides, Scientists, Institute for Public Information Workbook, 1970.

Peterle, Tony J. "Pyramiding Damage." *Environment,* August 1969.

Pollution by Pesticides. Conservation Foundation, 1969.

Pollution Caused Fish Kills, 1968. FWPCA, 1969.

Pottenger, F. M., Jr., and Krohn, B. "Poisoning from DDT and Other Chlorinated Hydrocarbon Pesticides—Pathogenesis, Diagnosis and Treatment." *Chemicals in Food and Cosmetics,* Pt. 2, November 23, 1951.

"Public Health Aspects of Pesticides." Hearings, New York State Assembly Study Committee on Health, Subcommittee on Environmental Health, August 1969.

"Rachel Carson Fund, The. A Report of Progress." National Audubon Society, 1969.

"Ravaged Summer." *Barrons,* July 6, 1970.

Reeder, Norm. "New Ways to Clean Pesticides out of Livestock." *Farm Journal,* June 1968.

Report from the Select Committee to Investigate the Use of Chemicals in Foods and Cosmetics, No. 2356, 1950.

Risebrough, Robert. "More Letters in the Winds." *Environment*, January–February 1970.

Rollins, Robert Z. "Drift of Pesticides." California Department of Agriculture. Presented, Tenth Annual Convention of the Agricultural Aircraft Assn., Inc., Palm Springs, Calif., January 15, 1960.

Schroeder, Henry A. "Metals in the Air." *Environment*, October 1971.

Shea, Kevin. "Blunted Weapons." *Environment*, January–February 1970.

Sheldrick, Michael G. "EDF vs. DDT: Scientists Fight Pollution." *Scientific Research*, August 4, 1969.

"Shell's No Pest Strip." *Consumer Reports*, November 1970.

Sloan, M. J. "Economic Entomology—From a Commercial Standpoint." Talk, Entomology Dept., Seminar, University of California, Berkeley, February 2, 1970.

Sloan, M. J. "Pesticides in Our Environment." Talk, Oklahoma Agricultural Chemical Assn., January 26, 1971.

Snyder, Jean. "What You'd Better Know About the Meat You're Eating." *Today's Health*, December 1971.

"Stricter Controls Urged to Keep Carcinogens Out of Food." *Medical World News*, February 10, 1967.

Stringer, John C. "Pesticide Contamination of Food and Drugs During Shipment." *FDA Papers*, May 4, 1968.

"U.S. Ecological Policy Devastates Vietnam." *Environmental Action*, June 4, 1970.

van den Bosch, Robert. "Pesticides: Prescribing for the Ecosystem." *Environment*, April 1970.

van den Bosch, Robert. "Statement on DDT." University of California, Berkeley, undated.

Walter, Eugene M., and Schurter, Dale L. "The Growing Crisis of Pesticides in Agriculture." *The Plain Truth*, May 1970.

Warshofsky, Fred. "Pesticides, Handle with Care." *Reader's Digest*, May 1966.

Wasserman, M., et al. "Department of Occupational Health, Hebrew University, Hadassah Medical School, Jerusalem." *Pesticides Monitoring Journal*, September 1967.

Whiteside, Thomas. "Department of Amplification." *The New Yorker*, June 20, 1970, and July 29, 1970.

"Who's for DDT?" *Time*, November 22, 1971.

Wright, T. D., and Deway, J. E. "Hard Pesticides—Can We Afford Not to Use Them?" *Conservation Circular*, N.Y. State College of Agriculture, Cornell University, Spring 1969.

Wurster, Charles F. "Aldrin and Dieldrin." *Environment*, October 1971.

Wurster, Charles F. "Chlorinated Hydrocarbon Insecticides and the World Ecosystem." Paper (undated).

Wurster, Charles F. "DDT Goes to Trial in Madison." *Bioscience*, Vol. 19, No. 9. September 1969.

Wurster, Charles F. "DDT in Human Milk." State University of New York at Stony Brook (Report) 1970.

Wurster, Charles F. "DDT in Mother's Milk." *Saturday Review*, May 2, 1970.

Wurster, Charles F. "Dutch Elm Disease Control." Environmental Defense Fund (Report) April 1968.

Wurster, Charles F. "Federal Government Regulation of Economic Poisons." Testimony, U.S. Senate Subcommittee on Energy, Natural Resources and the Environment, May 26, 1970.

Wurster, Charles F., and Wingate, David. "DDT Reduces Photosynthesis by Marine Phytoplankton." *Science,* March 29, 1968.

Wurster, Charles F., and Wingate, David. "DDT Residues and Declining Reproduction in the Bermuda Petrel." *Science,* March 1, 1968.

POPULATION

"Africa's Population Problem." Victor–Bostrom Fund, Spring, 1969.

Ardrey, Robert. "Control of Population." *Life,* 1970.

Boffey, Philip M. "Japan: A Crowded Nation Wants to Boost Its Birthrate." *Science,* February 13, 1970.

Brooks, Robert R. R. "People Versus Food." *Saturday Review,* September 5, 1970.

"Do We Need a National Population Policy?" *The Conservation Foundation Letter,* July 1970.

Draper, William H. "Is Zero Population Growth the Answer?" Speech at testimonial dinner, Washington, D.C., December 2, 1969.

Draper, William H. "The Pill and the Population Crisis." Testimony before Monopoly Subcommittee of the Senate Small Business Committee, March 3, 1970.

Ehrlich, Paul R. "Conservation and Voluntary Sterilization." National Conference on Conservation and Voluntary Sterilization—A New Alliance for Progress. New York City, October 1, 1967.

Ehrlich, Paul R. "Our Environmental Crisis." *The Plain Truth,* July 1970.

Ehrlich, Paul R. "The Population Explosion: Facts and Fiction." *Sierra Club Bulletin,* October 1968.

Ehrlich, Paul R. "Voluntary Sterilization, Conservation and Man's Survival." Assn. for Voluntary Sterilization, Inc., New York City, October 1, 1969.

Ehrlich, Paul R. "World Population: Is the Battle Lost?" *Reader's Digest,* February 1969.

Ehrlich, Paul R., and Holdren, John P. "Who Makes the Babies?" *Saturday Review,* February 6, 1971.

"Established Population Growth Commission." Message for the President of the U.S. Relative to Population Growth. House of Representatives, July 21, 1969.

"Famine Stalks the Earth." Hugh Moore Fund.

Glass, D. "Mankind Refuses to Leave Its Destiny to Chance." *World Health,* November 1965.

"Governments Act on Population Crisis." Victor–Bostrom Fund for the International Planned Parenthood Federation. Winter 1969–70.

"If All the Arable Land Was Properly Cultivated." *World Health,* November 1965.

Krutch, Joseph Wood. "A Naturalist Looks at Overpopulation." *Connecticut Conservation Reporter,* January 1968.

"Man Was a Rare Creature 500,000 Years Ago." *World Health,* November 1965.

McNaughton, S. J. "Rising Population: Its Effect on the Environment." *The Conservationist,* June–July 1970.

"1970 World Population Date Sheet." Population Reference Bureau, Inc. Washington, D.C., April 1970.

Nixon, Richard M. President's Message on Population. July 18, 1969.

"One Man's Answer to Overpopulation." *Life,* March 6, 1970.

Piotnow, Phyllis. Testimony for the Subcommittee of the Senate Small Business Committee, March 3, 1970.

"Population." Population Crisis Committee, March 1970.

"Population Crisis." Population Crisis Committee, May–June 1970.

"The Population Explosion: Facts and Fiction." *The Northwest Technocrat,* 1969.

"Population: The Biggest Boom." *Senior Scholastic,* October 27, 1969.

Spengler, Joseph J. "Population Problem: In Search of a Solution." *Science,* December 5, 1965.

Tydings, Sen. Joseph D. "Family Planning: A Basic Human Right." *Congressional Record,* May 8, 1969.

Udall, Rep. Morris, Jr. "Our Spaceship Earth—Standing Room Only." *Congressman's Report,* July 30, 1969.

Victor–Bostrom Fund Report for the International Planned Parenthood Federation. Spring 1970.

Weissman, Steve. "Why the Population Bomb Is a Rockefeller Baby." *Ramparts,* May 1970.

Williams, Joanne W. "Population vs. Wilderness." *Sierra Club Bulletin,* September–October 1966.

"Whatever Your Cause, It's a Lost Cause Unless We Control Population." Hugh Moore Fund (Leaflet) 1969.

Wolfers, D. "Problems of Expanding Populations. *Nature,* February 1970.

Wrong, Dennis H. "Portrait of a Decade." *New York Times Magazine,* August 2, 1970.

SOIL AND NUTRITION

Adeney, Martin. "Food Yields From Salty Lands." Scientists' Institute for Public Information Workbook, 1970.

Albrecht, William A. "Diagnoses or Post-Mortems?" *Natural Food and Farming,* September 1959.

Balfour, Eva. "The Haughley Experiment." The Soil Assn. (G.B.), (Reprint) undated.

Bear, Firman E. "The Inorganic Side of Life." *Normal Agriculture,* June 1953.

Boerma, A. H. "The State of Food and Agriculture 1969." Scientists' Institute for Public Information Workbook, 1970.

Braine, Gunnar M. "Sediment Is Your Problem." Soil Census Service, USDA, March 1958.

Commoner, Barry. *Nutrients Become Pollutants.* Scientists' Institute for Public Information Workbook, 1970.

Commoner, Barry. "Soil and Fresh Water: Damaged Global Fabric." *Environment,* April 1970.

Commoner, Barry. "Threats to the Integrity of the Nitrogen Cycle; Nitrogen Compounds in Soil, Water, Atmosphere and Precipitation." Global Effects of Environmental Pollution Symposium. Dallas, Texas, December 26, 1968.

Crimshaw, Nevil S. "Infant Malnutrition and Adult Learning." Scientists' Institute for Public Information Workbook, 1970.

Drew, Elizabeth B. "Hunger and Malnutrition in the U.S.—Going Hungry in America." Scientists' Institute for Public Information Workbook, 1970.

"International Food and Nutritional Problems: Urgent Needs—Inadequate Activities." Scientists' Institute for Public Information Workbook, 1970.

Jennings, Dana. "The National Grasslands." *Dakota Farmer,* May 1970.

McCarrison, Sir Robert. "Nutrition and Health." Lecture to London School Children, May 1937. Reprint, Lee Foundation for Nutritional Research.

Mead, Margaret. "Hunger, Food and the Environment." Scientists' Institute for Public Information Workbook, 1970.

Pottenger, Francis M., Jr. "The Effects of Heat-Processed Foods and Metabolized Vitamin D Milk on the Dentofacial Structures of Experimental Animals." *American Journal of Orthodontics and Oral Surgery,* August 1946.

Pramer, David. "The Soil Transforms." *Environment,* May 1971.

Revelle, Roger. "In the Shape of a Loaf of Bread." Scientists' Institute for Public Information, 1970.

"Soil Conservation at Home." USDA Agricultural Information Bulletin No. 244. Soil Conservation Service, May 1962. Revised February 1968.

Walter, Eugene M., and Shurter, Dale L. "Sick Soil—A Basic Cause of Poor Health." *The Plain Truth,* June–July 1970.

Walter, Eugene M., and Hogbert, Gene H. "Arid Lands—Can They Be Reclaimed in Time?" *The Plain Truth,* October 1969.

SOLID WASTE

Alexander, Tom. "Where Will We Put All That Garbage?" *Forbes,* October 1967.

Black, Ralph J., Muhick, Anton J., Klee, Albert J., Hickman, H. Lanier, Sr., and Vaughan, Richard D. "1968 National Survey of Community Solid Waste Practice—an Interim Report." Presented at the 1968 Annual Meeting of the Institute for Solid Wastes of the American Public Health Works Assn., Miami Beach, Fla., October 24, 1968.

Burkness, Dan. "Solid Waste: A National Disgrace." *Environmental Action,* August 15, 1970.

"Cleaning and Collection." *Sweep,* Autumn 1969.

"The Crisis of Survival." *Progressive Magazine,* April 1970.

Eisenbud, Merril. Third National Incinerator Conference. Speech, American Society of Mechanical Engineers. New York, May 6, 1969.

"Engineering and the Urban Crisis." *Design Perspective,* March 16, 1970.

"Fact Sheet," New York City Dept. of Sanitation, January 1970.

Feldman, M. M. "Municipal Solid Waste." *Transactions, New York Academy of Sciences,* February 5, 1969.

Fleming, Thomas. "Engulfed in Garbage." *This Week.* (Magazine defunct; tear sheet undated.)

Grinstead, Robert B. "No Deposit, No Return." *Environment,* November 1969.

Handley, Arthur. "Meeting the Solid Waste Problem." *The Conservationist,* April–May 1970.

Interim Report—1968 National Survey of Community Solid Waste Practices. USDHE&W, October 24, 1968.

Katz, Linda. "Nation's Clean Garbage Disposal Costs Soar." *Environmental Action,* April 9, 1970.

Kearing, Samuel J., Jr. "The Politics of Garbage." *New York Magazine,* April 13, 1970.

Lucia, Frank J. "Land Pollution. A Heap of Trouble." *Senior Scholastic,* October 27, 1969.

Lucia, Frank J. "On Your Mark, Get Set, SNOW." *Sweep,* Autumn 1969.

Nelson, Sen. Gaylord. "Introduction of the Packaging Pollution Control Act of 1970." *Congressional Record,* April 1, 1970.

"New Total Destruction System for Solid Waste." Doremus & Co. New York, N.Y., November 11, 1969.

Outwater, Eric B. "The Disposal Crisis—Our Effluent Society." *National Review,* February 24, 1970.

Powledge, Fred. "Can Kretchmer Make a Clean Sweep?" *New York Magazine,* March 22, 1971.

"Progressing Backwards." *Environmental Action,* November 28, 1970.

Rodda, J. D. "Sanitation Sailor Who Never Goes to Sea." *Sweep,* Spring 1959.

Rogus, Casmir A. Incineration in New York City. *Sweep,* Autumn 1959.

Roosevelt, James. "Guest Editor." *Sweep,* Summer 1965.

Ross, Clifford B. "The Rubbish Dilemma: Time to Face the Consequences." *New Jersey Business,* January 1969.

Seymour, Whitney North, Jr. "80,000,000 Bottles a Day." *The Conservationist,* February–March 1968.

Seymour, Whitney North, Jr. "320 Billion Pounds of Garbage a Year." Position Paper, 1968.

"Solid Wastes: A List of Available Literature." USDHE&W, Public Health Service, October–December 1968.

"Solid Waste Management: The Federal Role." USDHE&W, Bureau of Solid Waste Management, 1969.

"Some Facts About the World's Biggest Housekeeping Job." City of New York, Dept. of Sanitation (Pamphlet) undated.

"State/Interstate Solid Waste Planning Grants and Agencies." U.S. Public Health Service, January 1969.

"Summaries of Solid Waste Program." *Contracts,* July 1, 1965. USDHE&W, June 30, 1968.

"Summaries of Solid Waste Research and Training Grants." Consumer Protection and Environmental Health Service. USDHE&W, Public Health Service Publication No. 1596, 1968.

"Summary Report of Advanced Waste Treatment in Water Pollution Control." Research Series, U.S. Dept. of the Interior, FWPCA, July 1964–July 1967.

Teller, A. J. "New Economic Value Criterion for Recovery of a Habitable Environment." Statement, Joint Congressional Conference on New Priorities. *The Environment,* October 24, 1969.

Teller, A. J. "Preservation of Natural Resources Depends on Social Adequacy of Economics." (Speech) undated.

Teller, A. J. "Should All Our Environmental Waste Be Economic Waste?" *Professional Engineer,* February 1970.

Teller, A. J. Testimony, Senate Subcommittee on Air and Water Pollution, March 19, 1970.

Teller, A. J. "We Are Discovering the Obvious." Statement, Fourth Annual Conference of Regional Councils, March 18, 1970.

Trippett, Frank. "The Epic of Garbage." *Look,* November 4, 1969.

Vaughan, Richard D. "Solid Waste Management: The Federal Role. Symposium, Environmental Equilibrium Criteria, Cost, Cooperation." National Pollution Control Conference and Exposition, Houston, Texas. USDHE&W, April 22, 1969.

Walford, Lionel A. "Artificial Fishing Reef Study—Possible Uses of Waste Materials." Sandy Hook Marine Laboratory. September 10, 1966.

"Waste Disposal." *Sweep,* Winter 1962.

"We Can Tame the Garbage Monster." *Prevention,* January 1970.

"Where Will We Stash the Trash?" *Nation's Business,* September 1968.

"You Can Move to a Cleaner Block Without Leaving Home." *Operation Better Block,* New York, N.Y.

UNITED NATIONS

Astrom, Sverker (Swedish Ambassador). "Problems of the Human Environment." Statement in the Second Committee, November 10, 1969.

Astrom, Sverker. Preparatory Committee for the U.N. Conference on the Human Environment, first session. Statement, March 10, 1970.

Astrom, Sverker. Preparatory Committee for U.N. Conference on the Human Environment. Remarks at meeting of Sub-group No. 4, March 16, 1970.

Astrom, Sverker. "Problems of the Human Environment." Statement in the General Assembly, December 15, 1969.

Problems of the Human Environment. Report of the Secretary-General, May 26, 1969.

Report of the Preparatory Committee for the United Nations Conference on the Human Environment, April 6, 1970.

Resolution Adopted by the General Assembly. U.N. Conference on the Human Environment, January 8, 1971.

Sweden's Reply to the U.N. Enquiry in Connection with the Preparations for the U.N. Conference on the Human Environment. Royal Ministry for Agriculture, not dated.

WATER

"Advances in Waste-Water Treatment. Pilot Plant, Lebanon, Ohio." FWPCA, 1969.

"Another Wash Day Miracle: Eutrophication." *Environmental Action*, March 12, 1970.

"Anticipated Capital Needs for Sewage Facilities in New Jersey." N.J. State Dept. of Health, February 3, 1969.

Argenio, Modesto. "Lake Erie." *Sierra Club Bulletin*, March 1970.

Arnold, Dean E. "Lake Erie Alive but Changing." *The Conservationist*, December–January 1970–71.

Berg, George G. *Water Pollution*. Scientists' Institute for Public Information Workbook, 1970.

"Bowery Bay Pollution Control Project Brief of Treatment Facilities." Dept. of Public Works, N.Y.C.

"Brief Outline of the Inadequate Economic Justification of the Cross-Florida Canal Project." (Pamphlet) undated.

Brodine, Cynthia, and Berg, George G. "Water Crisis—The Rochester Area." *Scientist and Citizen*, March 1967.

Bylinsky, Gene. "The Limited War on Water Pollution." *Fortune*, February 1970.

"Citizens Against Water Pollution." New Jersey. Pamphlet.

Citizens for the Conservation of Florida's Natural and Economic Resources, Inc., March 1965.

"Clean Water. It's Up to You." Izaak Walton League (Booklet) 1969.

Clement, Roland C. "National Water Policy." Testimony, National Water Commission, Washington, D.C., November 21, 1969.

Commoner, Barry. "Lake Erie. Aging or Ill." *Scientist and Citizen*, December 1968.

"Cost of Clean Water and Its Economic Impact, The." Vols. 1, 2, 3. FWPCA, 1969.

Coughlin, Robert L. "Comparative Waste Loads." (Memo, U.S. Govt.) April 3, 1968.

Cunningham, John T. "A Foul Deed." *Newark Sunday News*, September 28, 1969.

Cunningham, John T. "New Jersey's Chance to Cry Foul." *New Jersey Business*, 1969.

"Detergent Pollution Control Act of 1970, The." *Congressional Record*, February 26, 1970.

"Disposal of Wastes from Water Treatment Plants." FWPCA, August 1969.

Drew, Elizabeth B. "Dam Outrage—The Story of the Army Engineers." *The Atlantic Monthly*, 1970.

Ehrlich, Henry. "The Ladies Save the Lakes." *Look,* April 21, 1970.

Epstein, Samuel. "NTA." *Environment,* September 1970.

"Facts About New York Harbor." *Argonaut,* February 1971.

"Federal Grants for Clean Water." FWPCA, January 1969.

"Federal Water Pollution Control Act—Oil Pollution Act." FWPCA, 1966.

Greene, Wade. "What Happened to the Attempts to Clean Up the Majestic, the Polluted Hudson?" *New York Times Magazine,* May 3, 1970.

Groopman, Abraham. "Effects of the Northeast Water Crisis on the New York City Water System." *Journal of the American Water Works Assn.,* January 1968.

Gruchow, Nancy. "Detergents: Side Effects of the Washday Miracle." *Environmental Action,* January 9, 1970.

Hughes, Gov. Richard J. "New Jersey's Approaches to Control of Water Pollution." *Public Health News,* November 1963.

"Hunt's Point Pollution Control Plant." N.Y.C. Dept. of Public Works.

"Improving Your Potomac." FWPCA, 1969.

"It's a Dirty Shame." League of Women Voters of New Jersey.

Kandle, Roscoe P. Conference in the Matter of Pollution of the Navigable Waters of Eastern New Jersey from Shark River to Cape May and Their Tributaries. Atlantic City, N.J., November 1–2, 1967.

Kandle, Roscoe P. "Regulations Concerning Classification of the Surface Waters of the Passaic River Basin." N.J. Dept. of Health, September 11, 1967.

Kardos, Louis T. "A New Prospect." *Environment,* March 1970.

Kilgore, Margaret. "Waste Not, Want Not." *The Binnacle,* May 11, 1968.

Kohn, Sherwood Davidson. "Warning: The Green Slime Is Here." *New York Times Magazine,* March 22, 1970.

"Lake Erie." *Sierra Club Bulletin,* March 1970.

Lamb, James C. "A Plan for Ending Lake Erie Pollution." *Public Works Magazine,* June 1969.

Langewiesche, Wolfgang. "The Great American River Cleanup." *Reader's Digest,* May 1969.

Levy, Gerald F., and Folsted, John William. "Swimmer's Itch." *Environment,* December 1969.

Lindsay, Sally. "How Safe Is the Nation's Drinking Water?" *Saturday Review,* May 2, 1970.

"Machine Livens Dead Rivers, A." *Report from Rutgers,* April 1970.

"Manpower and Training Needs in Water Pollution Control." Report of the Dept. of the Interior, FWPCA, to U.S. Congress, August 2, 1967.

Marine, Gene. "California's Water Plan: The Most Expensive Faucet in the World." *Ramparts,* May 1970.

McCloskey, Michael. "Water Projects—A Changing Perspective." *Sierra Club Bulletin,* July 1970.

"Needed: Clean Water. Problems of Pollution." Channing L. Bete Co., Inc. FWPCA, 1967.

Nelson, Sen. Gaylord. "Introduction of National Preservation Act of 1970." *Congressional Record,* February 16, 1970.

Nelson, Sen. Gaylord. "Water Pollution Control." *Congressional Record,* November 18, 1969.

New Jersey Effluents, New Jersey Water Pollution Control Assn., Vol. 2, No. 2, April 1969; Vol. 3, No. 1, October 1969.

"New York City Water Pollution Control Program Summary." May 1969.

"An Old Law and a New Court Decision." *Conservation Foundation Newsletter,* August 1970.

"Omnibus Bill Funding Dam Engineers." *Environmental Action,* October 3, 1970.

"The Operation of a Typical N.Y.C. Water Pollution Control Plant." N.Y.C. Dept. of Public Works (Fact sheet) undated.

Penman, H. L. "The Water Cycle." *Scientific American,* September 1970.

"A Primer on Waste Water Treatment." FWPCA, October 1969.

"Pollution Watch: The 1899 Refuse Act." *Textile Industries,* August 1971.

"Procedure for Evaluation of Water and Related Land Resources Projects." Water Resources Council, Washington, D.C., June 1969.

"Regulations Concerning Treatment of Wastewaters, Domestic and Industrial, Separately or in Combination, Discharged into the Waters of the Passaic River Basin Including the Newark Bay," 1967. New Jersey State Department of Health, February 1967.

"Regulations Establishing Classification for Waters." New Jersey Dept. of Health, February 28, 1967.

"Report of the Committee on Water Quality Criteria." FWPCA, April 1, 1968.

"Report on the Quality of the Interstate Waters of the Lower Passaic River and Upper and Lower Bays of New York Harbor." FWPCA, November 1969.

Research at Rensselaer. "Water Pollution." Rensselaer Polytechnic Institute, August 1969.

"Rutgers Takes a Hard Look at Environmental Problems." *Report from Rutgers,* October 1970.

Schaeffer, John R. "Reviving the Great Lakes." *Saturday Review,* November 7, 1970.

Schofield, Carl L. "Water Chemistry and Lake Productivity." *The Conservationist,* April–May 1970.

Schrag, Peter. "Life of a Dying Lake." *Saturday Review,* September 20, 1969.

Seeger, Pete. "Voyage to Save the Dying Hudson." *Look,* August 26, 1969.

Shaw, Robert S., and Legesser, Ernest R. "Water Pollution Control in New Jersey." (Statement) 1967.

Shea, Kevin P. "Dead Stream." *Environment,* July–August 1970.

"Showdown for Water." FWPCA (Booklet) October 1968.

"So You'd Like to Do Something About Water Pollution." League of Women Voters of the U.S. March 1969.

"Steps Toward Clean Water." Report to the Committee on Public Works of the U.S. Senate, from the Subcommittee on Air and Water Pollution, January 1966.

"Tallman's Island Pollution Control Project Extension." Dept. of Public Works, N.Y.C.

"Ten Years of Cleaning New Jersey's Streams and Waters," New Jersey State Department of Health, Summary Report of Activity, 1948–1958.

"They Love the Oyster." *The Lamp,* Spring 1960.

Torpey, Wilbur N., and Chasick, A. H. "Principles of Activated Sludge Operation." *Sewage and Industrial Wastes,* November 1965.

Torpey, Wilbur N., and Melbinger, Norman R. "Reduction of Digested Sludge Volume by Controlled Circulation." *Journal of Water Pollution Control Federation,* September 1967.

Udall, Stewart L. "Comeback for the Willamette River." *National Parks Magazine,* March 1968.

"U.S. Industry Makes Amends." *Newsweek,* January 26, 1970.

"Wastes from Watercraft." Report of the Dept. of the Interior, FWPCA, to the U.S. Congress, August 7, 1967.

"Watch on the Raritan Monitors Water Quality." *Public Works Magazine,* May 1964.

"Water," Worthington Corp. (Booklet) 1966.

"Water Pollution. The Blighted Great Lakes." *Life,* August 23, 1968.

"Water Pollution Control Research Development, Demonstration, and Training Projects." FWPCA, 1968.

"Water Pollution: A Deadly Tidal Wave." *Senior Scholastic,* October 27, 1969.

"Water Pollution Group." Rochester Committee for Scientific Information, 1964–69. June 1969.

"Water Pollution . . . a Pressing Problem." *Fast Facts,* Hospital Service Plan of N.J., October 1969.

The Water Quality Act of 1965. USDHE&W, November 1965.

"Water Quality Standards. Better Water for America." FWPCA (Pamphlet) March 1969.

Webster, Dwight A. "Temperature and Related Factors in Lakes." *The Conservationist,* December–January 1969–70.

Webster, Dwight A. "Water Quality Standards." FWPCA, March 1967.

"Weyerhaeuser Scientists Explore the Secrets of Clean, Clear Water." Weyerhaeuser Co. Pamphlet (undated).

"What You Can Do About Water Pollution." FWPCA.

"The Wonder of Water." Soil Conservation Society of America. (Undated.)

WILDLIFE

"Alaska's Wildlife." *Audubon,* July 1969.

"Background Information on the Canadian–Norwegian Slaughter of Seals." Committee for Humane Legislation, Inc., March 6, 1971.

"Bird Preservation. Notes and Comments." *The New Yorker,* March 14, 1970.

"The Brown Pelican. A Vanishing American." *Environment Southwest,* June 1969.

Charles, Gordon. "The State vs. the Predator." *Audubon,* November–December 1966.

Clement, Roland C. "Instincts, Laws and Ducks." *Transactions,* 34th North American Wildlife and Natural Resources Conference, March 1969.

Clement, Roland C. "Vanishing Wildlife and the Utah Environment." Speech, University of Utah Interdisciplinary Conference on the Future of Utah's Environment. Salt Lake City, October 17, 1968.

Dwyer, Tim. "Americans Wipe Out 40 Animal Species." *Environmental Action,* June 11, 1970.

"Extinction of the Brown Pelican Threatened by the Chemical Effect of DDT on Its Eggs." American Museum of Natural History, September 24, 1969.

Fisher, Jr. "African Wildlife: Man's Threatened Legacy." *National Geographic,* February 1972.

Forbes, James E. "Environmental Deterioration and Declining Species." *The Conservationist,* August–September 1970.

"Furs Look Better on Their Original Owners," Muir and Friends, Friends of the Earth and John Muir Institute for Environmental Studies, January 1970.

Gilbert, Bil. "Hunting Is a Dirty Business." *Saturday Evening Post,* October 21, 1967.

Harris, Fred R. Testimony before Commerce Subcommittee on Oceans and Atmosphere, February 15, 1972.

Herrington, Alice. Statement before Commerce Subcommittee on Oceans and Atmosphere, February 15, 1972.

James, Charles. "Death on the Road." *The Conservationist,* April–May 1970.

McCutchen, Steve. "Atomic Blast." *Audubon,* November–December 1965.

McIntyre, Joan. "Wildlife." *Earth Times,* May 1970.

"Migration to the Moon." *Connecticut Conservation Reporter,* January 1968.

Olla, Bori L. "Daily and Seasonal Rhythms of Activity in the Bluefish, Pomatomus Saltatrix." Sandy Hook Marine Laboratory. International Game and Fish Conference, San Juan, Puerto Rico, November 18, 1967.

Payne, Roger. "Among Wild Whales." *New York Zoological Society Newsletter,* November 1968.

Peterson, Roger Tory. World Wildlife Fund. Washington, D.C. Letter (undated).

Regenstein, Lewis. "Sea Brutality Decimates Seals, Whales, Polar Bears. *Environmental Action,* July 10, 1971.

Vandevere, Judson E., and Mattison, James A., Jr., "Sea Otters." *Sierra Club Bulletin,* October 1970.

"Wildlife Rescue Mission." *Weyerhaeuser Magazine,* April 1969.

Wood, Nancy. "The Wild Horse." *Audubon,* November 1969.

Zimmerman, D. R., "Death Comes to the Peregrine Falcon." *New York Times Magazine,* August 9, 1970.

Zimmerman, David R. "Last Hope for the Ospreys of Long Island Sound." *New York Times Magazine,* December 1971.

Notes

[1] Scientists at Michigan State University have isolated viruses from drinking water, apparently for the first time in the United States, according to *The New York Times*, January 5, 1972. The scientists said the action points to a serious health hazard that exists in many water supplies. The virus was described as a Type II polio virus of the kind used in oral polio vaccines; the viruses were taken from a 100-foot-deep well certified as safe by the State Health Department. They were believed to have been excreted by recently immunized persons and to have seeped from the sewer system through the ground to contaminate the water supply. The scientists said that if polio viruses can get from sewage into ground water, other types of virus that cause disease can also get into the water.

[2] The "general quality of water is deteriorating so rapidly, because of the new and complex pollutants that are dissolved and thus undetectable by conventional means, that water researchers are cautious about giving flat assurances of a completely pure future water supply." This is the view of water men and the American Waterworks Association research arm of the nation's water suppliers, reports *The New York Times*, April 9, 1970. Water men are quoted as saying that present filtration methods are "rudimentary." A New Jersey official is quoted as saying that "conventional water treatment . . . won't remove anything that's dissolved." Such substances were said to include chlorides in runoffs from salted roadways, and petroleum pollutants from tank trucks and traffic.

[3] Californians have been warned by their State Environmental Quality Study Council that the state "is losing the battle against pollution on all fronts," according to *The New York Times*, March 26, 1972. The 17-member council noted local progress against pollution. "But all in all, the report said, 'the situation has not improved significantly' since it issued its 1970 report, which warned, 'If the present course is continued our posterity will inherit a vast wasteland.' "

[4] California in 1966 became the first state to require by law the control of emissions of hydrocarbons and carbon monoxide from automobile exhausts. The control devices worked well enough, "but the sad fact is that air pollution in California—particularly here in the Los Angeles area—is worse than ever," according to Tom Wicker in *The New York Times*, March 5, 1972. He quotes Dr. James M. Pitts, director of the Air

Pollution Research Center for the University of California, as explaining that the devices effectively reduced hydrocarbons and carbon monoxide, but they increased auto engine temperatures; the result was greater emissions of nontoxic nitrogen oxides, which mixed with the air to produce the toxic nitrogen dioxide, PAN and ozone. Nitrogen dioxide has doubled, reaching levels more than ten times that supposedly permitted by state standards. Smokestacks that appear to be emitting nothing are also creating the menacing nitrogen dioxide in the air above. A study commission found that 10 million people in the Los Angeles area live in what amounts to a "health crisis." Funding for research to combat the problem was said to amount to "peanuts."

5 New York City's top environmental official, Jerome Kretchmer, revealed that the waste problem is so desperate that landfills no longer are being made level, as planned, but that garbage is being piled higher into hills, according to *The New York Times*, March 5, 1970. "No decision on the height of the hills has been made," the article states. Henry L. Diamond, state commissioner of environmental conservation, said serious spillage and leaching problems already exist, and piling the refuse higher will make those problems more severe.

6 New statistics reveal "a growing epidemic among younger men of heart attacks, a disease once thought to be confined to the elderly," according to *The New York Times*, April 9, 1972. "Since 1950, the heart attack death rate among men from age 25 to 44 years has risen 14 percent, from 45.7 to 52 per 100,000, while among men 45 to 64 years, the rate has risen 4 percent from 574.9 to 597.8 per 100,000."

A Korean War study—autopsies on young men (averaging 22 years of age) killed in combat—disclosed a high incidence of heart disease: ". . . accumulation of fatty deposits in the arterial walls that ultimately leads to its occlusion (blockage) was present in over 77 percent of these young men. In 10 percent the process had already occluded over 70 percent of the opening of one or more of the major arteries." The men showed external evidence of good health and represented "a group of Americans presumably in better than average health."

7 Representative Colmer, 82, announced on March 6 that he would not seek another term in Congress. His retirement, at the end of the current session, would open the way for the House Democratic leadership to obtain a firmer grip on the flow of key liberal legislation to the floor.

8 Nicholas Johnson, renegade member of the Federal Communications Commission and persistent critic of television, told a consumer group that television was the "foremost enemy of intelligent consumerism," according to *The New York Times*, March 5, 1972. Before the New York Consumer Assembly, "he said the medium ducked controversial issues in its programing, promoted unhealthful diets with its advertising, was 'the number one pusher of the drug life' with its commercials for tranquilizers, sleeping pills and the like, and contributed to the American ethic that people had to keep buying 'things' to be happy.

" 'For a medium that is so powerful to abuse its power is sad,' he said. 'For a medium that is statutorily required to operate in the public interest, it is close to criminal. . . . Television is the business of gathering you (in front of the set) and selling you like cattle to the advertisers.'" He added: " 'The problem in America is not that the top 100 corporation presidents are violating the law, though God knows they are, the problem is they're writing the laws.' "

9 Even when out driving in the country there is no escape from the assault of advertising. An average of three billboards profanes every mile

of highway, and in some areas there are 50 times that number, all urging the captive motorist to buy something, and marring the landscape. In an effort to fight this visual pollution, Congress passed the Highway Beautification Act of 1965, providing for the removal of highway signs. Originally the measure called for removal of all billboards "visible from the highway." But the powerful signboard lobby got the final measure so weakened that it provides only for removal of signs closer than 660 feet from the roadway. The act is so full of loopholes, so poorly enforced, that there are now an estimated 250,000 more signs than when the act was passed—an overall total of more than a million of the eyesores violating the land.

Under the act, the government pays 75 percent of the cost for removal of the things—that means the taxpayer pays. Recently the government took a more vigorous stand, threatening to withhold 10 percent of highway construction programs; all but three states have complied. Transportation Secretary Volpe has crusaded to get rid of the signs, but as of early 1972 only one token signboard had been removed, in Maine, while others were still springing up to add new blight.

10 The continuing pollution from power plants was revealed following a year-long independent study by the Council on Economic Priorities, reported in *The New York Times*, April 16, 1972. The study by the non-profit organization covered 124 major fossil fuel plants—which burn coal, oil or gas—operated by 15 utilities that supply 25 percent of the nation's power. The article states that:

—Only 45 of the plants (36 percent of the total) adequately controlled particulate emissions, or soot.

—Fifty-seven of the plants (46 percent) have continued to burn high-sulfur oil or coal—a major cause of sulfur dioxide gas.

—Sixty-six of the plants (81 percent) had no controls of any kind of emission of nitrogen oxides, gases that are known to induce emphysema and are the main ingredients of smog.

The report added: "In 1970, advertising and sales expenditures for the 15 companies totaled $126.9 million, 1.9 percent of their combined gross operating revenues. The 15-company research spending, only a small part of which goes to develop pollution-control techniques and more advanced generating methods, totaled $21.4 million in 1970—one-sixth the amount spent for advertising and sales."

Index

ABOUT THE AUTHOR

WILLIAM LONGGOOD was born and educated in Missouri. He attended elementary and high school in Kansas City and then the University of Missouri. Before becoming a free-lance writer, he was, for seventeen years, a reporter-feature writer for the New York *World-Telegram and Sun*. He also served as a Middle East correspondent for the Scripps-Howard newspapers. His newspaper work has been recognized with many awards, including a Pulitzer Prize. Mr. Longgood has written seven books, including the first major work on food chemicals, *The Poisons in Your Food*, published by Simon and Schuster in 1960, as well as many magazine articles. He also teaches writing courses at The New School for Social Research.